COMPUTATIONAL MODELLING OF MASONRY, BRICKWORK AND BLOCKWORK STRUCTURES

COMPUTATIONAL MODELLING OF MASONRY, BRICKWORK AND BLOCKWORK STRUCTURES

Edited by
J.W. Bull

SAXE-COBURG
PUBLICATIONS

published 2001 by
Saxe-Coburg Publications
Dun Eaglais, Station Brae, Kippen
Stirling, FK8 3DY, UK

Saxe-Coburg Publications is an imprint of Civil-Comp Ltd

ISBN 1-874672-04-0

British Library Cataloguing in Publication Data
A catalogue record for this book is available from the British Library

Chapter 9 of this book contains material from: Genna F., Di Pasque M., Veroli M., Ronca P., "Numerical analysis of old masonry buildings: a comparison among constitutive models", Engineering Structures, vol. 20, pages 37-53, 1997. This material is reprinted with kind permission of Elsevier Science Ltd., Oxford, England.

Printed in Great Britain by Bell & Bain Ltd, Glasgow

Contents

Preface

This book is aimed at: design engineers who need to know the latest advances in masonry, brickwork and blockwork; architects wanting to develop building shape still further; engineering consultants who must ensure designs are safe; academics who research into masonry and postgraduates approaching masonry for research purposes.

The use of masonry, brickwork and blockwork for building and civil engineering structures has a long history going back to ancient times. For many hundreds of years, the main masonry material has been stone, with clay becoming the main brickwork material. With the introduction of other materials including concrete, the term blockwork is now also used.

As masonry and brickwork structures have a long history, design standards have generally accepted the empirical design requirements that place 'natural' height restrictions on masonry and brickwork structures. Further, brickwork and blockwork are often still seen as just infill material for steel framed or reinforced concrete buildings.

With the advent of reinforcing or post tensioning brickwork, increasingly complex structures became a possibility. Further there is an increasing requirement to assess the strength of existing masonry structures and to determine the most suitable means of improving masonry performance.

Designers, engineers and research workers want to exploit masonry to its full potential. To determine how masonry reacts to extreme conditions, there is a need for the computational modelling of a whole range of masonry, brickwork and blockwork materials and structures, including the composite action of the brick/joint interface.

Although much of the information on masonry structures is available from publications in journals and at conferences, masonry research is widely scattered throughout the world. In recognition of this fact, the expert chapter authors in this book have been drawn together from around the world. Their up-to-date knowledge and expertise in masonry structures is drawn upon to determine the present state of the computational modelling of masonry, brickwork and blockwork structures and to point to future developments.

This book brings together a wide range of masonry, brickwork and blockwork disciplines and shows where computational modelling has been used successfully. The ten chapters have been divided into five topic areas: damage and failure models; vibration and earthquake effects; settlement; fire; and historic buildings. Each topic area can be described briefly as follows:

Damage and failure models

Chapter one reviews the literature regarding the homogeneous and heterogeneous constitutive laws for masonry and describes two models based on damage mechanics. In the first model, masonry is considered as a homogeneous, orthotropic material, but in the second model the overall mechanical properties of masonry are determined based on the properties of the components. The numerical results from the two models are in good agreement with experimental data.

Chapter two considers the formulation of an elastic-plastic joint element to determine the structural behaviour and characteristics of masonry structures. The chapter then goes on to investigate the effectiveness of the joint element.

Earthquake and vibration effects

Chapter three refers to the computational modelling of masonry structures subject to earthquake and vibration effects. After a general survey covering the effect of earthquake and vibration loads on civil engineering structures, the chapter then focuses on the modelling of the structure, the material and load simulation. Two case studies are included.

Chapter four studies the dynamic behaviour of masonry bell towers. Measurements of the transient response during full-circle ringing of a single heavy bell gave the natural frequency, mode shape, damping, and peak response for each tower. Finite element analyses of the towers for frequency, mode shape and transient response correlated well with the observed behaviour. Very large factors of safety were found in the bell towers.

Settlement

Chapter five describes the development and use of a complex three-dimensional finite element model to study the effect of constructing a tunnel beneath an existing masonry building. Results are presented for simple arrangements of masonry facades as well as more complex buildings for which field data are available.

Fire

Chapter six provides insight into factors affecting the thermo-structural behaviour of masonry firewalls. The elevated temperature material properties of some masonry materials are discussed. This has facilitated a greater understanding of the structural response of masonry in high temperature environments. A computational model has been used to successfully simulate the experimental data from fire testing of masonry walls.

Historic buildings

Chapter seven applies discontinuous deformation analysis, within the context of the computational modelling, to the assessment of masonry arch bridges. The method was tested on a series of benchmark problems related to the stability of a masonry arch under its own weight. The analysis predictions were compared with nonlinear finite element analysis and discrete element analysis predictions. Satisfactory comparisons of failure mode and formation of hinges were obtained.

Chapter eight reviews the basic problems associated with the computational modelling of masonry arches. The chapter discusses the basic characteristics of masonry materials and describes the development of a 'rigid block' mechanism method of analysis, which allows estimates of the ultimate load carrying capacity of arches to be obtained. As not all constituent blocks in a masonry arch are rigid, issues relevant to the non-linear elastic methods of analysis are also discussed.

Chapter nine looks at the conservation and rehabilitation of historic masonry structures and uses computational modelling to study the safety and stability of structural elements in old "difficult" masonry. The results show that masonry structures can reach unpredictable limit states. Furthermore, every technological and structural detail must be taken into account, and with computational modelling the results of a variety of different assumed structural models and analysis methods have to be compared with each other.

Chapter ten considers computational techniques for structural assessment of historic masonry structures and focuses on establishing feasible states of equilibrium that encourage interaction between the engineer and the computer. The computational models are based on finite element concepts developed with characteristics relating to equilibrium elements. The techniques involve the determination of thrust lines for skeletal structures using spreadsheets, the optimization of general thrust lines by limit analyses and the use of finite element models that can simulate cracking. The techniques are illustrated by case studies using historic structures.

My thanks go to my family for their support, to the chapter authors for their contributions and to Saxe-Coburg Publications for their help and guidance especially Barry Topping, Jelle Muylle, and Rosemary Brodie.

John W. Bull

©2001, Saxe-Coburg Publications, Stirling, Scotland
J.W. Bull, (Editor), Computational Modelling of Masonry, Brickwork and Blockwork Structures,
Saxe-Coburg Publications, Stirling, Scotland 1-26.

Chapter 1

Damage and Failure Models

E. Papa

Department of Structural Engineering, Polytechnic of Milan, Italy

Abstract

After having concisely reviewed parts of the available literature regarding homogeneous and heterogeneous constitutive laws for masonry, two models based on Damage Mechanics and suitable for the analysis of masonry structures subjected to plane stress conditions are presented. In the first, masonry is considered as a homogeneous, orthotropic material. A limited domain, endowed with a suitable hardening/softening law, is defined according to piecewise linear functions; material damage is taken into account as a function of inelastic deformation. In the second model the overall mechanical properties of masonry are determined based on the properties of the components, for which a unilateral damage model is assumed. The numerical results given by these two models appear to be in good agreement with the experimental data related to masonry panels and buildings.

1.1 Introduction

Several lines of research have been developed to analyse masonry structures. These include no-tension constitutive laws [1–3], time-independent incremental (nonholonomic) elastic/plastic models [4–6], and the use of limit analysis concepts [7, 8]. During the last decades material models based on damage mechanics concepts have been developed and adapted to masonry, possibly in combination with plasticity laws.

In any case, mechanically speaking, a correct modelling for masonry cannot ignore the fact that it is an aggregate of bricks with interposed mortar joints. Consequently, brick masonry is a heterogeneous material with orthotropic mechanical characteristics which depend not only on the properties of each phase (brick and mortar), but also on their interaction.

Starting from these considerations, models for masonry can be divided into two categories: heterogeneous models and homogeneous models. In the first model, bricks and mortar are separately discretised, whereas the second is based on the definition of equivalent homogeneous material, namely a fictitious material having mechanical properties that are equivalent to the overall, average properties of the given, non-homogeneous material.

Within each category, the damage and failure models can be based on a discrete or smeared crack approach. In the discrete approach, crack growth induces a material discontinuity: to properly model the process of crack closure and opening due to cyclic loads, a new structure geometry has to be described and a unilateral constraint introduced. A drawback of these numerical models comes from the large computational effort necessary to analyse real masonry structures. The smeared crack approach considers the cracked material as an equivalent continuum, over which cracks are not exactly localised but continuously distributed. Consequently, this approach can be used with satisfactory results when the material degradation is due to a system of cracks spread over a defined zone. In this sense it is possible to define the equivalent continuum as a damaged continuum.

In the following sections, after a brief review of several homogeneous and heterogeneous constitutive laws, two models conceived specifically for masonry are presented.

1.2 Heterogeneous models

A masonry structure can be analysed by discretising bricks and mortar joints separately. For each phase a suitable constitutive law is introduced in order to reproduce not only its own behaviour, but also the interaction between the two material components. In a finite element modelisation, two approaches can be followed: in the first, both bricks and mortar joints are discretised by using continuum finite elements of suitable size, whereas, in the second, an interface element is used to model the behaviour of mortar joints.

Since the first approach requires a large computational effort to analyse real masonry structures, it is used to simulate tests on small specimens in order to determine accurately the stress distribution in the masonry material.

The interface element of the second approach reproduces essentially the interaction between two adjoining bricks. In this way, the discretisation of mortar joints does not require further degrees of freedom to be introduced. Obviously, each brick must always be modelled through at least one finite element. Also in this case, the numerical problem quickly grows in complexity as the structure size increases.

One of the first heterogeneous models was proposed by Page [9] where masonry is discretised as a continuum of bricks which behave elastically, connected by mortar joints having poor mechanical properties. Indeed, the behaviour of the joints is assumed to be brittle, in the presence of tensile stresses, whereas a bilinear failure criterion is introduced when the joints are subjected to compressive stresses. Moreover, the normal stiffness is kept constant; whereas the shear modulus changes depending on the value of the normal stress. Assuming elastic behaviour for bricks, the computational effort is limited, but the lack of any failure criterion does not allow some of the important experimental phenomena to be taken into account, such as, for example crushing in compression. Consequently, this model can be successfully applied for the analysis of restricted, but important, categories of structures, for example shear walls.

In [10, 11], Lourenço et al. propose for masonry an interface model derived from the Mohr-Coulomb criterion: the compressive and tensile stresses are bounded by introducing two suitable surfaces, called cap surface and tension cut-off, respectively. The Mohr criterion and the tension cut-off vary according to a hardening-softening law, whereas the behaviour of the cap surface is not coupled with the other two. Indeed, the sliding and opening mechanisms are localised within the mortar joints; on the other hand the collapse for compression involves the masonry as a whole.

Lotfi and Shing [12] introduce an interface model having a hyperbolic yield function, which is a generalisation of the Mohr-Coulomb criterion. In comparison with Lourenço's model, these authors introduce a yield surface with a continuous gradient, obtaining in this way a unique analytical expression for the constitutive law of joints. The compressive behaviour of bricks is governed by the von Mises criterion, whereas the tensile failure is reproduced by introducing a Rankine-type tension cut-off. The interface model is able to simulate both shear dilatancy and joint compaction, that is the increase in normal stress due to sliding and the decrease in sliding phenomenon due to increase of compressive stresses. These two characteristics are reproduced by introducing a non-associated flow rule in which shear dilatancy vanishes when the joint is completely damaged. The mechanical characteristics of the interface evolve according to softening laws depending on the interaction between compressive plastic strains and the shear and tensile strengths of the joints.

In [13] Gambarotta and Lagomarsino study the behaviour of mortar joints by using damage mechanics concepts. The damage of joints is measured by an internal variable, associated with the strain energy release rate, to which the inelastic strain components (sliding and normal strain) are related. When the strain energy release rate becomes equal to the joint toughness that is assumed to depend on the damage itself, the joint damage rate evolves. Starting from these assumptions, the proposed model assumes brick masonry walls consist of continuum elements that behave elastically until a suitable failure criterion is activated, connected by interface elements which represent mortar joints.

1.3 Homogeneous models

The homogeneous models are based on the definition of an equivalent continuum, the characteristics of which can be obtained either by introducing suitable constitutive laws able to represent the directional behaviour of masonry, or starting from the properties of the masonry components, i.e., mortar and bricks, and using a suitable homogenisation procedure [14–16].

The existing macro-models, based on experimental tests performed only on assemblages of sufficient size under homogeneous stress, may be divided into two main categories. Within the first category fall those models which are essentially derived by models for concrete through slight modifications: see, for example, Chen [17], Ganju [18], Shing et al. [19]. Indeed, these constitutive laws are suitable for fully grouted reinforced concrete masonry, but they may fail in representing the behaviour of unreinforced brick masonry, which is particularly anisotropic owing to the presence of mortar joints.

The second category is of those models which have been developed specifically for masonry, taking *a priori* into account the anisotropy induced by the mortar joints. In the simple monotonic model proposed by Samarashinghe et al. [20], the behaviour of masonry is assumed as linear-elastic and brittle. When the failure surface, derived by Samarashinghe and Hendry [21], in the tension-compression range, has been reached, the residual stiffness and strength of the corresponding element is supposed to vanish. Since no complete failure criterion was available at that time, no strength limitation was introduced in the biaxial compression and biaxial tension ranges.

A more elaborated monotonic model has been proposed by Page et al. [22] using a complete idealised failure surface derived by Dhanasekar et al. [23] in the plane of the stress components referred to the joint's orientation. In this model, the plastic strain components in each direction are related only to the corresponding stresses by a power law. The local behaviour is assumed to be elasto-plastic and brittle; only if one of the principal stresses is tensile, does it become elastic-brittle. By introducing this model into a finite element code, the two sources of non-linearity, plasticity and failure, are treated separately, maintaining one constant when the other evolves.

In the model proposed by Contro and Sacchi [24], an anisotropic failure criterion with zero-strength in tension is assumed. The behaviour in biaxial compression is reproduced introducing a suitable work-hardening law. Considering different orientations of the principal stresses to the bed joints, the failure domain is bounded by six straight lines in the compression-compression range.

By interpreting brick masonry as a composite material, its overall mechanical properties are determined as functions of the properties of the components. This goal is obtained by using homogenisation techniques, which provide the stiffness matrix of an idealised homogeneous material having mechanical characteristics equivalent to the heterogeneous material of the whole. Obviously, application of this procedure requires the introduction of suitable constitutive laws for each component of masonry. With the models based on the definition of an equivalent continuum, real masonry walls can be analysed with an acceptable computational effort obviously lower than the effort required by the application of heterogeneous models.

The homogenisation procedure can be applied only to determine the equivalent material parameters or included in the constitutive law. In the first case, the analyses are certainly simpler, but it is not possible to realise whether failure takes place in the mortar or the bricks. On the contrary, in the second case the computational effort is greater since the behaviour of the homogeneous material must always be defined taking account of the mechanical characteristic of its components.

Several authors employed this procedure and obtained interesting results. For example, to simulate the lateral response of in-plane loaded brick masonry shear walls Gambarotta and Lagomarsino [25] propose a continuum model in which masonry is considered as an equivalent stratified medium. The layers represent the mortar bed joints and the brick units, respectively. The constitutive equations are obtained through a homogenisation procedure involving the damage model described in the previous section for the mortar joints. A similar damage law is assumed for the brick layers: another internal variable is introduced which represents crushing damage. The damage variation in mortar and bricks is obtained through the solution of a Linear Complementarity Problem. Using different damage indicators, it is possible to distinguish between the different failure modes within the material element. In particular, this model permits one to evaluate the dissipated energy due to the cyclic displacement: its knowledge is very important to be able to evaluate the seismic vulnerability of masonry buildings. The simulations of the experimental results obtained for rectangular slender and squat shear walls and for a large-scale brick masonry shear wall show the capabilities and the validity limits of this model.

In the model proposed by Pietruszczak et al. [26] the horizontal mortar joints are assumed as planes of weakness, whereas the vertical joints are treated as parallel distributions of inclusions having lower mechanical properties than the remaining continuum. Vertical and horizontal joints are separately taken into account, through a double homogenisation of the material. The material collapse can be due to brick crushing or joint failure for opening and/or sliding. In the first case, a brittle elastic mechanism in the vertical joint-brick system is introduced, while in the second case it is simulated by an elasto-plastic criterion. The numerical simulations show that the presence of horizontal mortar joints lead to a significant directionality in the material response. Moreover, these results point out that the model is particularly fit for studying the failure behaviour of masonry subjected to biaxial stresses.

1.4 An elastic-plastic damage model for masonry

To simulate the behaviour of masonry subjected to cyclic loads, an elastic-plastic model with damage has been developed and implemented in a finite element computer package for non-linear analyses.

In this model, masonry is assumed as an orthotropic continuum subjected to plane stress conditions. A piecewise-linear surface (with the same characteristics as a classical yield surface in plasticity [27]) that bounds the elastic domain is defined in the stress space $\sigma_{11}-\sigma_{22}-\sigma_{12}$ (with σ_{22} normal to the mortar beds); it is symmetrical with respect to the $\sigma_{11} - \sigma_{22}$ plane [28]. The shear stress σ_{12} can never exceed the values corresponding to the dark areas. The entire surface is formed by eighteen planes (see Figure 1.1). Their intersections with the $\sigma_{22} - \sigma_{12}$ plane are the segments α and β of Figure 1.2, which correspond to the grey and the darker surfaces, respectively. Plane

α is introduced in order to limit the maximum shear stress in the mortar joints [29]. It can also be inferred from Figure 1.1 that the compressive strength predicted by the model is much higher than the tensile strength.

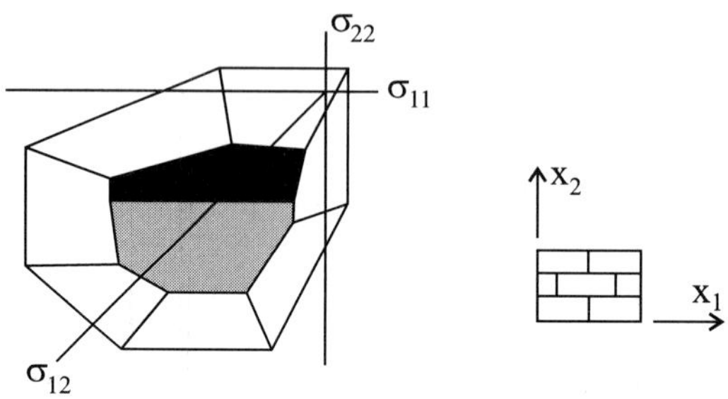

Figure 1.1: Piecewise-linear surface.

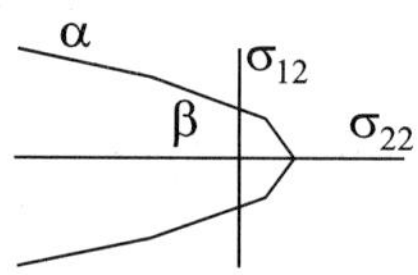

Figure 1.2: Section in the plane $\sigma_{11} = 0$.

Using the model for the solution of a structural problem, the load history is divided into a convenient number of time steps, during which increments of the external actions are applied. The response to each step is found by means of the following iterative scheme:

a) stresses are predicted by assuming a linear elastic response to the strain increments;

b) no inelastic strains are assumed if the predicted stress point falls within the current elastic domain;

c) if the predicted stress point lies beyond the current piecewise linear surface, inelastic strain increments are computed on the basis of an associate flow rule, by using the theoretical background typical of classical plasticity. In a first step, they are determined as if they were traditional plastic strains. However, at the end of each time step they are split into two contributions [30]: a non recoverable (plastic) strain and a damage-induced strain: the damage processes are quantified by a convenient decrease in stiffness, as discussed below.

Owing to the piecewise linear pattern of the yield surface and the normality rule, the inelastic strain increments can be obtained by properly combining plastic multipliers. Such multipliers, in turn, are related to the projections of the inelastic strain increments along axes normal to the yield planes. Imposing a piecewise linear law for the hardening/softening rule (cf. Figure 1.3), the vector of inelastic strain increments $\Delta\underline{\varepsilon}^i$ can be expressed as

$$\Delta\underline{\varepsilon}^i = \underline{n}\Delta\underline{\mu} \tag{1.1}$$

$$\text{with } \underline{n} = \begin{bmatrix}\underline{n}^1 & \underline{n}^2 & \dots & \underline{n}^Y\end{bmatrix} \tag{1.2}$$

where $\underline{n}^k$ represents the unit outward vector normal to the k-th plane ($k = 1, ..., Y$ with $Y = 18$ in this case). The plot of Figure 1.3 corresponds to the traditional stress-strain curve that would be obtained with a fictitious elastic modulus $E = \{\underline{n}^k\}^\top \underline{D}\underline{n}^k$ (where $\underline{D}$ denotes the material stiffness matrix) and with a uniaxial stress applied along the direction normal to the k-th plane. The r_j^k coefficients represent the distance between the k-th plane and the origin at different hardening states. The vector $\Delta\underline{\mu}$ collects the relevant (eighteen) scalar quantities $\Delta\mu^k$. Since a piecewise linear hardening/softening rule is characterised by several critical stresses beyond which an inelastic mode becomes active, each multiplier $\Delta\mu^k$ actually consists of different contributions. For instance, in the case of Figure 1.3 we have

$$\Delta\mu^k = \sum_{j=1,5} \Delta\lambda_j^k \tag{1.3}$$

where $\Delta\lambda_j^k$ are the plastic multipliers.

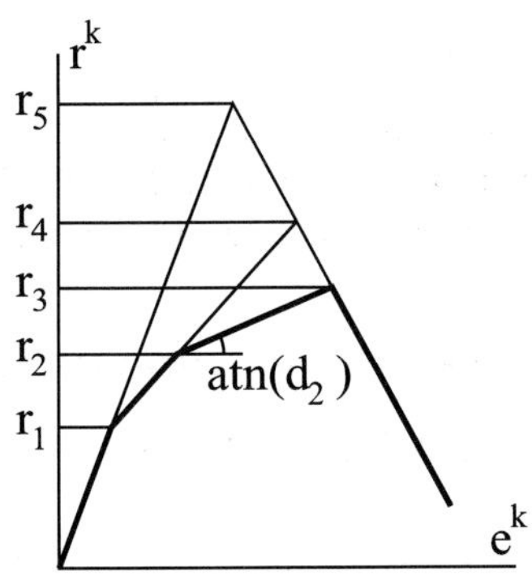

Figure 1.3: Plane hardening law.

For the sake of simplicity, suppose that only the k-th mode is activated. The non-negative contributions $\Delta\lambda_j^k$ must satisfy the following Linear Complementarity Problem [LCP] (written in matrix form):

$$\underline{\Phi}^k = [\underline{N}^k]^\top \{\underline{\sigma}^0 + \underline{D}\Delta\underline{\varepsilon}\} - [\underline{N}^k]^\top \underline{D}\underline{N}^k\Delta\underline{\lambda}^k - \underline{H}^k\Delta\underline{\lambda}^k - \underline{r}^k \leq \underline{0} \tag{1.4}$$

under the conditions $\Delta\underline{\lambda}^k \geq \underline{0}$; $\left\{\underline{\Phi}^k\right\}^\top \Delta\underline{\lambda}^k = 0$ and with

$$\left[\underline{N}^k\right] = \left[\; \left\{\underline{n}^k\right\}^\top \; \left\{\underline{n}^k\right\}^\top \; \left\{\underline{n}^k\right\}^\top \; \left\{\underline{n}^k\right\}^\top \; \left\{\underline{n}^k\right\}^\top \;\right] \tag{1.5}$$

$$\Delta\underline{\lambda}^k = \left\{\; \Delta\lambda_1^k \;\; \Delta\lambda_2^k \;\; \Delta\lambda_3^k \;\; \Delta\lambda_4^k \;\; \Delta\lambda_5^k \;\right\}^\top \tag{1.6}$$

$$\underline{r}^k = \left\{\; r_1^k \;\; r_2^k \;\; r_3^k \;\; r_4^k \;\; r_5^k \;\right\}^\top \tag{1.7}$$

$$\underline{H}^k = \begin{bmatrix} h_1^k & 0 & 0 & 0 & 0 \\ 0 & h_2^k & 0 & 0 & 0 \\ 0 & 0 & h_3^k & h_4^k & h_5^k \\ 0 & 0 & h_4^k & h_4^k & h_5^k \\ 0 & 0 & h_5^k & h_5^k & h_5^k \end{bmatrix} \tag{1.8}$$

The symbol $\underline{\sigma}^0$ in Equation (1.4) denotes the stress at the beginning of the current time step, whereas the hardening/softening parameters h_j^k are related to the slopes of the $r^k - e^k$ plot (e.g. d_2 - see Figure 1.3). The above equations have been given in the incremental form, typical of non-linear problems. Therefore, each finite increment of the external loads implies an increment $\Delta\underline{\varepsilon}$ of the total strain vector at a certain location (stress point) where the constitutive law is to be enforced. This law is satisfied by finding convenient increments $\Delta\underline{\lambda}^k$ ($k = 1, \ldots, Y$), such that $\underline{\Phi}^k \leq \underline{0}$, as required by the inequality (1.4), $\Delta\underline{\lambda}^k \geq \underline{0}$ and $\left\{\underline{\Phi}^k\right\}^\top \Delta\underline{\lambda}^k = 0$. Indeed, the j-th multiplier concerned with the k-th plane $\Delta\lambda_j^k$ ($j = 1, \ldots, 5$ in the case of Figure 1.3) has to be zero if $\Phi_j^k < 0$.

Considering the plot of Figure 1.3, the rank of this LCP problem is initially equal to five. During the analysis the rank decreases to one when the plane considered is totally collapsed to the origin of the axes, or to a minimum distance if defined.

Considering all the Y planes of the piecewise linear yield surface, the actual stress increment is obtained from this equation:

$$\Delta\underline{\sigma} = \underline{D}\Delta\underline{\varepsilon} - \underline{D}\left\{\sum_{k=1,Y} \underline{N}^k \Delta\underline{\lambda}^k\right\} = \underline{D}\Delta\underline{\varepsilon} - \underline{D}\underline{N}\Delta\underline{\lambda} \tag{1.9}$$

where $\underline{N} = \left[\; \underline{N}^1 \;\; \underline{N}^2 \;\; \ldots \;\; \underline{N}^Y \;\right]$, and $\Delta\underline{\lambda}$ collects the Y subvectors $\Delta\underline{\lambda}^k$.

Therefore, at each stress point the inelastic strains corresponding to a given incremental total strain $\Delta\underline{\varepsilon}$ can be found by solving the LCP

$$\underline{\Phi} = \underline{N}^\top \left\{\underline{\sigma}^0 + \underline{D}\Delta\underline{\varepsilon}\right\} - \underline{N}^\top \underline{D}\underline{N}\Delta\underline{\lambda} - \underline{H}\Delta\underline{\lambda} - \underline{r} \leq \underline{0} \tag{1.10}$$

$$\underline{\Phi}^\top \Delta\underline{\lambda} = 0 \tag{1.11}$$

$$\Delta\underline{\lambda} \geq 0 \tag{1.12}$$

where $\underline{H} = \text{diag}\left[\underline{H}^k\right]$ and $\underline{r}$ collects the subvectors $\underline{r}^k$ ($k = 1, \ldots, Y$). The solution in terms of $\Delta\underline{\lambda}$ is iterative [31].

1.4.1 Damage evolution law

In order to take damage effects into account, the vector of the inelastic strains $\Delta\underline{\varepsilon}^i = \underline{N}\Delta\underline{\lambda}$ (determined at the end of each time step due to a given increment of the external loads) can be subdivided into two contributions: $\Delta\underline{\varepsilon}^p$ and $\Delta\underline{\varepsilon}^d$. The first contribution is the plastic unrecoverable strains, whereas the second one represents strains related to damage phenomena that are supposed to be reversible. Accordingly, in this model damage is quantified by updating the material stiffness and compliance matrices. The partition is obtained introducing a parameter η (with $0 \leq \eta \leq 1$) in the following coaxial laws:

$$\Delta\underline{\varepsilon}^d = \eta\Delta\underline{\varepsilon}^i \quad (1.13)$$

$$\Delta\underline{\varepsilon}^p = (1-\eta)\Delta\underline{\varepsilon}^i \quad (1.14)$$

At the end of each time interval and at each stress point, the total strain is obtained by the following expression:

$$\underline{\varepsilon} = \underline{C}^0\underline{\sigma} + \underline{\varepsilon}^p + \Delta\underline{\varepsilon}^d = \underline{\varepsilon}^p + \left[\underline{C}^0 + \Delta\underline{C}\right]\underline{\sigma} = \underline{\varepsilon}^p + \underline{C}^0\underline{\sigma} + \Delta\underline{C}\underline{\sigma} \quad (1.15)$$

where $\underline{C}^0$ (written for an orthotropic material and referred to axes $x_1 - x_2$ introduced in Figure 1.1) is the initial compliance matrix at the beginning of each step, whereas $\underline{\varepsilon}$, $\underline{\varepsilon}^p$ and $\underline{\sigma}$ are the vectors that collect the total strains, the plastic strains and the stresses at the end of the current step, respectively. The increment $\Delta\underline{C}$ can be split into the contributions $\Delta\underline{C}_N$ and $\Delta\underline{C}_S$ [32]:

$$\underline{\varepsilon} = \underline{\varepsilon}^p + \underline{C}^0\underline{\sigma} + \Delta\underline{C}\underline{\sigma} = \underline{\varepsilon}^p + \underline{C}^0\underline{\sigma} + \Delta\underline{C}_N\underline{\sigma} + \Delta\underline{C}_S\underline{\sigma} \quad (1.16)$$

Indeed, by introducing the constant terms a, b, c and d, the following pattern can be assumed for these matrices:

$$\Delta\underline{C}_N = \begin{bmatrix} a & b & 0 \\ b & c & 0 \\ 0 & 0 & 0 \end{bmatrix} \quad (1.17)$$

$$\Delta\underline{C}_S = \begin{bmatrix} 0 & 0 & 0 \\ 0 & 0 & 0 \\ 0 & 0 & d \end{bmatrix} \quad (1.18)$$

Hence, the damage strain rate is equal to:

$$\Delta\underline{\varepsilon}^d = \Delta\underline{\varepsilon}^N + \Delta\underline{\varepsilon}^S = \rho^N\left|\underline{\sigma}^N\right|\underline{n}^N + \rho^S\left|\underline{\sigma}^S\right|\underline{n}^S = \Delta\underline{C}_N\left|\underline{\sigma}^N\right|\underline{m}^N + \Delta\underline{C}_S\left|\underline{\sigma}^S\right|\underline{m}^S \quad (1.19)$$

In the above equations $\underline{\sigma}^N$ is a vector whose non-zero terms are the normal stresses σ_{11} and σ_{22}, while the only non-zero entry of $\underline{\sigma}^S$ is the shear stress σ_{12}. The vectors $\underline{\varepsilon}^N$ and $\underline{\varepsilon}^S$ are featured by a similar pattern. The symbol $|\underline{x}|$ denotes the Euclidean norm of $\underline{x}$, while $\underline{n}^N$, $\underline{n}^S$, $\underline{m}^N$, $\underline{m}^S$ are unit vectors that give the directions of $\Delta\underline{\varepsilon}^N$, $\Delta\underline{\varepsilon}^S$, $\underline{\sigma}^N$, $\underline{\sigma}^S$ in the spaces $\varepsilon_{11} - \varepsilon_{22} - \gamma_{12}$ and $\sigma_{11} - \sigma_{22} - \sigma_{12}$, respectively. Assuming ρ^N and ρ^S equal to the ratios $\frac{|\Delta\underline{\varepsilon}^N|}{|\underline{\sigma}^N|}$ and $\frac{|\Delta\underline{\varepsilon}^S|}{|\underline{\sigma}^S|}$, the matrix compliance increments are:

$$\Delta\underline{C}_N = \frac{\rho^N}{\{\underline{n}^N\}^\top\underline{m}^N}\left[\underline{n}^N\{\underline{n}^N\}^\top\right] \quad (1.20)$$

$$\Delta\underline{C}_S = \frac{\rho^S}{\{\underline{n}^S\}^\top\underline{m}^S}\left[\underline{n}^S\{\underline{n}^S\}^\top\right] \quad (1.21)$$

From a numerical point of view, the contour lines of damage will not be drawn referring to the compliance increment, but will introduce a fictitious internal variable which gives the position of the stress point in the $r^k - e^k$ plot. Referring to Figure 1.3, its value varies between 1 (elastic behaviour) and 6 (complete failure).

1.4.2 Numerical analyses

Before simulating the behaviour of real masonry walls subjected to horizontal cyclic loads, the reliability of the model was checked by performing some simple analyses.

Several cyclic displacement histories were imposed on a single finite element: Figure 1.5 shows the numerical response of the model subjected to the load history of Figure 1.4. It is possible to note that its different compression and tensile strength have been reproduced.

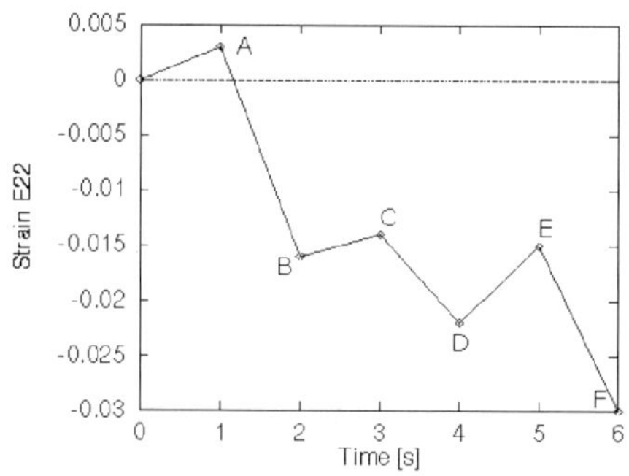

Figure 1.4: Time history.

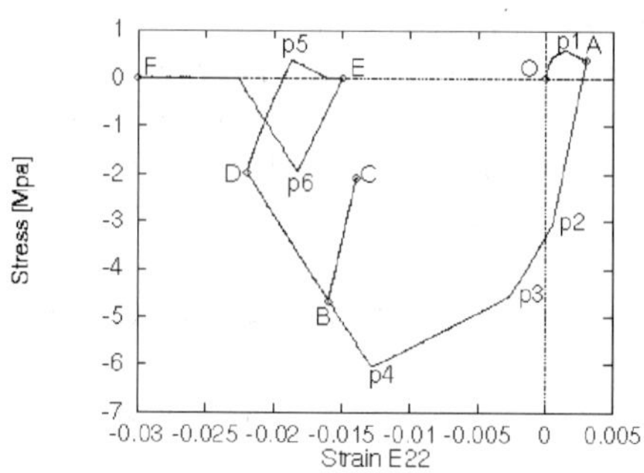

Figure 1.5: Stress-strain curve.

Figure 1.6 shows the response due to an increasing level of alternate shear strain obtained by considering both the yield surface of Figure 1.1 (Figure 1.6a) and by making the plane α inactive (Figure 1.6b). In the latter case, the maximum shear stress always increases, towards high unrealistic values, while the shape of curve in Figure 1.6a reproduces a realistic behaviour.

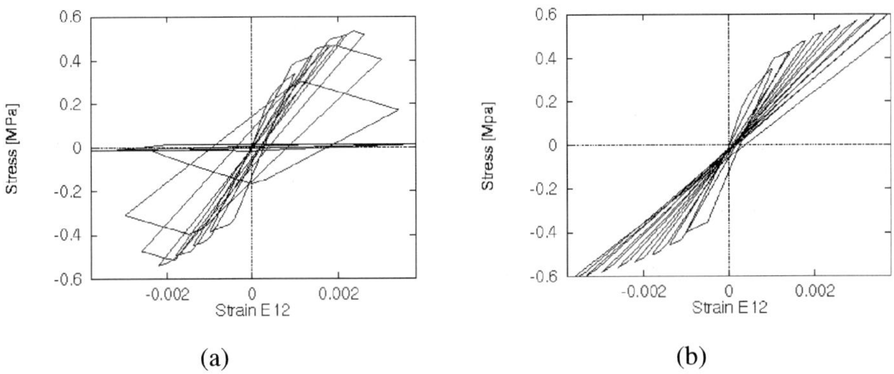

Figure 1.6: Shear stress-strain curve with (a) and without (b) plane α.

After these qualitatively encouraging results, the model was used to simulate shear tests carried out on two geometrically different kinds of masonry walls: a slender panel (1 m in width and 2 m in height) and a squat panel (of the same width, but 1.35 m in height). A constant vertical load is applied and a cyclic displacement at the top imposed. The experimental responses are obviously different for the two walls [33]. The horizontal force vs. horizontal displacement plot for the squat panel shows a degradation in stiffness and strength in conjunction with significant dissipation. On the other hand, the same curve for the slender panel is characterised by smaller dissipation, due to a lower reduction in stiffness, and a lack of softening behaviour. Indeed, the collapse for the squat panel depends on the growth of shear damage in the central zone of the panel, whereas the failure mechanism for the slender panel is due to a rocking behaviour. Figures 1.7a,b and 1.8a,b show a good agreement between the experimental (a) and the numerical (b) responses, both in terms of failure loads and dissipation.

Moreover the distributions of shear damage depicted in Figures 1.9a,b reproduce the failure mechanism, such as the presence of significant tensile and compressive damage at the top and the bottom of the slender panel which confirm its bending behaviour (see Figure 1.10a,b).

This model was used to simulate the behaviour of a real masonry wall tested at the University of Pavia [34]. It is a two-story masonry building subjected to dead vertical loads and to two equal horizontal forces applied at each floor. These forces vary in a cyclic way and change in sign. Figures 1.11a,b show the total horizontal

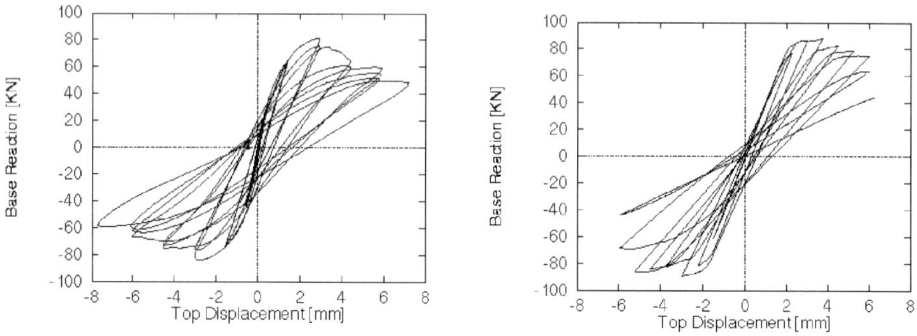

(a) (b)

Figure 1.7: Squat panel: experimental (a) and numerical (b) response.

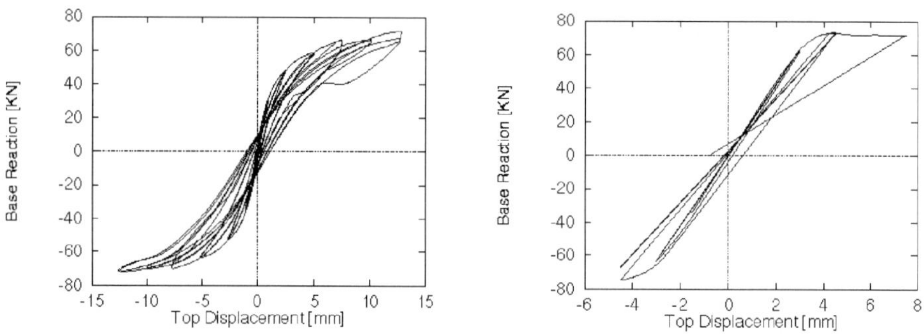

(a) (b)

Figure 1.8: Slender panel: experimental (a) and numerical (b) response

loads vs. horizontal displacement at the left corner of the second floor. In this case, the experimental and numerical responses are in good agreement. The experimental failure is due to the growth of shear cracks at the central zone of the ground floor, in addition to tensile cracks at the bottom of the wall. The numerical damage pattern at failure is depicted in Figures 1.12a,b. Note that significant shear and tensile damage is present in the central zone of the ground floor and at the bottom, respectively.

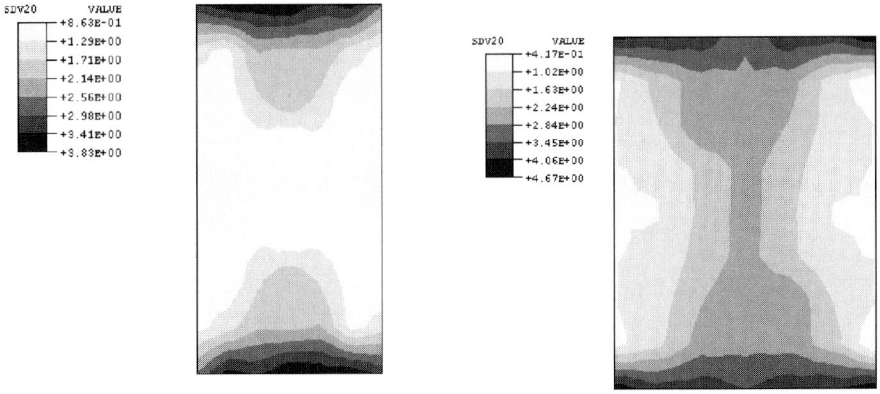

(a) (b)

Figure 1.9: Distribution of shear damage for the slender (a) and squat (b) panel.

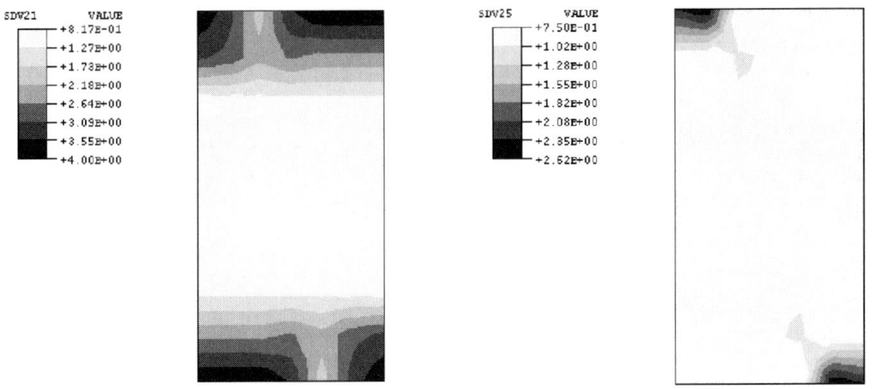

(a) (b)

Figure 1.10: Slender panel: distribution of tensile (a) and compressive (b) damage.

1.5 A unilateral damage model based on a homogenisation procedure

In this section, the constitutive law of masonry is obtained by introducing a damage model for bricks and mortar and then using a suitable homogenisation procedure.

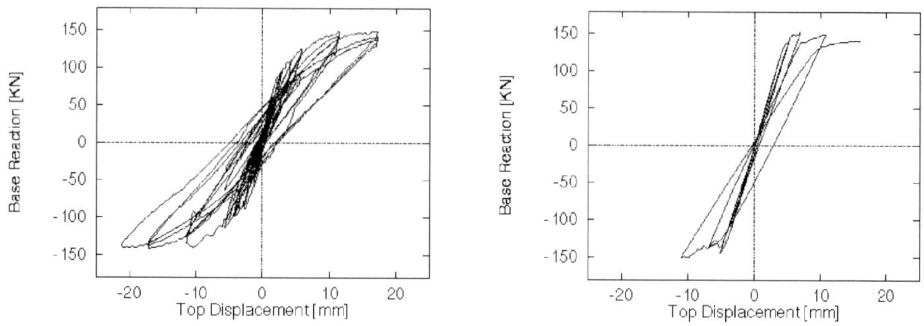

(a) (b)

Figure 1.11: Real masonry wall: experimental (a) and numerical (b) response.

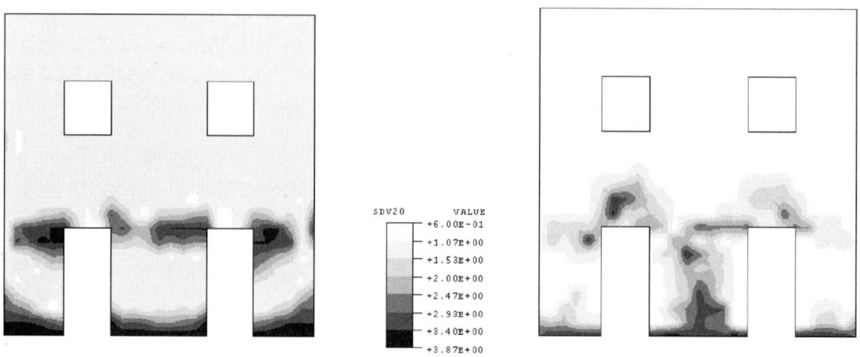

(a) (b)

Figure 1.12: Distribution of tensile (a) and shear (b) damage.

1.5.1 The damage model for bricks and mortar

The model conceived for brittle, isotropic materials is characterised by three internal variables D_T, D_C and D_S which describe the damage induced by tensile, compressive and shear stresses, respectively. The proposed model can satisfactorily reproduce the recovery of initial stiffness that occurs in these materials when a compressive load is applied after a first cycle in tension [35]. This phenomenon is called "unilateral effect", accordingly, the damage model described here will be defined as a unilateral damage model. Moreover, this model is also able to capture damage effects due to the presence of shear stresses of the same order of magnitude as the normal stresses.

The introduction of D_T, D_C and D_S permits taking into account the different behaviour in tension and compression typical of rock-like materials. These variables vary between 0 ("virgin" material) and 1 (completely fractured volume element).

Calling the stresses obtained from the strains through the undamaged elastic stiffness matrix "elastic stresses" and considering a suitable partition of the stress tensor, the evolution law of D_T is defined as a function of an equivalent elastic tensile stress

$$\hat{\sigma}^+ = \sqrt{\left\langle \sigma_{11}^{el} \right\rangle_+^2 + \left\langle \sigma_{22}^{el} \right\rangle_+^2 + \left\langle \sigma_{33}^{el} \right\rangle_+^2} \tag{1.22}$$

where the superscript "el" means elastic and the brackets $\langle * \rangle_+$ represent the positive part of $*$. In the same manner, the evolution laws of D_C and D_S are functions of an equivalent elastic compressive stress $\hat{\sigma}^-$ (defined similarly to $\hat{\sigma}^+$, see Eq. (1.22)) and an equivalent elastic shear stress $\hat{\tau}$, defined as:

$$\hat{\tau} = \sqrt{\left(\tau_{12}^{el}\right)^2 + \left(\tau_{23}^{el}\right)^2 + \left(\tau_{13}^{el}\right)^2} \tag{1.23}$$

To reproduce the unilateral effects, two independent damage variables D_{TS} and D_{CS} are defined as convenient weighted averages of the variables D_T, D_C and D_S; the weights depend upon the load condition. To be more specific, referring to a plane stress state ($\tau = \tau_{12}^{el}$), D_{TS} and D_{CS} are given by:

$$D_{TS} = \frac{\left|tr\underline{\sigma}^{el}\right|}{Z} D_T + \frac{|\tau|}{Z} D_S \quad \text{if } tr\underline{\sigma}^{el} \geq 0 \tag{1.24}$$

$$D_{CS} = \frac{\left|tr\underline{\sigma}^{el}\right|}{Z} D_C + \frac{|\tau|}{Z} D_S \quad \text{if } tr\underline{\sigma}^{el} \leq 0 \tag{1.25}$$

where

$$Z = \left|tr\underline{\sigma}^{el}\right| + |\tau| \tag{1.26}$$

It is possible to note that in uniaxial tension (compression) $D_{TS} = D_T$ $(D_{TS} = 0)$ and $D_{CS} = 0$ $(D_{CS} = D_C)$. If a pure shear test is considered, the situation is described by $D_{TS} = D_{CS} = D_S$.

Hence, the stress-strain relation can be written in the following form:

$$\underline{\varepsilon} = \underline{K}^{-1}\underline{\sigma}\left\{\frac{1}{(1 - D_{TS})}\frac{\langle tr\underline{\sigma}\rangle_+}{tr\underline{\sigma}} + \frac{1}{(1 - D_{CS})}\frac{\langle tr\underline{\sigma}\rangle_-}{tr\underline{\sigma}}\right\} \tag{1.27}$$

where $\underline{K}$ is the material stiffness matrix.

The evolution laws of the damage variables D_T, D_C and D_S are defined by introducing three threshold values $\hat{\sigma}_0^+$, $\hat{\sigma}_0^-$ and $\hat{\tau}_0$. When one of the equivalent elastic stresses attains its threshold value, an increment in the corresponding damage variable may develop according to the relevant evolution law. Then, it is possible to calculate the values of D_{TS} and D_{CS}, the rates of which can be only positive [36].

The evolution laws for D_T, D_C and D_S are governed by the following relationships:

$$D_T = 1 - \frac{\hat{\sigma}_0^+ \left(\hat{\sigma}_u^+ - \hat{\sigma}^+\right)}{\hat{\sigma}^+ \left(\hat{\sigma}_u^+ - \hat{\sigma}_0^+\right)} \tag{1.28}$$

$$D_C = \begin{cases} 1 - \frac{\hat{\sigma}_0^-}{\hat{\sigma}^-} & \text{if } \hat{\sigma}^- \le \hat{\sigma}_{u1}^- \\ 1 - \frac{\hat{\sigma}_0^- \left(\hat{\sigma}_{u2}^- - \hat{\sigma}^-\right)}{\hat{\sigma}^- \left(\hat{\sigma}_{u2}^- - \hat{\sigma}_{u1}^-\right)} & \text{if } \hat{\sigma}^- > \hat{\sigma}_{u1}^- \end{cases} \tag{1.29}$$

$$D_S = 1 - \frac{\hat{\tau}_0 \left(\hat{\tau}_u - \hat{\tau}\right)}{\hat{\tau} \left(\hat{\tau}_u - \hat{\tau}_0\right)} \tag{1.30}$$

where $\hat{\sigma}_0^+$,$\hat{\sigma}_u^+$, $\hat{\sigma}_0^-$, $\hat{\sigma}_{u1}^-$, $\hat{\sigma}_{u2}^-$, $\hat{\tau}_0$ and $\hat{\tau}_u$ are the threshold and limit values defining the response of the model when applied to the simulation of uniaxial tension-compression and pure shear tests. The relevant stress-strain curves are assumed to be piece-wise linear, as shown in Figures 1.13a,b respectively. In detail, the threshold values $\hat{\sigma}_0^+$, $\hat{\sigma}_0^-$ and $\hat{\tau}_0$ are the peak values in the relevant stress-strain curves. The corresponding elastic strain values are ε_0^+, ε_0^- and γ_0. $\hat{\sigma}_u^+$, $\hat{\sigma}_{u2}^-$ and $\hat{\tau}_u$ are stresses associated to the limit strains ε_u^+, ε_{u2}^- and γ_u through the elastic constitutive law. Finally, $\hat{\sigma}_{u1}^-$ is the stress corresponding to ε_{u1}^- , that is, the strain at which the softening branch in compression starts. As Figure 1.13 shows, the model does not allow the presence of permanent strains to be taken into account.

Using piece-wise linear stress-strain curves it is easily possible to limit mesh-dependence problems of the finite element analyses. Indeed, when a cohesive damage (such as the present one) is used to analyse a structural system discretised with a suitable number of finite elements, the structural response generally depends on the size of the finite elements employed. To circumvent this mesh-dependence problem, the fracture energy value must remain constant under changes in size of the finite elements [37,38]. By assuming that stresses, strains and displacements are continuous in each finite element, the fracture energy G_f can be obtained by multiplying a characteristic length l_C of any finite element by the material energy dissipated per unit volume g_f, which corresponds to the area subtended by the stress-strain plot of the material. Consequently, it must increase or decrease with decreasing or increasing element size to maintain the value of G_f constant [39]. Then, it is convenient to introduce the piece-wise linear stress-strain curves expressed as functions of a suitable parameter α, which represents the ratio between a material characteristic length l and the length l_C of the finite element. The evolution law for D_T can be written as:

$$D_T = 1 - \frac{\hat{\sigma}_0^+ \left(\alpha\hat{\sigma}_u^+ - \hat{\sigma}^+\right)}{\hat{\sigma}^+ \left(\alpha\hat{\sigma}_u^+ - \hat{\sigma}_0^+\right)} \tag{1.31}$$

$$\alpha = \frac{l}{l_C} \tag{1.32}$$

Since the definition of the fracture energy G_f is

$$G_f = l_C g_f \tag{1.33}$$

$$g_f = \int_0^\infty \sigma(\varepsilon)\, \mathrm{d}\varepsilon \tag{1.34}$$

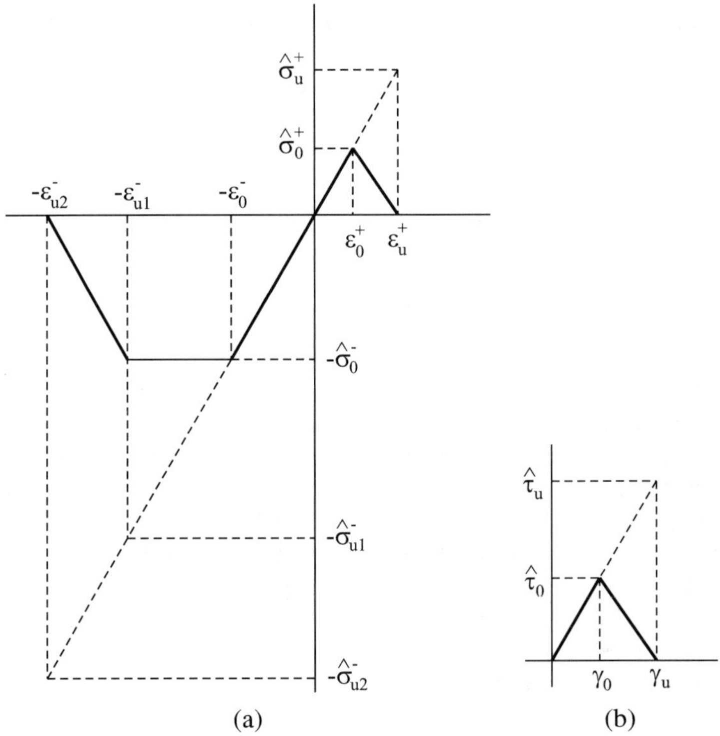

Figure 1.13: Stress-strain curves in uniaxial tension-compression (a) and shear (b)

the parameter α can be obtained by substituting the stress-strain law for uniaxial tension in equation (1.34). The parameter α becomes:

$$\alpha = \frac{l}{l_C} = \frac{2G_f E}{\hat{\sigma}_u^+ l_C \hat{\sigma}_0^+} \tag{1.35}$$

where E is the material elastic modulus and l_C is a function of the dimensions of the finite element. The value of l depends upon the size of the representative volume element, *i.e.*, the smallest volume element that has to be repeated periodically to reproduce the structure. To avoid the snap-back phenomenon, the value of the characteristic length l_C of the finite element must be always greater than the critical value l_{Cmin}. By imposing this condition of critical stability, l_{Cmin} is equal to

$$l_{Cmin} = \frac{2G_f E}{\left(\hat{\sigma}_0^+\right)^2} \tag{1.36}$$

With the same procedure, it is possible to write the evolution laws for D_C and D_S as a function of suitable parameters.

1.5.2 The homogenisation procedure

In principle, the damage model presented above can also be applied to the study of masonry structures, although it is not able to take into account the orthotropic behaviour which is typical of masonry. This problem can be circumvented by obtaining the constitutive law of masonry starting from the mechanical properties of its components, *i.e.* mortar and bricks. The aim is to provide the stiffness matrix of a fictitious homogeneous material (which is mechanically equivalent to masonry) through a homogenisation technique, considering mortar and bricks as homogeneous materials undergoing damage phenomena. These effects are described by using the unilateral damage model of the previous Section. Obviously, the model parameters will be different for each of the two materials.

The application of any homogenisation procedure requires the definition of the representative volume element (RVE) of masonry. A possible choice for the RVE is shown in Figure 1.14 where the dashed area indicates the representative volume element considered. By neglecting the vertical mortar joints [40], the nine members of the homogenised stiffness matrix are obtained by means of the method of layers [41]. This procedure is based on the hypothesis that the displacement field is periodic along the direction of layers (in our case, the axis x_2) and constant along the other two directions. Details on the application of this method are reported in Reference [42].

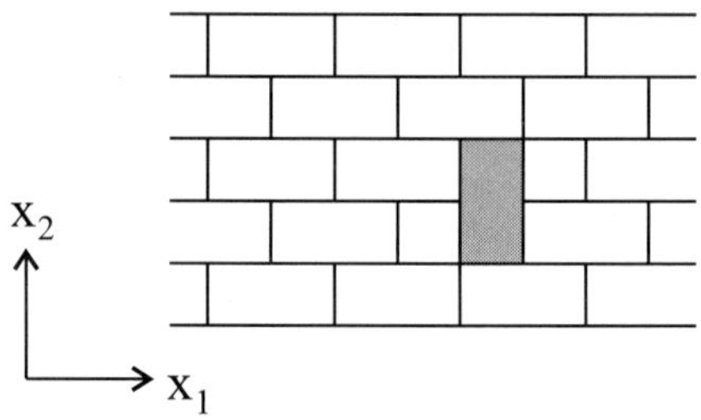

Figure 1.14: Representative volume element of masonry.

It is important to note that the terms of the stiffness matrix that do not refer to the axis of periodicity x_2 vary almost linearly with damage in mortar and bricks and vanish only when the mortar and bricks damage parameters are both equal to one. However, the components referred to the periodicity axes exhibit a non-linear dependence on the damage parameters, and their values are also equal to zero when either of the two damage parameters is equal to one.

Since the volume fractions of mortar and bricks are involved in the definition of the RVE, the homogenisation procedure allows the geometric features of masonry to be taken into account. It is possible to note, for example, that the stiffness matrix components decrease in value as the size of the mortar joints increase.

Starting from the components, the stiffness matrix of masonry under plane stress conditions can be obtained. If this material model is to be implemented into a finite element code for the analysis of in-plane loaded structural elements, it is important to underline that at every Gauss point the stresses in the mortar and in the bricks have to be computed on the basis of the overall stresses. These local stresses provide the damage parameters for each material component, thus allowing the evaluation of the damaged-elastic stiffness matrix of the masonry [43].

The considerable computational effort required by this numerical procedure justifies the assumption of neglecting the vertical mortar joints. If this were not the case, it would be necessary to apply the method of layers (and, consequently, to compute the local stresses) three times.

Finally, it is important to emphasise that by this model it is possible to determine if the local failure at every Gauss point is due to the mortar or to the bricks. Indeed, when the damage in the mortar or the bricks exceeds its critical value, the new stress increment is carried by the bricks or the mortar respectively and/or through a stress redistribution.

1.5.3 Numerical analyses

To check the mesh-independence of the numerical results of the unilateral damage model, a masonry panel was analysed using three different meshes: the number of elements of the finest mesh (mesh c) is four times the number of elements in the second mesh (mesh b), which, in turn, is four times the number of elements in the coarser mesh (mesh a). The panel is fixed at the bottom and subjected to a constant evenly distributed vertical load and to a horizontal increasing load applied at the top left corner. Figure 1.15 shows that the differences in the three plots of the horizontal force vs. the horizontal top displacement are negligible.

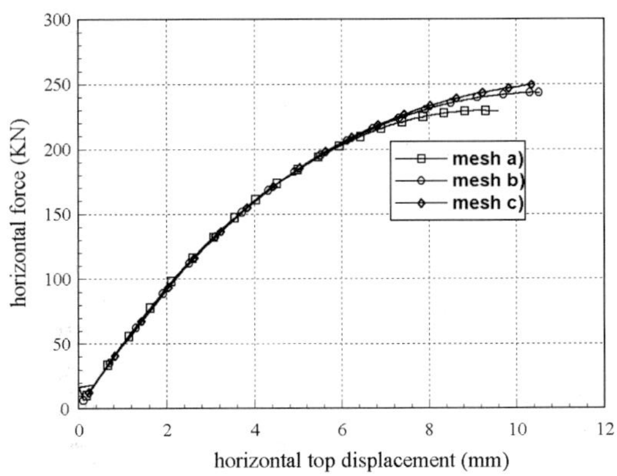

Figure 1.15: Horizontal force vs. horizontal top displacement of a masonry panel.

The same panel was analysed considering a cyclic history for the horizontal load applied at the top left corner. The horizontal force vs. the horizontal top displacement plotted in Figure 1.16 shows the capability of the model effectively to take the unilateral behaviour of masonry into account. Moreover this curve is qualitatively in good agreement with the experimental results reported in Figures 1.7 and 1.8a. Note that the bearing capacity in terms of the horizontal force and the relevant displacement are in good agreement with the experimental findings. However, the model is not able to capture the experimental hysteretic effect, since it does not take permanent strains and sliding at the brick-mortar interfaces into account.

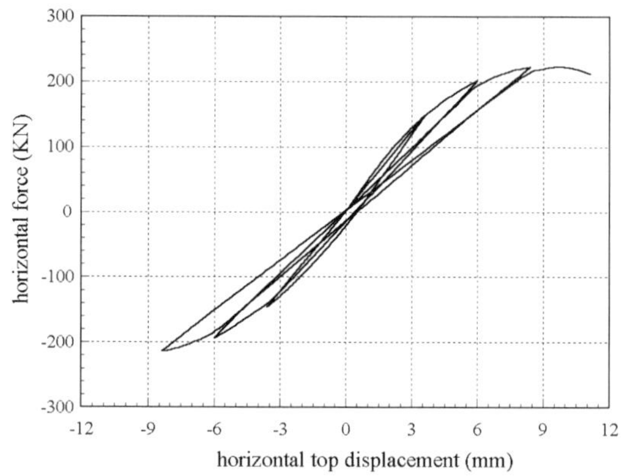

Figure 1.16: Horizontal cyclic force vs. horizontal top displacement.

Moreover, the experimental response of a miniaturised wall, which correspond to the real wall described in Section 1.4.2, was considered. This panel was built and tested at the Department of Structural Engineering of the Politecnico di Milano, by using a convenient test rig suitable for biaxial load conditions [44]. The panel size is $1/8$ of the real wall and is subjected to the same load conditions. Obviously, the value of the dead vertical loads is equal to $1/64$ of the value of the vertical loads applied to the real wall. To use the numerical model, the values of the material parameters were determined by using the experimental results from tests on specimens of mortar and bricks nominally identical to the ones used in building the wall [45]. Their values are reported in Reference [42]. Figures 1.17a and b show a comparison of the experimental and the numerical results in terms of total horizontal force vs. horizontal displacement at the left corner of the second floor. The computed failure load is almost identical to the ultimate experimental load, and the shape of the curves are quite similar.

Finally, Figures 1.18a, b, c and d show the contour lines of D_{TS} and D_{CS} for the mortar (a,b) and the bricks (c,d). Damage is greater in the mortar than in the bricks and, for both materials, the damage D_{TS} (which is much more widespread than the damage D_{CS}) is localised at the bottom of the wall and in the zones around the doors. These distributions reproduce the experimental collapse of the miniaturised wall. Indeed, in the experimental test on this wall, contrary to the real wall, shear

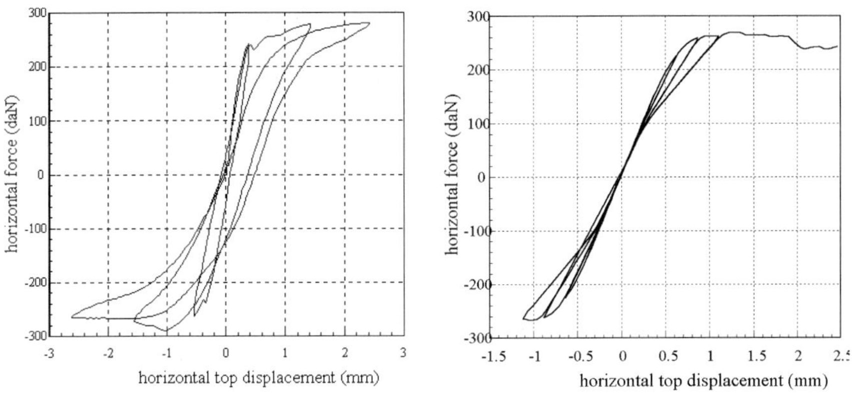

(a) (b)

Figure 1.17: Experimental data (a) and numerical results (b) of a miniaturised wall.

cracks are not present in the central zone of the ground floor and the collapse is due only to a bending mechanism [42].

1.6 Conclusions

The existing numerical models for masonry have been divided into two groups: the heterogeneous models and the homogeneous models. The heterogeneous models analyse the masonry walls discretising bricks and mortar separately through finite element and/or interface elements. A suitable constitutive law is then assumed for each component. In this way it is possible to take into account, with particular accuracy, the characteristics of mortar joints, which play a very important role in the global behaviour of masonry. The principal limitation of these models consists of the high computational effort they require, especially when a real wall or building must be analysed.

In the homogeneous models, masonry is assumed as a continuum material: its properties are either obtained by introducing suitable constitutive laws able to represent the directional behaviour of masonry, or by starting from the characteristics of the masonry components, *i.e.*, mortar and bricks, and using a suitable homogenisation procedure. In this way, real masonry buildings can be studied with reasonable computational effort. Moreover, considering the masonry as a composite material, it is possible to detect the occurrence of failure either in the bricks or in the mortar joints, although these components are not discretised separately.

After a brief review of these two categories of models, two different homogeneous models have been presented. Both have been conceived to study the behaviour of masonry panels and walls subjected to horizontal cyclic loads.

In the first model, masonry is considered as an orthotropic material for which a

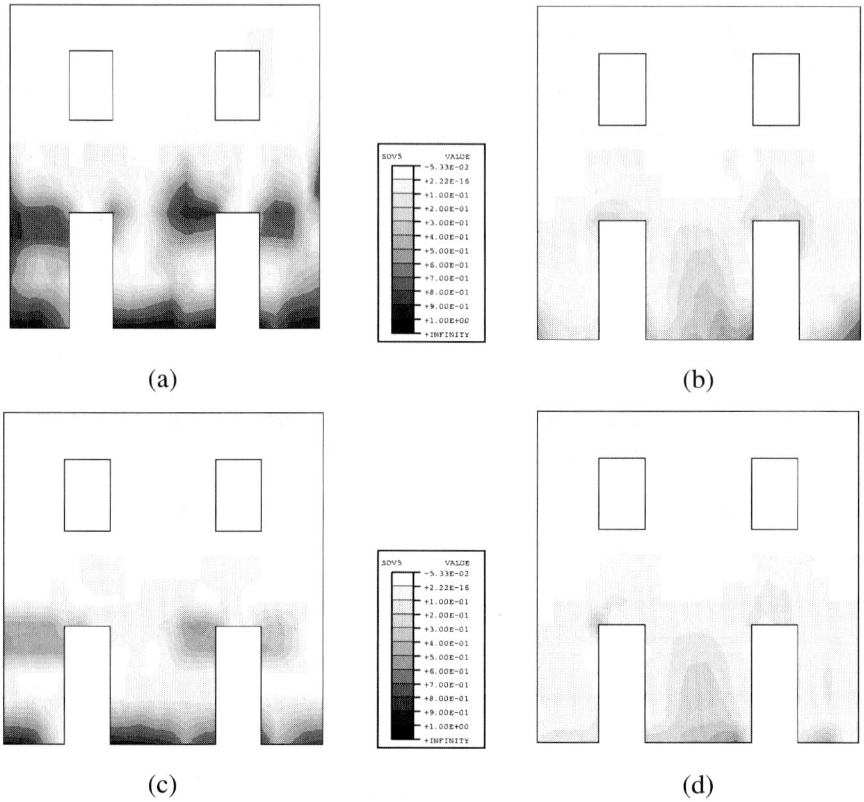

Figure 1.18: Distribution of D_{TS} (a) and D_{CS} (b) for the mortar and distribution of D_{TS} (c) and D_{CS} (d) for the bricks.

suitable piecewise linear yield surface is defined. For each plane, a hardening/softening law is introduced and the evaluation of stress in the masonry is performed by solving a Linear Complementarity Problem. The material damage is taken into account considering the change in the stiffness matrix as a function of the inelastic deformations. Several numerical analyses on masonry panels and walls have been performed and the results are in good agreement with the experimental results. Further developments for the model will be implementing more realistic damage laws according to the different damage mechanisms and considering the effects of the out-of-plane displacements on the damage distribution under seismic actions.

In the second model, the mechanical properties of masonry have been obtained starting from the characteristics of the material components, through a suitable homogenisation procedure. The mechanical behaviour of the bricks and the mortar joints is described through a unilateral damage model based on the introduction of three damage variables related to simple tension, uniaxial compression and shear stresses. These three variables are multiplied by convenient weights and linearly combined, yielding two independent damage variables: in this way it is possible to describe the behaviour of brittle materials subjected to alternating tensile-compressive cyclic loads. For this model, the computational effort is reasonable, provided the vertical mortar joints are neglected: consequently it is possible to simulate unreinforced masonry walls of any size. The simulations of tests on masonry panels and on miniaturised walls show a satisfactory agreement between the numerical and the experimental data. It is important to note that the results given by the model do not depend on the size of the mesh used to discretise the structure. Moreover, since the evaluation of the overall mechanical properties of masonry requires estimating stresses and damage in mortar and bricks, this model allows for it to be established in which of the components failure takes place. As regards possible future developments of the model, it seems important to take the presence of permanent strains into account and to reproduce the damage growth due to repeated cycles.

References

[1] Maier G. and Nappi A., "A theory of perfectly no-tension discretized structural system", Eng. Struct., 12, 229-233, 1990.

[2] Panzeca T. and Polizzotto C., "Constitutive equations for no-tension materials", Meccanica, 23, 88-93, 1988.

[3] Baratta A.,"Metodi di calcolo per strutture non reagenti a trazione", in "Proceedings of 7th National Congress AIMETA", Trieste, Italy, 275-286, 1984.

[4] Shing P.B., "In-plane resistance of reinforced masonry shear walls", J. of Struct. Eng. ASCE, 116, 619-640, 1990.

[5] Dhanasekar M., Kleeman P.W. and Page A.W., "Biaxial stress-strain relations for brick masonry", J. Struct. Div., 5, 1085-1100, 1985.

[6] Hendry A.W., "Structural brickwork", The Macmillan Press, Hong Kong, 1981.

[7] Biolzi L., "Evaluation of compressive strength of masonry walls by limit analysis", J. Struct. Eng., 114(10), 2179-2189, 1988.

[8] Franciosi V., "An approach to the limit design of masonry arches under seismic loads", in "Plasticity Today", Sawczuk A. and Bianchi G., (Editors), Elsevier Appl. Sc. Pubbl., London, 821-834, 1985.

[9] Page A.W., "Finite element model for masonry", J. Struct. Div. ASCE, 104, 1267-1285, 1978.

[10] Lourenço P.B., "Computational strategies for masonry structures", PhD. Thesis, Delft Univ. of Technology, 1996.

[11] Lourenço P.B. and Rots J.G., "Implementation of an interface cap model for the analysis of masonry structures", in "Computational Modelling of Concrete Structures", Mang H., Bicanic N. and de Borst R., (Editors), 123-134, 1994.

[12] Lotfi H.R. and Shing P.B., "Interface model applied to fracture of masonry structures", J. Struct. Engrg. ASCE, 120, 63-80, 1994.

[13] Gambarotta, L. and Lagomarsino, S., "Damage models for the seismic response of brick masonry shear walls. Part I: the mortar joint model and its applications", Earth. Eng. and Struct. Dynamics, 26, 423-439, 1997.

[14] Pande G.N., Liang J.X. and Middleton J., "Equivalent elastic moduli for brick masonry", Computer & Geotechnics, 8, 243-265, 1989.

[15] Middleton J., Pande G.N., Liang J.X. and Kralj B., "Some recent advances in computer methods in structural masonry", in "Computer Methods in Structural Masonry", Middleton J. and Pande G.N. (Editors), Books and Journals International, Swansea, U.K., 1-21, 1991.

[16] Maier G., Papa E. and Nappi A., "On damage and failure of brick masonry", in "Experimental and Numerical Methods in Earthquake Engineering", Kluwer, Dordrecht, 223-245, 1991.

[17] Chen W.F., "Plasticity in reinforced concrete", McGraw-Hill, 1982.

[18] Ganju T.N., "Non-linear finite element analysis of clay brick masonry", in "Proceedings of 6th Australian Conf. on the Mechanics of Structures and Materials", Univ. of Canterbury, 59-65, 1977.

[19] Shing P.B., Klamerus E., Spaeh H. and Noland J.L., "Seismic performance of reinforced masonry shear walls", in "Proceedings of 9th World Conf. on Earthquake Engineering", Japan, 103-108, 1988.

[20] Samarashinghe W., Page A.W. and Hendry A.W., "A finite element model for the in-plane behaviour of brickwork", Proc. of Inst. of Civ. Engrs., 72, 171-178, 1982.

[21] Samarashinghe W. and Hendry A.W., "The strength of brickwork under biaxial tensile and compressive stress", in "Proceedings of 7th Int. Symposium on Load Bearing Brickwork", London, 1980.

[22] Page A.W., Kleeman P.W. and Dhanasekar M., "An in-plane finite element model for brick masonry", in "New Analysis Techniques for Structural Masonry", Anand S.C. (Editor), ASCE Press, New York, 1-18, 1985.

[23] Dhanasekar M., Page A.W. and Kleeman P.W., "The failure of brick masonry under biaxial stresses", Proc. Inst. Civ. Engrs., 79, 295-313, 1985.

[24] Contro R. and Sacchi Landriani G., "Plane stress analysis for anisotropic no-tension materials with hardening compression", Eng. Frac. Mech., 21, 947-955, 1985.

[25] Gambarotta L. and Lagomarsino S., "Damage models for the seismic response of brick masonry shear walls. Part II: the continuum model and its applications", Earth. Eng. and Struct. Dynamics, 26, 441-462, 1997.

[26] Pietruszczak S. and Niu X., "A mathematical description of macroscopic behaviour of brick masonry", Int. J. of Solids & Struct., 29(5), 531-546, 1992.

[27] Maier G., "A matrix structural theory of piecewise linear elastoplasticity with

interacting yield planes", Meccanica, 5, 54-66, 1970.

[28] Nappi A., Facchin G. and Marcuzzi C., "Structural dynamics: convergence properties in the presence of damage and applications to masonry structure", Struct. Eng. & Mech., (to appear).

[29] Callerio A. and Papa E., "An elastic-plastic model with damage for cyclic analysis of masonry panels", in "Computer Methods in Structural Masonry (STRUMAS IV)", Pande G.N., Middleton J. and Kralj B. (Editors), E&F Spon, 19-26, 1998.

[30] Ortiz M., "A constitutive theory for the inelastic behaviour of concrete", J. Mech. Mat., 4, 67-93,1985.

[31] Mangasarian O.L., "Solutions of symmetric linear complementarity problems by iterative methods", J. of Optimisation Theory & Appl., 22, 465-485, 1977.

[32] Callerio A., "Comportamento sismico delle murature negli edifici monumentali: un metodo di analisi basato sulla meccanica del danneggiamento", PhD. Thesis, Dep. of Structural Engineering, Politecnico of Milan, Italy, 1998 (in Italian).

[33] Magenes G., "Comportamento sismico di murature di mattoni: resistenza e meccanismi di rottura di maschi murari", Ph.D. Thesis, Dept. of Structural Mechanics, Univ. of Pavia, 1992 (in Italian).

[34] Calvi G.M. and Magenes G., "Experimental research on response of URM building system", in "Guidelines for Seismic Evaluation and Rehabilitation of Unreinforced Masonry Buildings", Abrams D.P. et al. (Editors), State University of New York at Buffalo, 3/41-57, 1994.

[35] Mazars J., Berthaud Y. and Ramtani S., "The unilateral behaviour of damaged concrete", Eng. Fracture Mech., 35, 629-635, 1990.

[36] Lemaitre J. and Chaboche J.L., "Mécanique des matériaux solides", Dunod, Paris, 1985.

[37] Bazant Z.P. and Cedolin L., "Blunt crack band propagation in finite element analysis", J. Eng. Mech. ASCE, 105, 297-315, 1979.

[38] Bazant Z.P. and Cedolin L., "Fracture mechanics of reinforced concrete", J. Eng. Mech. ASCE, 106, 1287-1306, 1980.

[39] Koubaa B., Papa E. and Nappi A., "Local damage models suitable for preventing mesh-dependence effects", in "Computational Plasticity. Fundamentals and Applications", Owen D.R.J., Onate E., Hinton E., (Editors), Pineridge Press, 1131-1142, 1995.

[40] Papa E., "Sulla meccanica del danneggiamento con particolare riferimento alle murature", PhD. Thesis, Dept. Eng. Struct., Politecnico, Milan, 1990.

[41] Sanchez-Palencia E., "Non-homogeneous media and vibration theory", Springer-Verlag, Berlin, 1980.

[42] Papa E., "A unilateral damage model for masonry based on a homogenisation procedure", Mech. of Cohesive-Frictional Mat., 1, 349-366, 1996.

[43] Papa E. and Nappi A., "A numerical approach for the analysis of masonry struc-

tures", Masonry International, 7, 18-24, 1993.

[44] Papa E. and Nappi A., "A test rig for the application of biaxial cyclic loads to structural models", Materials and Structures, 28, 299-307, 1995.

[45] Binda L., Tiraboschi C., Mirabella Roberti G., Baronio G. and Cardani G., "Experimental and numerical investigation on a brick masonry building prototype", in "Report 5.0: Measuring masonry materials properties: detailed results from an extensive experimental research", Dept. of Struct. Eng., Politecnico of Milan, 1995.

©2001, Saxe-Coburg Publications, Stirling, Scotland
J.W. Bull, (Editor), Computational Modelling of Masonry, Brickwork and Blockwork Structures,
Saxe-Coburg Publications, Stirling, Scotland 27-52.

Chapter 2

Formulation of Elastic-plastic Joint Elements and their Application to Practical Structures

T. Aoki

School of Design and Architecture, Nagoya City University, Japan

Abstract

The first part of this chapter is devoted to the formulation of the elastic-plastic joint element to determine the structural behavior and characteristics of masonry structures[1]. In masonry structures, due to the material properties, only compressive stress is assumed to exist. In the second part, to ensure the effectiveness of the joint element, the case of plain concrete/mortar under combined stress and the case of soil where slippage occurs under a footing, are investigated.

2.1 Introduction

Stone and/or brick has been a construction material in Europe for over a thousand years. There are a great number of masonry structures, and eminent examples, especially in Italy – are the Pantheon in Rome (2nd century), Santa Maria del Fiore in Florence (15th century), and the San Pietro in Rome (16th century). These structures were constructed mainly as religious monuments and became some of the most important buildings representing the period, the city, the style, and the culture. Unfortunately, the stability of many of these structures is threatened by growing fractures. During the restoration work (1980 to 1995) of the dome of Santa Maria del Fiore,

[1]This chapter is developed from reference [1] (in Japanese).

the data concerning temperature, humidity, and displacement were measured both inside and outside this huge structure. A number of structural analyses based on the Finite Element Method were also carried out. The restoration work, however, was implemented from only an artistic point of view.

In general, when materials are gradually loaded beyond their elastic capacity, plastic deformation occurs and the materials separate into a few pieces. Slippage begins, fractures develop, and the discontinuous deformation begins, followed by collapse. The analytical method used to determine these phenomena is not unusual. There are several methods for analyzing and calculating these phenomena. As has been shown by the analysis of discontinuous rocks, the joint element is effectively modelled for analyzing structures composed of two different materials with very different strengths. The joint's behavior can be predicted in the follow situation:

1. A brick and mortar structure such as the Florentine Duomo [2–4], Hagia Sophia [5], and Isahaya arch bridge [6],

2. The interface between a soil and a concrete structure, and

3. A slippage plane within a soil.

Joint elements have been studied and developed in the field of Rock Engineering and Soil Mechanics [7–11]. These joint elements are often used to express non-linear behavior, as in the above three cases.

2.2 The Formulation of Elastic-plastic Joint Elements (Model I: Truss members breaking in tension)

Figure 2.1 shows a brick and mortar structure, where the joint element is the mortar between the bricks. The elastic-plastic joint element of the mortar truss members in a two-dimensional situation is illustrated schematically in Figure 2.2 as a composite model. By introducing a suitable number of members and assigning different material characteristics to each, a variety of sophisticated composite actions can be obtained. But such mechanical parameters must be determined experimentally. However, for reasons of representation, we have simplified the model (Figures 2.2 and 2.3). The joint element consists of three members forming a truss structure: one vertical and two diagonal members where φ_0 is an angle between the vertical and the diagonal members. Each member is assumed to have the same thickness, t, and the same height, h. However, each member can have different material properties: E_{sd}, E_{sv} the stiffness, σ_{td} , σ_{tv} the tensile strength, and σ_{cd}, σ_{cv} the compressive strength of the diagonal and the vertical members, respectively. The stress-strain relationships of all three members in tension and compression are shown in Figure 2.3. We see that beyond the proportional limits σ_{cd} and σ_{cv}, the stress-strain lines under compression are assumed straight lines parallel to the ε-axis, but beyond the proportional limits σ_{td} and σ_{tv}, the stress-strain lines under tension are assumed to soon become zero.

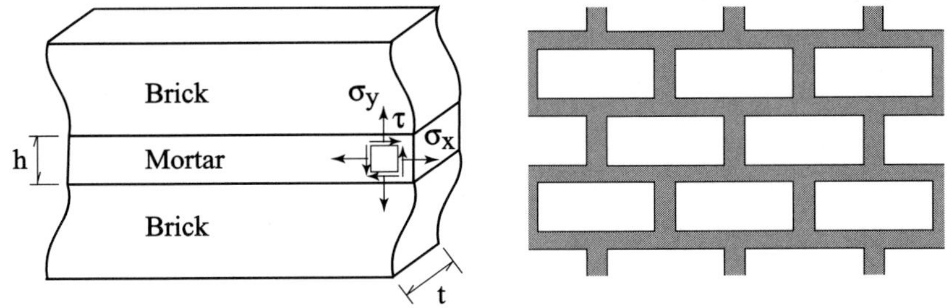

Figure 2.1: Joint element (mortar) and brick structures.

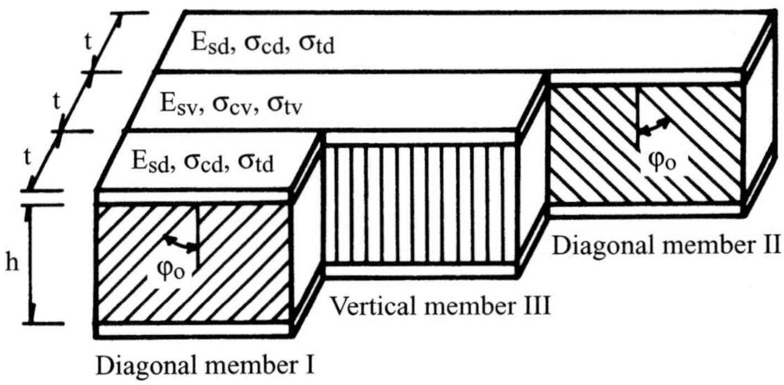

Figure 2.2: Elastic-plastic joint element composed of three truss-like members.

Figure 2.4 shows the vertical displacement, V, and the horizontal displacement, U, of the joint element when it is subjected to the normal stress σ and the shear stress τ. There is a different stress field (σ_I, σ_{II}, σ_{III}) in each member which contributes to the total stress field of σ and τ. That is, the total stress field is obtained by summing the different contributions of each member. Consequently, we can determine easily the stresses of truss members σ_I, σ_{II}, σ_{III}, and the relationship between the stresses σ and τ, by means of equilibriums which are given by Equation (2.1).

$$\begin{aligned} \sigma &= (\sigma_I + \sigma_{II}) \cos^2 \varphi_0 + \sigma_{III} \\ \tau &= (\sigma_I - \sigma_{II}) \sin \varphi_0 \cos \varphi_0 \end{aligned} \tag{2.1}$$

The corresponding strains of truss members $\varepsilon_I, \varepsilon_{II}, \varepsilon_{III}$ are obtained by Equation (2.2),

$$\begin{aligned} \varepsilon_I &= \sin \varphi_0 \cos \varphi_0 \gamma + \cos^2 \varphi_0 \varepsilon \\ \varepsilon_{II} &= -\sin \varphi_0 \cos \varphi_0 \gamma + \cos^2 \varphi_0 \varepsilon \\ \varepsilon_{III} &= \varepsilon \end{aligned} \tag{2.2}$$

where $\varepsilon = V/h$ and $\gamma = U/h$. Here, we assume that the joint element does not stretch in the x-direction in Figure 2.4.

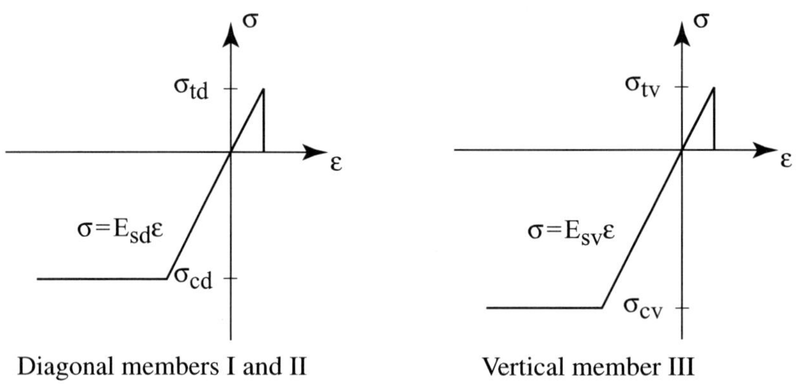

Figure 2.3: Stress-strain relationships of members.

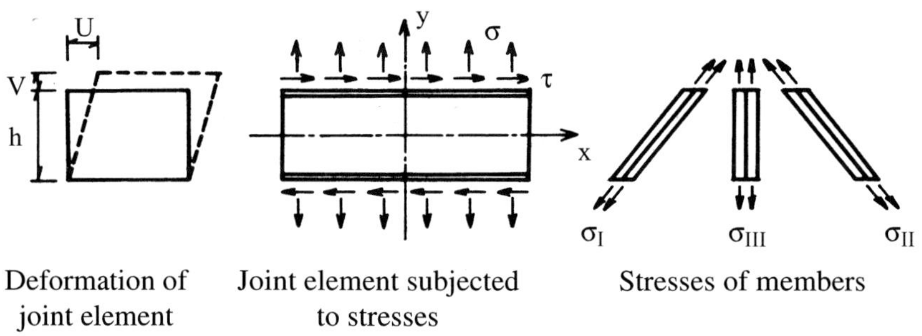

Figure 2.4: Joint element subjected to stresses

2.2.1 Angle φ_0 and Stiffness of E_{sd} and E_{sv}

Let us consider the angle φ_0 between the vertical and the diagonal members as well as the stiffness of the vertical member, E_{sv}, and of the two diagonal members, E_{sd}. The joint element is assumed to be an isotropic material with Young's modulus, E, and Poisson's ratio, ν. It is again assumed that the joint element does not stretch in the x-direction, which is to say $\varepsilon_x = 0$.

In the case of plane strain ($\varepsilon_z = 0$), (for example, a long column subjected to a uniform load in the normal direction) we have the stress-strain relationship under the conditions of $\varepsilon_x = 0$ and $\varepsilon_y = \varepsilon$ given by Equation (2.3).

$$\sigma = \frac{E(1-\nu)}{(1+\nu)(1-2\nu)}\varepsilon, \qquad \tau = \frac{E}{2(1+\nu)}\gamma \tag{2.3}$$

From Equations (2.1), (2.2) and (2.3), we find both the angle φ_0 and the stiffness of E_{sd} and E_{sv}, expressed by Equation (2.4).

$$\begin{aligned} \tan^2\varphi_0 &= \frac{\dfrac{1}{2(1+\nu)}}{\dfrac{(1-\nu)}{(1+\nu)(1-2\nu)} - v} \\ d &= \frac{E_{sd}}{E} = \frac{1}{(1+\nu)\sin^2(2\varphi_0)} \\ v &= \frac{E_{sv}}{E} \end{aligned} \tag{2.4}$$

In the case of plane stress ($\sigma_z = 0$), (for example, a thin membrane subjected to a load in its plane) we have the stress-strain relationship under the conditions of $\varepsilon_x = 0$ and $\varepsilon_y = \varepsilon$, as follows.

$$\sigma = \frac{E}{1-\nu^2}\varepsilon, \qquad \tau = \frac{E}{2(1+\nu)}\gamma \tag{2.5}$$

From Equations (2.1), (2.2), and (2.5), we find φ_0, E_{sd}, and E_{sv} given by Equation (2.6).

$$\begin{aligned} \tan^2\varphi_0 &= \frac{\dfrac{1}{2(1+\nu)}}{\dfrac{1}{(1-\nu^2)} - v} \\ d &= \frac{E_{sd}}{E} = \frac{1}{(1+\nu)\sin^2(2\varphi_0)} \\ v &= \frac{E_{sv}}{E} \end{aligned} \tag{2.6}$$

Supposing $\nu = 0.16$, Poisson's ratio, and $\varphi_0 = 45°$, the angle, in Equations (2.4) and (2.6), we arrive at the subsequent stiffness of members d and v.

$$d = \frac{E_{sd}}{E} = 0.862, \qquad v = \frac{E_{sv}}{E} = 0.634 \tag{2.7}$$

for the plane strain ($\varepsilon_z = 0$), and

$$d = \frac{E_{sd}}{E} = 0.862, \qquad v = \frac{E_{sv}}{E} = 0.595 \tag{2.8}$$

for the plane stress ($\sigma_z = 0$).

If the joint element is assumed to be an isotropic material with a Poisson's ratio of $\nu = 0$, both of the stress-strain relationships for the plane strain and for the plane stress become equal, as expressed by Equation (2.9).

$$\sigma = E\varepsilon, \qquad \tau = \frac{1}{2}E\gamma \tag{2.9}$$

From Equations (2.1), (2.2), and (2.9), we find φ_0, E_{sd}, and E_{sv} given by Equation (2.10).

$$\begin{aligned} \tan^2\varphi_0 &= \frac{\frac{1}{2}}{1-\nu} \\ d &= \frac{E_{sd}}{E} = \frac{1}{\sin^2(2\varphi_0)} \\ v &= \frac{E_{sv}}{E} \end{aligned} \tag{2.10}$$

Subsequently, we have the stiffness of members d and v by substitution as the angle $\varphi_0 = 45°$ into Equation (2.10).

$$d = \frac{E_{sd}}{E} = 1.0, \qquad v = \frac{E_{sv}}{E} = 0.5 \tag{2.11}$$

Also, we have the stiffness of the member d and the angle φ_0 by substitution the stiffness of the member v at $v = Esv/E = 0$ into Equation (2.10).

$$d = \frac{E_{sd}}{E} = 1.125, \qquad \varphi_0 \approx 35.26° \tag{2.12}$$

2.2.2 Yielding Condition for Plane Stress

Let us consider the yielding condition for plane stress. In the usual approach, the yielding condition of the joint element is determined by two factors: one takes into account the possible combinations of the order in which the members turn to the plastic phase, and the other gives the appropriate properties of the three members derived from experiments or theories. The appropriate properties include Young's modulus, compressive strength, and tensile strength. However, for reasons of representation, we consider the yielding condition, herein, to simulate the behavior of the thin layer of concrete and mortar, provided by Equation (2.35) [12]. This Equation was determined by the experiment of plain concrete under combined stress [12]. Table 2.1 specifies the characteristic values of the joint element to determine the yielding condition.

(a) When the conditions of $\sigma > 0$ and $\tau = 0$ are applied, the first vertical member III will fail in tension, followed by the diagonal members I and II. Member III will fail when the stresses reach (σ_t, τ).

$$\begin{aligned} \sigma_t &= \left(1 + \frac{2d}{v}\cos^4\varphi_0\right)\sigma_{tv} \\ \tau &= 0 \end{aligned} \tag{2.13}$$

Young's modulus		E	$= E$
Poisson's ratio		ν	$= 0.16$
Strength	Compressive	σ_c	$= \sigma_c$
	Tensile	σ_t	$= 0.1\sigma_c$
Angle		φ_0	$= 45°$
Young's modulus	diag. member	E_{sd}	$= 0.862E$
	vert. member	E_{sv}	$= 0.595E$
Compressive strength	diag. member	σ_{cd}	$= 0.25\sigma_c$
	vert. member	σ_{cv}	$= 0.75\sigma_c$
Tensile strength	diag. member	σ_{td}	$= 0.969\sigma_t$
	vert. member	σ_{tv}	$= 0.579\sigma_t$

Table 2.1: Material constants determining yielding condition.

At the point at which member III fails, members I and II will not yet have reached their failure points (σ_{td}). Consequently, when stress σ_I is $\sigma_I < \sigma_{td}$, and also stress σ_{II} is $\sigma_{II} < \sigma_{td}$, then σ_I is equal to:

$$\sigma_{td} > \frac{d}{v}\cos^2\varphi_0\sigma_{tv} \tag{2.14}$$

After member III fails, members I and II will fail when the stress reaches $\widetilde{\sigma_t}$, where $\widetilde{\sigma_t} < \sigma_t$, given by Equation (2.15).

$$\widetilde{\sigma_t} = 2\sigma_{td}\cos^2\varphi_0 < \sigma_t \tag{2.15}$$

In this case, the stress-strain relationship is as shown in Figure 2.5.

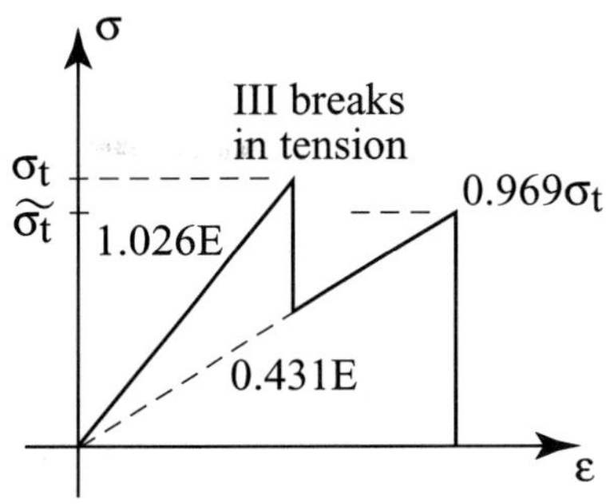

Figure 2.5: Stress-strain relationship in the case of $\sigma > 0$ and $\tau = 0$.

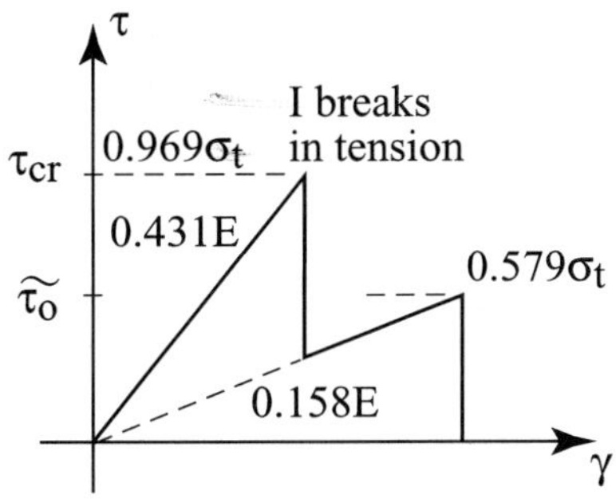

Figure 2.6: Stress-strain relationship in the case of $\sigma = 0$ and $\tau > 0$.

(b) When the conditions of $\sigma = 0$ and $\tau > 0$ are applied, the first diagonal member I will fail in tension, followed by vertical member III. Member I will fail when the stresses reach (σ, τ_{cr}).

$$\begin{aligned} \sigma &= 0 \\ \tau_{cr} &= 2\sigma_{td} \sin \varphi_0 \cos \varphi_0 \end{aligned} \tag{2.16}$$

After member I fails, member III will fail when the stress reaches $\widetilde{\tau_0}$, where $\widetilde{\tau_0} < \tau_{cr}$, as given by Equation (2.17).

$$\widetilde{\tau_0} = \sigma_{tv} \tan^2 \varphi_0 < \tau_{cr} \tag{2.17}$$

From Equations (2.16) and (2.17), we can obtain Equation (2.18).

$$\frac{\sigma_{tv}}{\sigma_{td}} < 2\cos^2 \varphi_0 \tag{2.18}$$

In this case, the stress-strain relationship is as shown in Figure 2.6.

(c) When the conditions of $\sigma < 0$ and $\tau = 0$ are applied, first diagonal members I and II will yield in compression, followed by vertical member III. When the stress-strain relationships of the members are as assumed in Figure 2.3, Equation (2.19) is obtained.

$$\begin{aligned} \sigma_c &= 2\sigma_{cd} \cos^2 \varphi_0 + \sigma_{cv} \\ \tau &= 0 \end{aligned} \tag{2.19}$$

In this case, the stress-strain relationship is as shown in Figure 2.7.

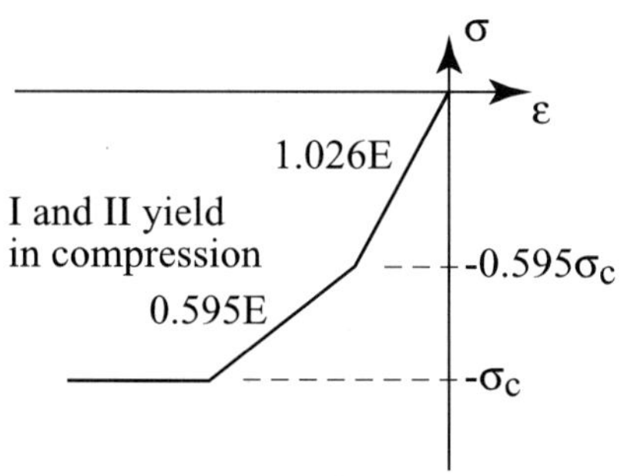

Figure 2.7: Stress-strain relationship in the case of $\sigma < 0$ and $\tau = 0$.

(d) When the conditions of $\sigma \neq 0$ and $\tau > 0$ are applied, the first diagonal member I will fail in tension. Member I will fail when the stress reaches τ as given by Equation (2.20).

$$\tau = 2\sigma_{td} \sin \varphi_0 \cos \varphi_0 - \frac{2\frac{d}{v} \sin \varphi_0 \cos^3 \varphi_0}{1 + 2\frac{d}{v} \cos^4 \varphi_0} \sigma \tag{2.20}$$

At the point at which member I fails, members II and III will not yet have reached their yielding and failure points. Consequently, when stress σ_{II} is $-\sigma_{cd} \leq \sigma_{II} \leq \sigma_{td}$, and stress σ_{III} is $-\sigma_{cv} \leq \sigma_{III} \leq \sigma_{tv}$, then:

$$\begin{aligned} -\sigma_{cd} \;&\leq\; \frac{-\tau}{2\sin\varphi_0\cos\varphi_0} + \frac{\frac{d}{v}\cos^2\varphi_0}{1+2\frac{d}{v}\cos^4\varphi_0}\sigma \leq \sigma_{td} \\ -\sigma_{cv} \;&\leq\; \frac{\sigma}{1+2\frac{d}{v}\cos^4\varphi_0} \leq \sigma_{tv} \end{aligned} \tag{2.21}$$

After member I fails, member III will fail when the stress reaches τ.

$$\tau = (\sigma_{tv} - \sigma)\tan\varphi_0 \tag{2.22}$$

At the point at which member I fails, member II will not yet have reached its yield point (σ_{cd}). Consequently, when stress σ_{II} is $-\sigma_{cd} \leq \sigma_{II}$, it is equal to:

$$-\sigma_{cd} \leq \frac{\sigma - \sigma_{tv}}{\cos^2\varphi_0} \tag{2.23}$$

After member I fails, member II will yield when the stress reaches τ, that is

$$\tau = \sigma_{cd}\sin\varphi_0\cos\varphi_0 \tag{2.24}$$

At the point at which member I fails, member III will not yet have reached its yield point (σ_{cv}) or its breaking point (σ_{tv}). Consequently, when stress σ_{III} is $-\sigma_{cv} \leq \sigma_{III} \leq \sigma_{tv}$, it is equal to:

$$-\sigma_{cv} \leq \sigma + \sigma_{cd}\cos^2\varphi_0 \leq \sigma_{tv} \tag{2.25}$$

In case (d), the maximum stresses (σ, τ) determined by Equations (2.22) to (2.25), must be below the yielding line determined by Equations (2.20) and (2.21).

(e) When the conditions of $\sigma < 0$ and $\tau > 0$ are applied, the first diagonal member II will yield in compression, and then diagonal member I will fail in tension. After member II yields, member I will fail when the stress reaches τ, where

$$\tau = (\sigma_{td} + \sigma_{cd})\sin\varphi_0\cos\varphi_0 \tag{2.26}$$

At the point at which member II yields, member III will not yet have reached its yield point (σ_{cv}) or its breaking point (σ_{tv}). Consequently, when stress σ_{III} is $-\sigma_{cv} \leq \sigma_{III} \leq \sigma_{tv}$, it is equal to:

$$-\sigma_{cv} \leq \sigma - (\sigma_{td} - \sigma_{cd})\cos^2\varphi_0 \leq \sigma_{tv} \tag{2.27}$$

(f) When the conditions of $\sigma < 0$ and $\tau > 0$ are applied, the first diagonal member II will yield in compression, and then vertical member III will yield in compression.

After member II yields, member III will yield when the stress reaches τ given by Equation (2.28).

$$\tau = 2\sigma_{cd} \sin\varphi_0 \cos\varphi_0 + \tan\varphi_0(\sigma + \sigma_{cv}) \tag{2.28}$$

At the point at which member II yields, member I will not yet have reached its yield point (σ_{cd}) or its breaking point (σ_{td}). Consequently, when stress σ_I is $-\sigma_{cd} \leq \sigma_I \leq \sigma_{td}$, it is equal to:

$$-\sigma_{cd} \leq \frac{\tau}{\sin\varphi_0 \cos\varphi_0} - \sigma_{cd} \leq \sigma_{td} \tag{2.29}$$

Here, $\tau < 0$ is omitted because the shape of the yielding condition is symmetrical about the σ-axis. Figure 2.8 shows the yielding condition when applied to the above six cases. (a) to (f) in Figure 2.8 correspond to the cases (a) to (f) of the yielding condition. The yielding condition of the joint element is represented by the solid line. Similarly, the yielding condition of Equation (2.35) is represented by the broken line.

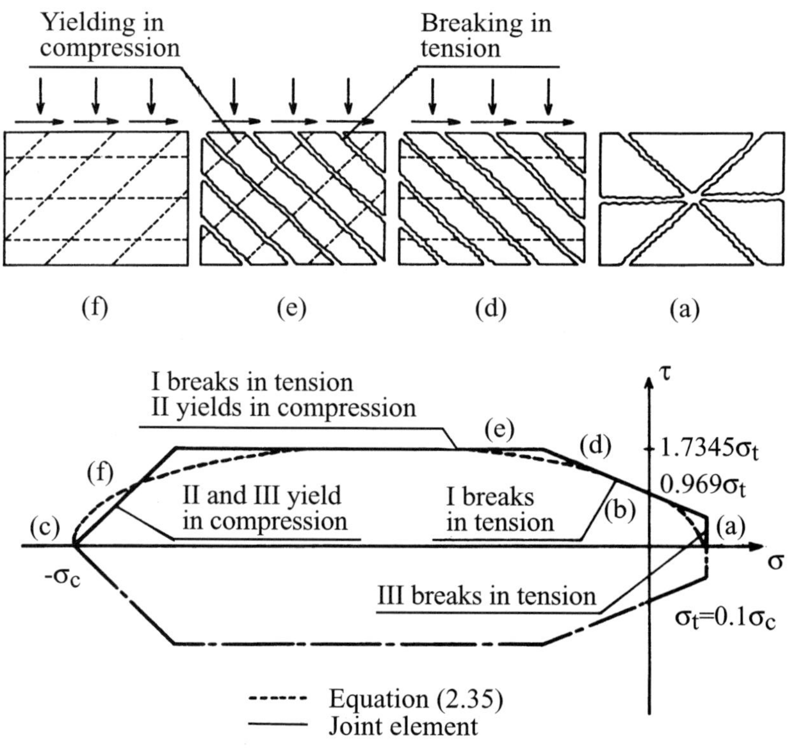

Figure 2.8: Yielding condition of concrete.

2.2.3 Formulation of Elastic-plastic Joint Element for the Finite Element Method

Let us consider the joint element with the length, $\widehat{ABC} = \ell_0$, and the height, h_0, shown in Figure 2.9. Here, we assume $h_0 \ll \ell_0$. Displacements on the upper surface (U_u, V_u) and on the lower surface (U_ℓ, V_ℓ) are expressed by Equation (2.30).

$$\begin{bmatrix} U_u & U_\ell \\ V_u & V_\ell \end{bmatrix} = \begin{bmatrix} NA(\xi) & 0 & NB(\xi) & 0 & NC(\xi) & 0 \\ 0 & NA(\xi) & 0 & NB(\xi) & 0 & NC(\xi) \end{bmatrix} \begin{bmatrix} U_4 & U_1 \\ V_4 & V_1 \\ U_5 & U_2 \\ V_5 & V_2 \\ U_6 & U_3 \\ V_6 & V_3 \end{bmatrix} \tag{2.30}$$

where $NA(\xi) = \xi(\xi-1)/2$, $NB(\xi) = 1-\xi^2$, $NC(\xi) = \xi(\xi+1)/2$, respectively.

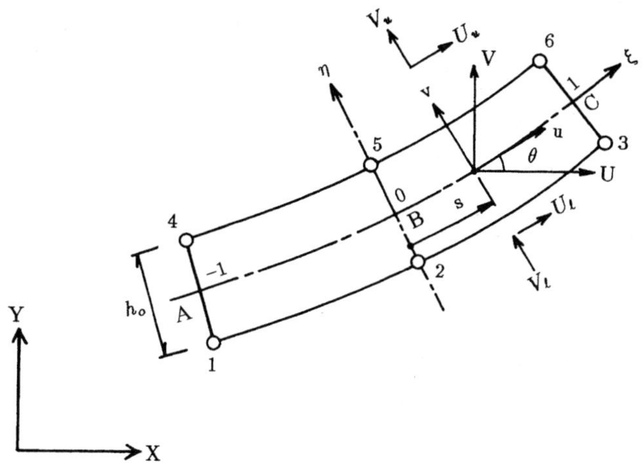

Figure 2.9: Joint element.

Displacements on the upper surface (u_u, v_u) and on the lower surface (u_ℓ, v_ℓ) in the $\xi - \eta$ coordinate are obtained by Equation (2.31).

$$\begin{bmatrix} u_u & u_\ell \\ v_u & v_\ell \end{bmatrix} = \begin{bmatrix} \cos\theta & -\sin\theta \\ \sin\theta & \cos\theta \end{bmatrix} \begin{bmatrix} U_u & U_\ell \\ V_u & V_\ell \end{bmatrix} \tag{2.31}$$

If the height h_0 is assumed to be sufficiently small, we have the strains of truss members ε_i, as follows.

$$\varepsilon_i = \frac{\cos\varphi_i}{h_0}\begin{bmatrix} -\sin\varphi_i & -\cos\varphi_i & \sin\varphi_i & \cos\varphi_i \end{bmatrix}\begin{bmatrix} U_\ell \\ V_\ell \\ U_u \\ V_u \end{bmatrix} \tag{2.32}$$

where $i = I, II, III$. The corresponding angles of truss members φ_I, φ_{II}, φ_{III} are $\varphi_I = \varphi_0$, $\varphi_{II} = -\varphi_0$, $\varphi_{III} = 0$, respectively. Consequently, when the incremental strain is $\Delta\varepsilon_i$ and the initial stress is σ_{0i} in each member, we have the incremental strain energy ΔU given by Equation (2.33).

$$\begin{aligned} \Delta U &= \sum_{i=1}^{n}\int \Delta U_i ds \\ &= \sum_{i=1}^{n}\int_{-1}^{1} t\left|\frac{ds}{d\xi}\right|\left[\sigma_{0i}\Delta\varepsilon_i + \frac{1}{2}E_{ti}(\Delta\varepsilon_i)^2\right] h_0 d\xi \end{aligned} \tag{2.33}$$

where n, is the number of truss members, t, the thickness of the joint element and E_{ti}, the tangent modulus of truss members, respectively.

2.3 An Analysis of Plane Concrete Under Combined Stress (Model II: Joint elements for thin layers of mortar)

We now discuss the joint element used to simulate the collapse behavior of thin layers of concrete and mortar between the bricks in masonry structures. The hysteretic characteristic of each member of the joint element is shown in Figure 2.10. We understand that up to the maximum stress in compression, the stress-strain line is assumed to be the line proposed by Equation (2.34) [13].

$$\sigma = \frac{E_0\varepsilon}{1 + [(E_0/E_s) - 2](\varepsilon/\varepsilon_c) + (\varepsilon/\varepsilon_c)^2} \tag{2.34}$$

where E_0 is the tangent modulus at the point $\sigma = 0$; f_c', the maximum compressive strength and ε_c, the strain at the point $\sigma = f_c'$; ε_u, the crashing strain in compression; σ and τ, the stress and the strain under the uniaxial compression; and $E_s = f_c'/\varepsilon_c$, respectively. Beyond the maximum stress in compression, the stress-strain line is assumed to be a straight line through the points (f_c', ε_c) and $(0.2f_c', 4\varepsilon_c)$. Material properties of the joint element and the characteristic values of the members of the joint element are listed in Table 2.2.

As shown in Figure 2.11, we applied the joint elements to the problem of proportional loading $(\lambda\sigma, \lambda\tau)$ where the joint elements are subjected to the uniform stresses σ and τ. The solution process is as follows. As shown in Figure 2.10, the stress-strain

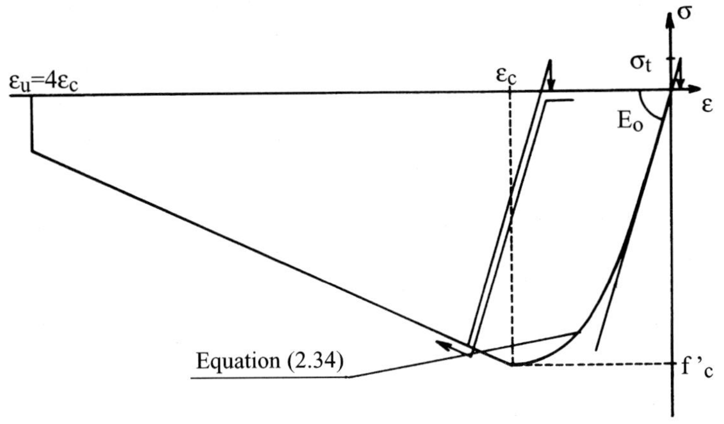

Figure 2.10: Stress-strain relationship assigned for the mortar.

Thickness		t	$= 0.5cm$
Height		h	$= 2.0cm$
Young's modulus		E	$= 2.06 \times 10^4 N/mm^2$
Poisson's ratio		ν	$= 0.16$
Strength	Compressive	σ_c	$= 20.6N/mm^2$
	Tensile	σ_t	$= 2.06N/mm^2$
Angle		φ_0	$= 45°$
Young's modulus	diag. member	E_{sd}	$= 0.862E$
	vert. member	E_{sv}	$= 0.595E$
Compressive strength	diag. member	σ_{cd}	$= 0.25\sigma_c$
	vert. member	σ_{cv}	$= 0.75\sigma_c$
Tensile strength	diag. member	σ_{td}	$= 0.945\sigma_t$
	vert. member	σ_{tv}	$= 0.580\sigma_t$
Strain		ε_c	$= -0.0025$

Table 2.2: Material constants simulating thin concrete/mortar layer.

curve is non-linear with regard to the applied load level. Therefore, the structural behavior is also non-linear with regard to the applied load level, so that an iteration process for the convergence of the non-linear solution is necessary to solve the problem of each increment of the load. As for the iteration schemes for the solution of the non-linear finite element equations, we use an improved version of the Newton-Raphson method. The stress and the strain terms are calculated separately for each member. Although hysteretic characteristics of each member are assumed as shown in Figure 2.10, the joint element exhibits the yielding condition as shown in Figure 2.12(a). The solid line is determined by the experiment of plain concrete under combined stress given in Equation (2.35) [12]. The dots represent the results of our analysis [1].

$$\frac{\tau}{\sigma_c} = \sqrt{-0.104\left(\frac{\sigma}{\sigma_c}\right)^2 + 0.095\left(\frac{\sigma}{\sigma_c}\right) + 0.0089} \qquad (2.35)$$

Figure 2.12(b) shows the stress-strain relationship where the joint elements are subjected to the uniaxial uniform stress σ. Equation (2.34) is expressed by the solid line [13]. The dots again represent the results of our analysis.

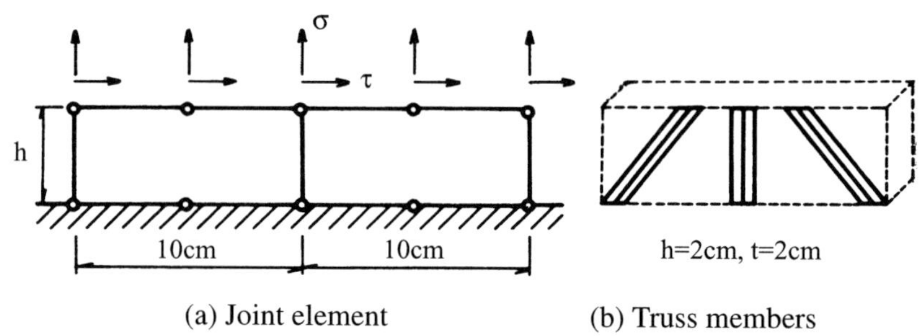

Figure 2.11: Analytical model of thin concrete and mortar.

2.4 Slippage under a Footing

2.4.1 Joint Elements for Soil (Model III: Mohr-Coulomb yielding condition)

Let us discuss the joint element used to simulate the slippage of soil based on the assumption that when the shear failure of a soil occurs, the state of the stresses can be determined by the Mohr-Coulomb yielding condition as expressed by Equation (2.36).

$$\tau = c - \sigma \tan \phi \qquad (2.36)$$

where σ is the normal stress on every section through a mass of cohesive soil, τ, the corresponding shear resistance per unit area, c, the cohesion, and ϕ, the angle of internal friction. As shown in Figure 2.13(a), the joint element consists of two diagonal members.

Assuming from Equations (2.20) and (2.36) that only shearing collapse occurs, we

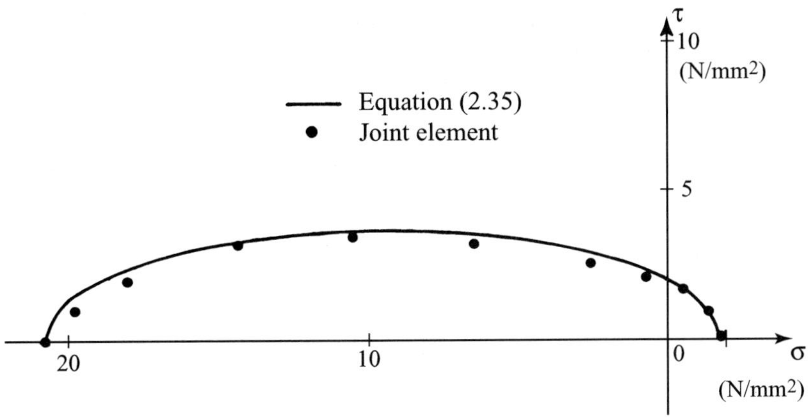

(a) Yielding condition of joint element

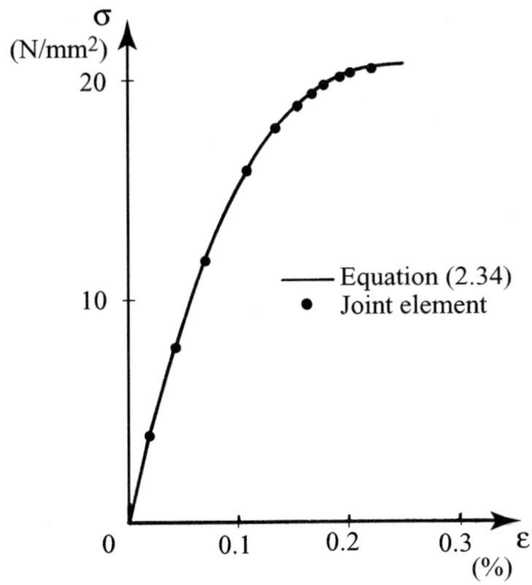

(b) Stress-strain relationship

Figure 2.12: Results of analyses of thin concrete and mortar.

obtain the yielding condition given by Equation (2.37).

$$\begin{aligned} c &= 2\sin\varphi_0\cos\varphi_0\sigma_{td} \\ \tan\varphi_0 &= \frac{2\dfrac{d}{v}\sin\varphi_0\cos^3\varphi_0}{1+2\dfrac{d}{v}\cos^4\varphi_0} \end{aligned} \tag{2.37}$$

Substituting Equation (2.2) into Equation (2.1), and assuming plane strain ($\varepsilon_z = 0$) is

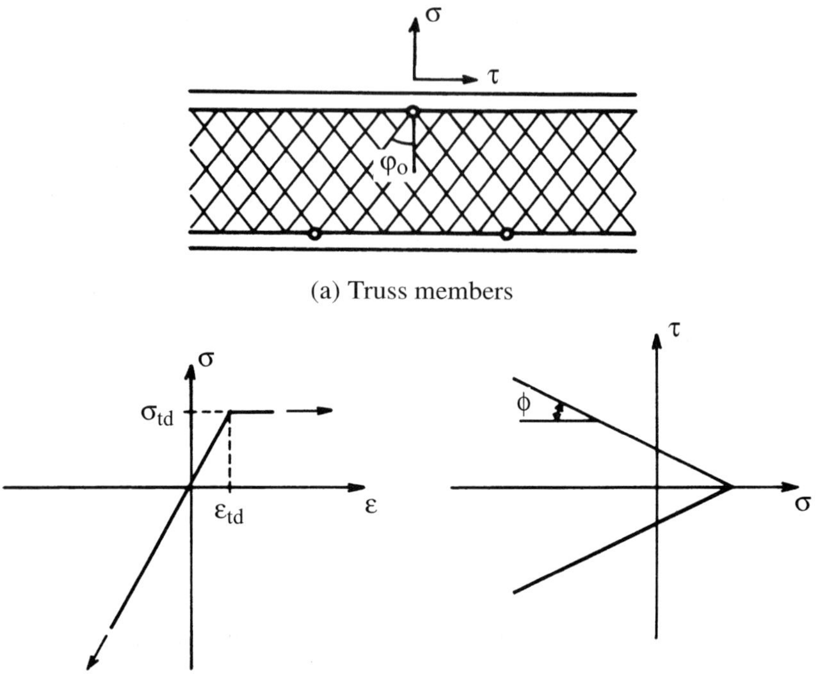

(b) Stress-strain relationship (c) Yielding condition

Figure 2.13: Joint element simulating soil.

in the soil, we obtain Equation (2.38) from Equation (2.37).

$$\tan\phi\tan\varphi_0 = \frac{1-2\nu}{2(1-\nu)} \tag{2.38}$$

From Equations (2.37) and (2.38), we obtain the tensile strength σ_{td} of the diagonal members and the angle φ_0, given by Equation (2.39).

$$\begin{aligned} \sigma_{td} &= \frac{c}{2\sin\varphi_0\cos\varphi_0} \\ \tan\varphi_0 &= \frac{1-2\nu}{2(1-\nu)}\cot\phi \end{aligned} \tag{2.39}$$

In this case, member III is not used, so that $v = E_{sv}/E = 0$ is assumed. Consequently, from Equation (2.37), we obtain Equation (2.40).

$$\tan\phi = \tan\varphi_0 \tag{2.40}$$

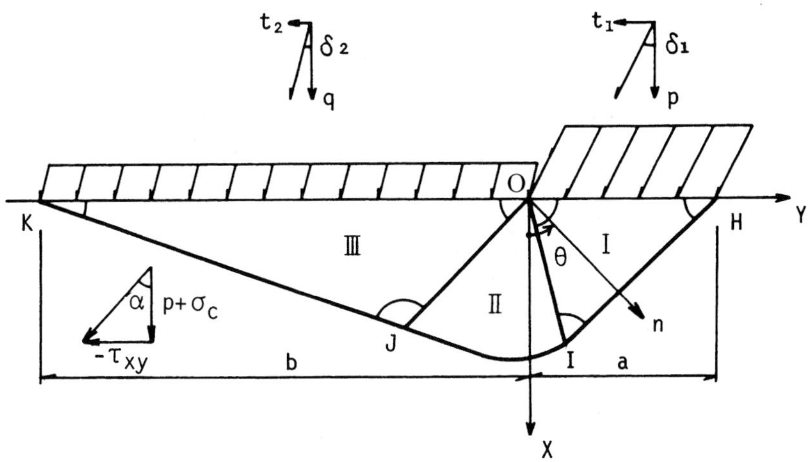

Figure 2.14: Zones at the limit equilibrium state.

Supposing the angle of internal friction $\phi = 30^\circ$ and Poisson's ratio $\nu = 0.25$ into Equation (2.4), we obtain Equation (2.41).

$$d = E_{sd}/E = 1.067 \tag{2.41}$$

We assume herein that only shear fracture occurs, so that only the tensile strength σ_{td} is necessary to determine the yielding condition. The stress-strain relationship of members I and II is shown in Figure 2.13(b).

Assuming $\varphi_0 = \phi$, $\sigma_{td} = c/(2\sin\phi\cos\phi)$, $\sigma_{cd} = \infty$, and $d = 1.067$, we obtain the yielding condition as shown in Figure 2.13(c).

2.4.2 Analysis of Slippage Occurs under a Footing

Let us apply the joint element to the problem in order to estimate the critical load of soil when it is subjected to the uniformly distributed pressure with the angle $\theta = 20.105^\circ$ under a footing as shown in Figure 2.14, where θ is the angle between the vertical axis and the axis in the direction of the uniformly distributed pressure. Figure 2.14 shows an analytical problem which Reissner solved and the equilibrium critical load of this problem which he derived.

2.4.2.1 A Solution Based on Limit Equilibrium Method

Let us consider a problem of limit equilibrium based on the assumption that the unit weight of the soil is equal to zero when the normal stress and the shear stress are subjected to the surface of the soil. The problem of computing the uniform vertical loads p and q when subjected to the surface of the soil are indicated in Figure 2.14.

Zone I is an active Rankine state. If we assume Equations (2.42) to (2.44),

$$\sigma_c = c\tan\phi \tag{2.42}$$

$$\delta_1 = \tan^{-1}\left(\frac{t_1}{p+\sigma_c}\right) \tag{2.43}$$

$$\delta_2 = \tan^{-1}\left(\frac{t_2}{q+\sigma_c}\right) \tag{2.44}$$

we obtain the boundary conditions as follows:

$$\sigma_x = p \tag{2.45}$$

$$\tau_{xy} = -(p+\sigma_c)\tan\delta_1 \tag{2.46}$$

From the relationship between the slope of the surface of the soil and in the direction of the external load, we obtain

$$\theta = 0, \qquad (\varepsilon = 0) \tag{2.47}$$

$$\alpha = \delta_1, \qquad n = 0 \tag{2.48}$$

Consequently, we get Equation (2.49).

$$\begin{aligned} 2\delta_I &= -\delta_1 - \sin^{-1}\left(\frac{\sin\delta_1}{\sin\phi}\right) \\ \sigma_I &= \frac{p+\sigma_c}{\cos\delta_1\cos^2\phi}\left(\cos\delta_1 - \sqrt{\sin^2\phi - \sin^2\delta_1}\right) \end{aligned} \tag{2.49}$$

On the other hand, Zone III is a passive Rankine state. Supposing $\sigma_x = q$, $\tau_{xy} = -(q+\sigma_c)\tan\delta_2$, $\theta = 0$, $\alpha = \delta_2$, and $n = 0$, we obtain Equation (2.50).

$$\begin{aligned} 2\delta_{III} &= -\delta_2 - \pi + \sin^{-1}\left(\frac{\sin\delta_2}{\sin\phi}\right) \\ \sigma_{III} &= \frac{q+\sigma_c}{\cos\delta_2\cos^2\phi}\left(\cos\delta_2 + \sqrt{\sin^2\phi - \sin^2\delta_2}\right) \end{aligned} \tag{2.50}$$

If δ_2 is given, we obtain the load strength q in Equation (2.51).

$$q = -\sigma_c + (p+\sigma_c)\left(\frac{\cos\delta_2\cos\delta_1 - \sqrt{\sin^2\phi - \sin^2\delta_1}}{\cos\delta_1\cos\delta_2 + \sqrt{\sin^2\phi - \sin^2\delta_2}}\right)$$

$$\exp\left[-\left\{\pi - \delta_1 + \delta_2 - \sin^{-1}\left(\frac{\sin\delta_1}{\sin\phi}\right) - \sin^{-1}\left(\frac{\sin\delta_2}{\sin\phi}\right)\right\}\tan\phi\right] \tag{2.51}$$

Equation (2.51) is the relationship between the vertical loads p and q at the point O in Figure 2.14, where both p and q are boundary values. The shear forces at the state of limit equilibrium are t_1 and t_2 which are subjected to the surface of zones I and III. Assuming $q = 0$, $t_2 = 0$, $\delta_1 = 18°$, $\delta_2 = 0$, $c = 0.0245 N/mm^2$, $\phi = 30°$, we obtain

$$\sigma_c = c \tan\phi = 0.04246 N/mm^2 \tag{2.52}$$

Consequently, from Equation (2.51), we obtain

$$p = 0.3357 N/mm^2 \tag{2.53}$$

From Equation (2.43), we obtain

$$t_1 = 0.1229 N/mm^2 \tag{2.54}$$

From the above, we obtain the critical load (P_{cr}) against the total diagonal loads.

$$\begin{aligned} P_{cr} &= 400(mm) \times 10(mm) \times \sqrt{(0.3357)^2 + (0.1229)^2}(N/mm^2) \\ &= 1430.0(N) \end{aligned} \tag{2.55}$$

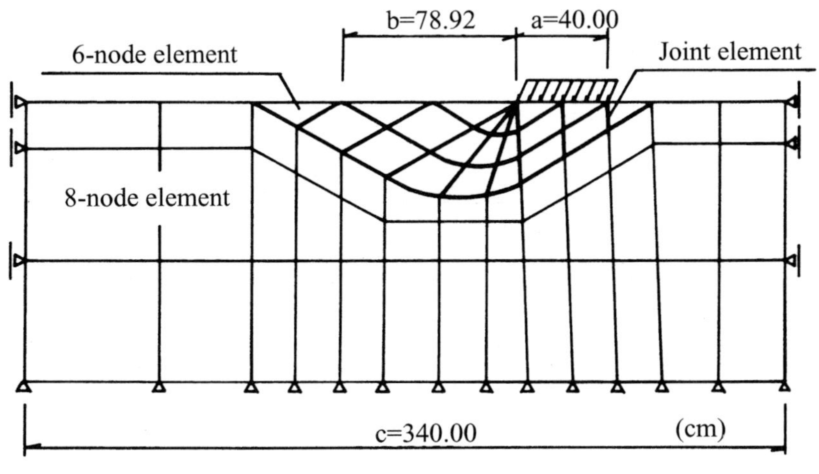

Figure 2.15: Analytical model of soil.

2.4.2.2 Analytical Model

A complete solution is obtained on the assumption that the soil has a unit weight equal to zero. Therefore, for the purpose of comparing it with the theory, the unit weight of the soil is assumed to be zero and the soil is also assumed to be homogeneous, without taking into consideration the influence of the pore-water pressure. Figure 2.15 shows an analytical model which has elastic-plastic joint elements and finite elements distributed along the slippage lines determined by the limit equilibrium method. Plane strain ($\varepsilon = 0$) is, herein, assumed in the soil to deal with this problem as a two-dimensional problem. The yielding condition is determined by the Mohr-Coulomb yielding condition. As boundary conditions of the displacements, the horizontal displacements are fixed on both sides, and both horizontal and vertical displacements are fixed at the bottom of the model. Table 2.3 specifies the material properties of the soil and the characteristic values of the members of the joint element.

Thickness		t	=	$1.0cm$
Height		h	=	$1.0cm$
Young's modulus		E	=	$24.5N/mm^2$
Poisson's ratio		ν	=	0.25
Weight per unit volume		γ	=	$0.0g/cm^3$
Cohesion		c	=	$0.0245N/mm^2$
Angle of internal friction		ϕ	=	$30.0°$
Angle		φ_0	=	$30°$
Young's modulus	diag. member	E_{sd}	=	$1.067E$
	vert. member	E_{sv}	=	$0.000E$
Strength	Compressive	σ_c	=	∞
	Tensile	σ_t	=	$0.0283N/mm^2$
Strain		ε_{td}	=	0.001084

Table 2.3: Material constants simulating soil.

2.4.2.3 Results of the Analysis

Figure 2.16 shows the development of the deformation at various stages of loading. The solid lines in Figure 2.17 show the development of the yielding zones at various stages of loading. Here the load parameter λ is the ratio of the applied load to the theoretical critical load. Normal and shear stresses at the stage of yielding are shown in Figure 2.18.

Figure 2.19 shows the relationships between the diagonal load which multiplied the diagonal stress by the width of the footing, and the horizontal displacements of points D and E which are along the slippage lines. Beyond the diagonal load of about $1422.0N$, the non-linear behavior begins ($\leftarrow$ mark in Figure 2.19). The value of this load corresponds to the value of the load required to establish a state of limit equilibrium, $1430.0N$, determined by the limit equilibrium method.

Regarding the points A, B, and C, each of which is located along the slippage lines, Figure 2.20 shows the stress-strain relationships and Figure 2.21 the shear stress-strain relationships. The yielding points of both stresses are very close to the stresses

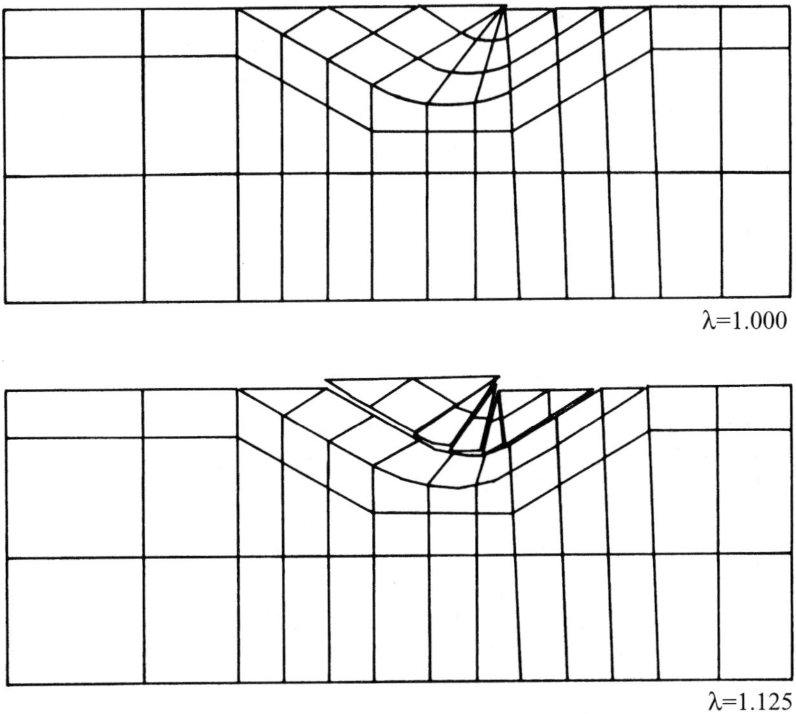

Figure 2.16: Development of the deformation at various stage of loading.

determined by the limit equilibrium method (Table 2.4). For points A and B beyond the elastic limit, the stresses gradually decrease.

2.5 Conclusion

In the case of plain concrete under combined stress, the yielding condition determined by the joint element is consistent with the curve determined through experiment. According to the analysis of the stresses and the critical load in the soil where slippage occurs under a footing, the analytical results correspond well to the theoretical values derived by Reissner (Table 2.4).

The main advantage of this joint element is that it allows the description of the material behavior and the yielding condition by use of components which individually exhibit a simple stress-strain relationship. In addition, this joint element is much simpler than others in expressing a discontinuous plane in the plastic phase based on a mathematical formulation. However, the state of deformations such as dilatancy is a future problem to be solved when shear failure of the soil occurs.

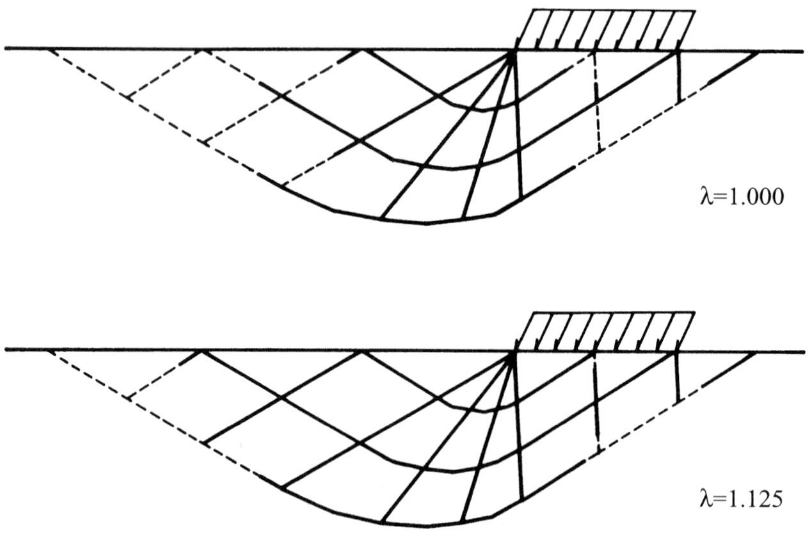

Figure 2.17: Development of the yielding zones at various stages of loading.

Zone	Angle	Theory		Analyses	
	θ	σ	τ	σ	τ
I		0.1794	0.1281	0.1778	0.1273
II	-0.0402	0.1692	0.1222	0.1628	0.1187
	-0.8238	0.1377	0.1040	0.1187	0.0933
	-1.6073	0.1109	0.0886	0.1324	0.1010
	-2.0643	0.0972	0.0806	0.1040	0.0843
	-2.8479	0.0764	0.0685	0.0636	0.0613
	-3.6314	0.0586	0.0583	0.0706	0.0653
	-4.0874	0.0496	0.0532	0.0576	0.0578
	-4.8719	0.0359	0.0452	0.0267	0.0399
	-5.6555	0.0242	0.0385	0.0324	0.0432
III		0.0212	0.0368	0.0213	0.0350

Table 2.4: Theoretical values and analytical results (N/mm^2).

Thus, the two-dimensional elastic-plastic joint element presented here is efficient for use in the Finite Element Method analysis to describe a discontinuous plane in the plastic phase. More elaborate material behavior can be modelled by a model which includes several members.

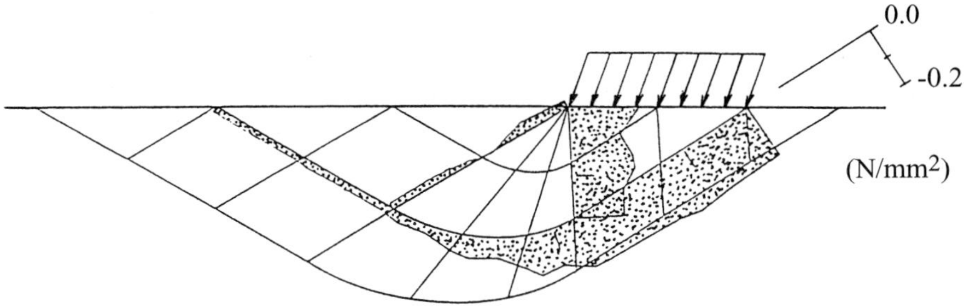

(a) Normal stress at the stage of yielding (λ=1.0)

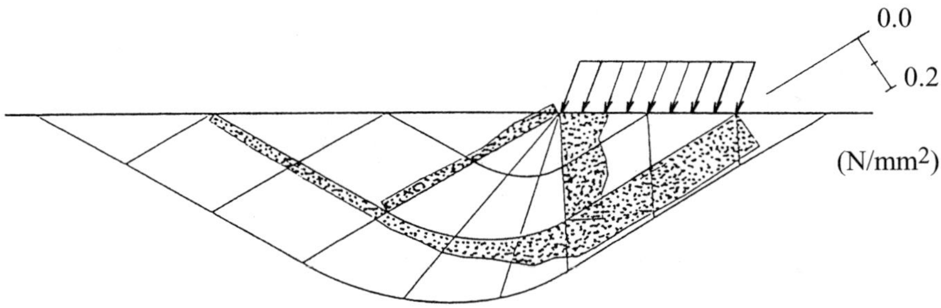

(b) Shear stress at the stage of yielding (λ=1.0)

Figure 2.18: Normal and shear stresses at the stage of yielding ($\lambda = 1.0$).

References

[1] Kato, S., Hidaka, K., Aoki, T., "A Study on the Formulation of a Elastic-plastic Joint Element by Truss Elements – An Application of the Theory of Effective Strength", Trans. of A.I.J., No.370, 50-59, 1986. (in Japanese).

[2] Kato, S., Hidaka, K., Aoki, T., "Analytical Studies of the Historical Masonry Structure – The Florentine Dome of Santa Maria del Fiore", Proceedings of IASS International Symposium on Shells, Membranes and Space Frames (Published by Elsevier Science Publishers B.V., 1986), Vol.1, Osaka, 225-232, 1986.

[3] Kato, S., Hidaka, K., Aoki, T., "Structural Role of the Wooden Ring of the Dome of Santa Maria del Fiore in Florence", Proceedings of IASS-MSU International Symposium on DOMES from Antiquity to the Present, Istanbul, 261-270, 1988.

[4] Aoki, T., Hidaka, K., Kato, S., "Structural Stability and Profile in the Dome of Santa Maria del Fiore, Florence", Proceedings of STREMA International Symposium on Structural Repair and Maintenance of Historical Buildings (Published by Birkhauser Verlag), Florence, 211-220, 1989.

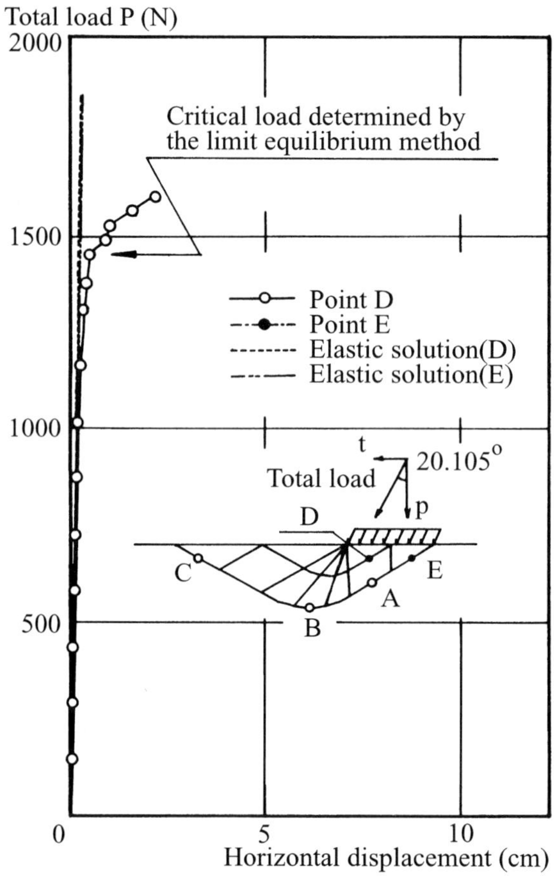

Figure 2.19: Load-displacement relationships.

[5] Hidaka, K., Aoki, T., Kato, S., "Structural Stability and Profile in the Dome of Hagia Sophia, Istanbul", Proceedings of STREMA International Symposium on Structural Repair and Maintenance of Historical Buildings (Published by Birkhauser Verlag), Florence, 267-276, 1989.

[6] Aoki, T., "Structural Characteristics of Isahaya arch bridge –1– Structural role of the sludge foundation", Proc. of Annual Convention of A.I.J., Kyushu Branch, 321-324, 1992. (in Japanese).

[7] Goodman, R.E., Taylor, R.L., Brekke, T.L., "A Model for the Mechanics of Jointed Rock", ASCE SM 3, Vol.94, 637-659, 1968.

[8] Goodman, R.E., "Methods of Geological Engineering in Discontinuous Rocks", West Publishing Co., St. Paul, 1975.

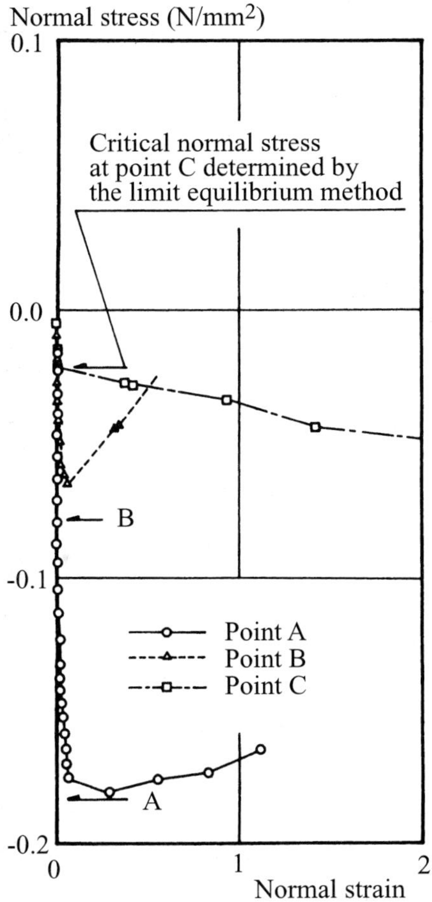

Figure 2.20: Normal stress-strain relationships.

[9] Ghaboussi, J., Wilson, E.L., Isenberg, J., "Finite Element for Rock Joints and Interfaces", ASCE SM 10, Vol.99, 833-848, 1973.

[10] Heuze, F.E., Barbour, T.G., "New Models for Rock Joints and Interfaces", ASCE GT 5, Vol.108, 757-776, 1982.

[11] Gexiurun, "Non-Linear Analysis of a Joint Element and its Application in Rock Engineering", Int. Numerical and Analytical Methods in Geomechanics, Vol.5, 229-245, 1981.

[12] Tsuboi, Y., Suenaga, Y., "Experimental Study on Failure of Plain Concrete under Combined Stress (Part3)", Trans. of A.I.J., No.64, 25-36, 1960.

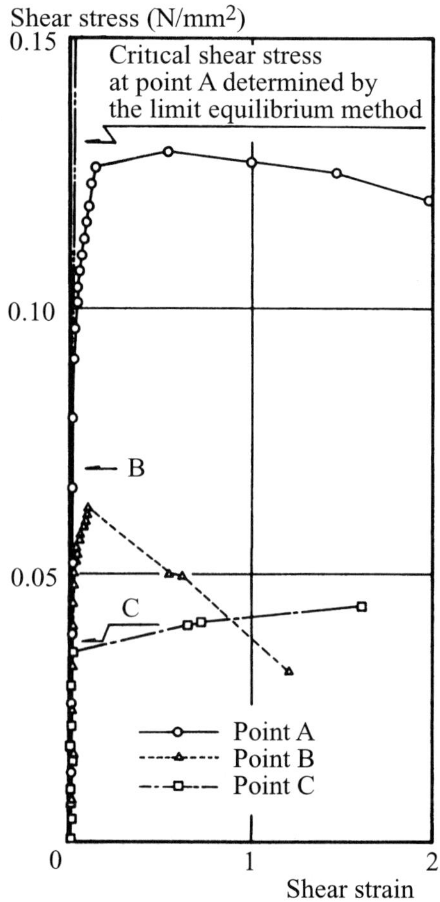

Figure 2.21: Shear stress-strain relationships.

[13] Chen, W.F., "Plasticity in Reinforced Concrete", McGraw-Hill Book Co., 1982.

©2001, Saxe-Coburg Publications, Stirling, Scotland
J.W. Bull, (Editor), Computational Modelling of Masonry, Brickwork and Blockwork Structures,
Saxe-Coburg Publications, Stirling, Scotland 53-77.

Chapter 3

Earthquake and Vibration Effects

C.A. Syrmakezis and A.A. Sophocleous

Institute of Structural Analysis & Aseismic Research, Faculty of Civil Engineering, National Technical University of Athens, Greece

Abstract

This chapter refers to the problems of the computational modelling of masonry structures due to earthquake and vibration effects. After a general survey covering the effect of earthquake and vibration loads on civil engineering structures, the investigation focuses on masonry structures. Modelling refers to structural, material and action simulation. Two case studies illustrate the proposed general methodology.

3.1 Introduction

Time varying loads during the analysis and design process have to be taken into consideration. Earthquake and vibration have a significant effect on all civil engineering structures. Loads encountered in civil engineering structures can be divided in two main categories: static and dynamic.

The appearance of inertia and damping forces in the structure during a dynamic loading is the most characteristic distinction between static loading and dynamic loading effects.

Inertia and damping forces appear as a result of the accelerations and velocities induced in the structure and have to be taken into account in the analysis and design procedure. Their magnitudes depend on the type of excitation and the dynamic properties of the structure.

The term "dynamic" refers to "time-varying" effects. Opposite to a static action, dynamic action is a loading with magnitude, direction and point of application varying with time t as indicatively shown in Figure 3.1. The structural response due to a dynamic loading is also time-varying.

Vibrations induced in a structure by man, machine, water, wind, traffic, blast, etc. have various effects on structures; mainly on its serviceability and load bearing capacity.

Earthquake loading, critical for the safety of the structure, is one of the most important dynamic loadings. Earthquake generation mechanism is briefly described by the geotectonic activity in the earth's crust and the upper mantle. For a long period of time elastic strain energy is accumulated in a fault zone, until the rock strength in a critical direction is reached. During the rupture, a substantial part of the released energy radiates as kinetic energy in the form of seismic waves. These waves induce abrupt horizontal and vertical motions in structures. When the structure responds to the imposed ground motions during a strong earthquake, its maximum response acceleration may be several times the maximum ground acceleration, and depends on the structure's mass, stiffness, strength and energy dissipation capacity. This base acceleration generates inertial forces in the structure creating a forced vibration response (Figure 3.2). When the seismic excitation stops, the movement of the structure goes on through a free vibration response, although dynamic loading does not now exist.

Figure 3.3 summarises the various types of static-dynamic loading.

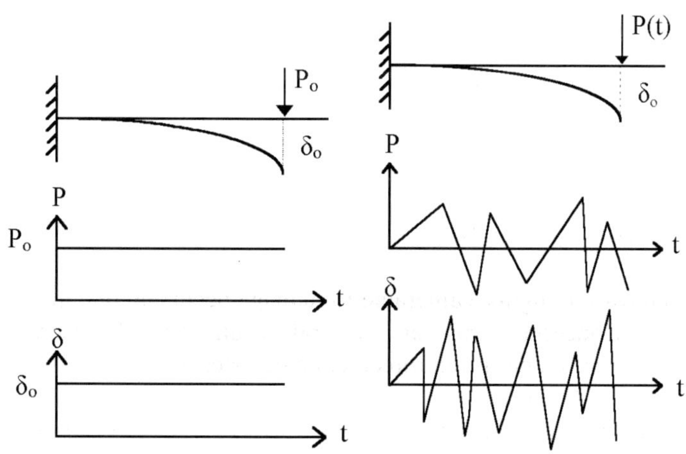

Figure 3.1: Static (a) and Dynamic (b) action.

The possible damage due to seismic or other types of vibrations are:

- loss of the load-bearing capacity of the structure
- local (irreversible) plastic zones of the bearing structural elements
- destruction of secondary elements

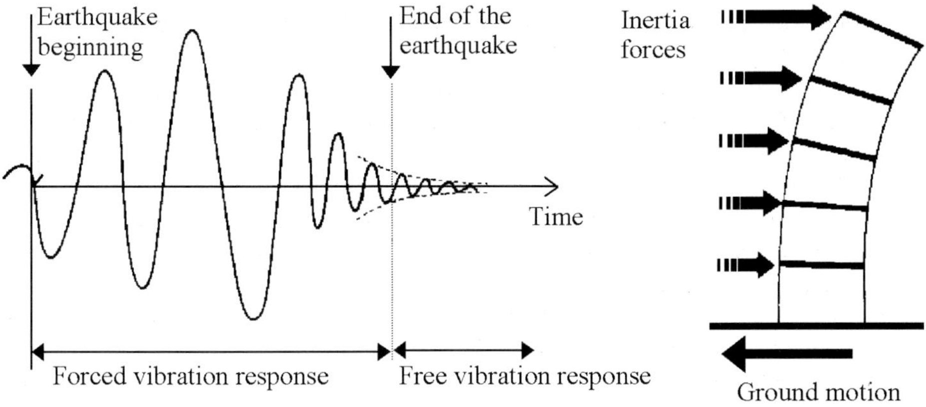

Figure 3.2: The earthquake excitation.

- excessive deformations
- deterioration due to fatigue
- intolerable vibration velocities and accelerations
- excessive noise.

3.2 Masonry structures

Masonry structures are highly vulnerable to damage occurring due to an earthquake ground motion. Concerning this case, several peculiarities of masonry structures, compared to the other modern (reinforced concrete, steel) structures, have to be considered for design and redesign purposes [1–3].

The first point refers to the distribution of masses and consequently, the distribution of inertia forces along the height of the structure. In this type of structure masonry masses have to be considered distributed along the height of the vertical bearing elements, while at the floor levels masses are relatively small. This fact is opposed to the principle governing the other type of structures (reinforced concrete, steel), where masses can generally be considered lumped at the floor levels, while masses of vertical elements can generally be ignored.

A second point is the rather low stress capacity of masonry structures, due to its low shear and tensile strengths, as well as its brittle mode of failure. It is to be noted that, although in the design for vertical loads compressive strength is the predominant factor, for the earthquake resistant design tensile and shear strengths govern the behaviour of a masonry structure.

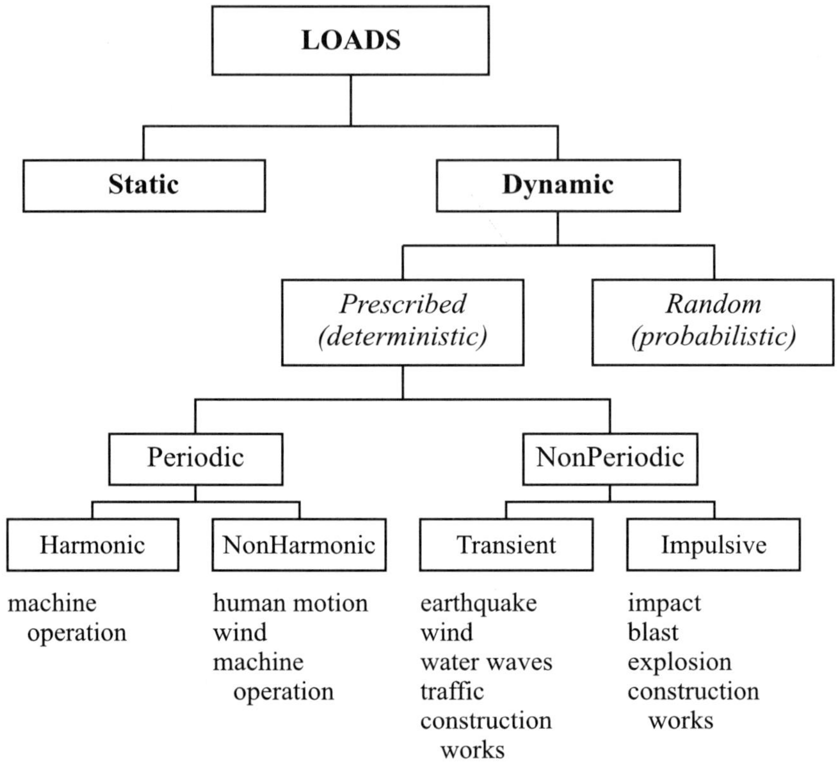

Figure 3.3: Various types of static-dynamic loading.

Masonry stress capacity is generally expressed through the stress capacity of units (blocks) and mortar. In Figure 3.4 horizontal and vertical mechanisms of tensile strength are shown. It can be easily concluded that the level of the block and mortar tensile strength strongly affects the tensile resistance of the whole masonry.

A third point is the type and quality of connections among various bearing elements of the structure. Connections between the vertical walls, the horizontal floors and the roof, must be carefully investigated. This investigation must ensure the safe propagation of loads through the elements of the structural system, until its final transmission to the supporting foundation.

The geometric characteristics of the walls play an important role in the masonry structure's response (type and extent of failures).

Bending failure could be observed for high values of the height to length ratio $(h_w/l_w > 2)$, while for low values of the ratio $(h_w/l_w < 1)$ shear failure appears. For the case of low friction coefficient values, sliding of the wall may be observed. In Figure 3.5 these possible failure mechanisms are shown.

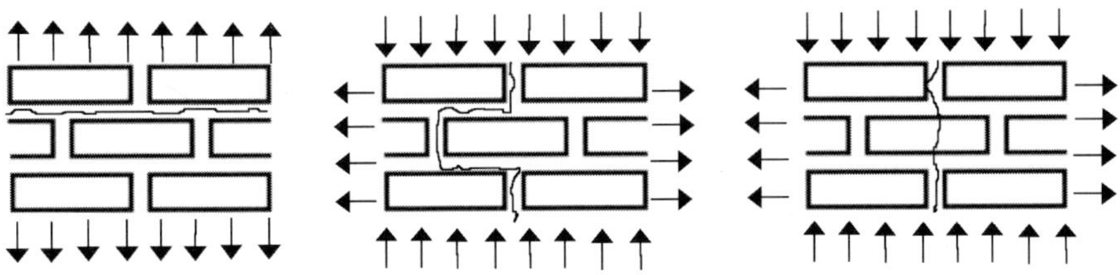

Figure 3.4: Horizontal and Vertical mechanisms of tensile strength.

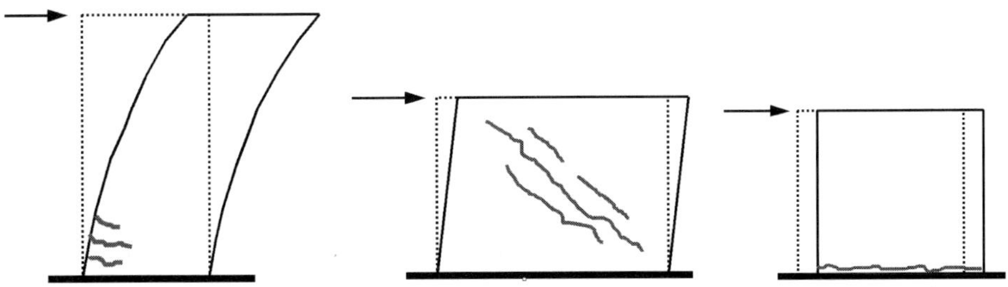

Figure 3.5: Failure mechanisms for different wall geometries.

3.3 Methods of analysis

Analysis of a masonry structure is a fundamental step either for the design of a new structure, or for the redesign of an existing structure.

During the design (or redesign) procedure a suitable method of analysis must be applied for the estimation of the response of the structure. The selection criteria are based on three factors: the type of the external actions, the material behaviour, and the type of the structural model selected. All three factors have to be properly simulated for the analysis.

Depending on the type of the external actions the analysis procedure could be either static or dynamic. Dynamic analysis is suitable for the cases of irregular structural configuration in plan and/or in height. In these cases significant differences between static and dynamic analysis force level and their distribution along the height of the structure can be observed. Dynamic analysis also provides useful information concerning the dynamic characteristics of all structures, such as the modes of vibration and the natural periods of vibrations, leading to a better understanding of the dynamic response of the structure. On the other hand, for the case of rather regular structures with limited height, equivalent static analysis can also be applied successfully.

The real stress-strain relationship of masonry is non-linear, so a non-linear analysis method should be more suitable. Such a method, if applied, leads to quite accurate results, for the case of reinforced masonry, where the stress-strain relationship is almost the same for all areas of the structure, and significant ductility exists. For unreinforced masonry which exhibits a brittle behaviour, non-linear analysis does not always lead to reliable results due to the uncertainties in the material response. The use of elastic analysis is then preferable.

Finally depending on the structural model of the structure, 3-D, 2-D or 1-D analysis can be performed. For masonry structures constructed in seismic areas a three-dimension analysis is preferred, leading to a more reliable consideration of the real response of the structure and especially of the real and accidental torsional effects. Two-dimension analysis is used mainly for structures which are rather small and regular in plan and in elevation. In these cases the torsional effects are limited and two dimensional analysis of separate bearing elements can be applied with relatively good approximation. One dimensional analysis is rarely used, only for rough preliminary calculations of single elements, e.g. single piers.

3.4 Masonry computational models

For the analysis and design of a masonry structure a proper computational model has to be created. This model has to provide satisfactory mathematical simulation of the real structure response on three levels:

a) Structure

b) Actions

c) Materials

Due to the uncertainties mentioned above, creation of such a model for a masonry structure is difficult. Assumptions to be made for the synthesis of the model have to be chosen very carefully, with special attention to their impact on the whole model, as well as to the possibility of feedback actions on other assumptions. Critical decisions to be taken fall on the shoulders of the designer.

To assist in that effort, the availability of as much data and information as possible is necessary. For example, in-situ or laboratory test results concerning the behaviour of the materials of the structure are very helpful for the final decisions concerning the materials model. On the other hand for the actions simulation, the accidental nature of some types of loading (concerning their values, directions, duration etc., e.g. earthquake), lead to additional difficulties.

Parametric studies give a solution for a successful confrontation to the difficulties. Several computer runs of the mathematical model with various values for the parameters can provide useful information concerning the identification of the sensitivity of the design parameter values leading to an optimum final design.

To illustrate the above arguments, some general outlines are presented in the next three paragraphs.

3.5 Structural Simulation

Structural simulation is the first and major step of the analysis procedure of a masonry structure. An appropriate and efficient mathematical model of the structure must be developed, representing the behaviour of the real structure at a maximum possible level.

Only structural elements with significant stiffness and/or ductility should be taken into account. Non-structural brittle elements can be neglected.

Since "exact solutions" (i.e. solutions describing the behaviour of the structure as continuous function over the structure under given loading and boundary conditions) are feasible only for a limited class of structural problems, modern computer techniques using the finite element method are the most versatile and effective approach.

A 3-D or 2-D finite element model seems to be generally the most suitable for the analysis of such structures.

Model reliability can be improved through specific simulation parameters such as:

- stiffness of the foundation area
- rigidity degree of connections between bearing walls
- rotation capacity of wooden floors
- prestressing elements
- etc.

3.6 Simulation of Actions

All types of actions induced on the structure must be properly implemented in the model during the analysis procedure. For common actions (dead and live loads) both their values and their direction of application are generally well defined through codes of practice. However, for accidental actions, and mainly for earthquake actions, there are different approaches for the ways they can be taken into consideration.

On a probabilistic basis, a series of artificial accelerograms, developed on the basis of the available data of previous seismic events in the region, are considered as the best simulation of the next expected earthquake ground motion to attack the structure.

For the evaluation of seismic loads during the analysis procedure, modern aseismic codes allow the use of a design spectrum, as an envelope of a number of different accelerograms related to real earthquakes expected to occur at the place where the structure is located. As an example, the well known and widely applied EC8 spectrum is presented in Figure 3.6.

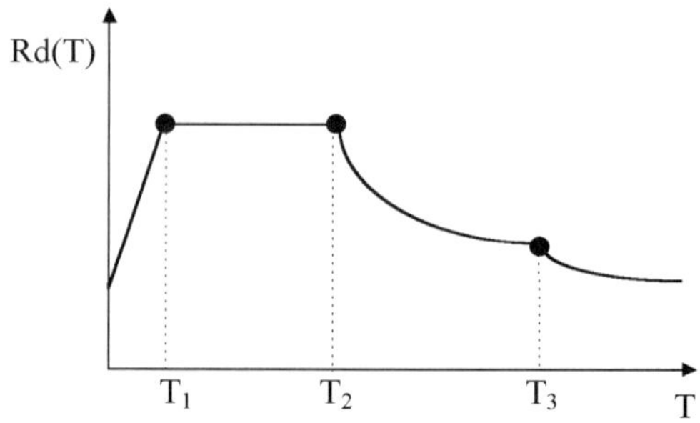

Figure 3.6: EC8 Design Spectrum.

3.7 Simulation of Materials Characteristics

3.7.1 General

The materials characteristics are critical input data. Namely, the compressive-tensile strengths of the materials, the modulus of elasticity, Poisson ratio, and shear modulus are of primary importance. For the estimation of those parameters, a combination of analytical or semi-empirical methods and experimental data have to be used.

3.7.2 Modulus of elasticity - Poisson ratio

Concerning the modulus of elasticity E_w, various formulae have been suggested. In all cases E_w is directly related to the compressive strength f_{wc}:

$$E_w = \alpha \cdot f_{wc} \tag{3.1}$$

Indicatively the following values are proposed:

EC6 [4]

If the modulus of elasticity is not evaluated through experimental tests (EN 1052-1), then $\alpha = 1000$ and $E_w = 1000 f_{wc}$. This value can be taken into account for the limit state analysis. If the serviceability limit state is investigated then a reduction factor of 0.6 must be applied to the above value, i.e. $E_w = 600 f_{wc}$.

Hendry [5]

- For the tangential modulus of elasticity: $E_w = 2 f_{wc}/\varepsilon_u$, where $\varepsilon_u = 2.5 \sim 3.5^0/_{00}$.
- For the section modulus of elasticity: $E_w = (400 \sim 1000) f_{wc}$.

CIB [6]

For the section modulus of elasticity: $E_w = 1000 f_{wc}$.
Poisson ratio $\nu = 0.15$.

Mauerwerk [7]
For the tangential modulus of elasticity: $E_w = (500 \sim 8000) f_w c$.
Poisson ratio $\nu = 0.10 - 0.20$.

Chuxian [8]
For the section modulus of elasticity: $E_w = 115 f_{wc}^{3/2}$.

3.7.3 Shear Modulus

The shear modulus can be evaluated according to the following well known relation:

$$G_w = \frac{E_w}{2 \cdot (1 + \nu)} \tag{3.2}$$

or by taking $G_w = 0.40 E_w$ according to EC6 requirements.

3.7.4 Compressive - Tensile Strength

For the determination of the masonry compressive and tensile strength, several semi-empirical expressions exist. Among them the following formulae are the most common:

EC6 [4]
Several relations are given in EC6 for the evaluation of the characteristic compressive strength of the masonry depending on the type of mortar, the filling (or not) of the vertical joints etc. The most general relation (mainly for new masonry) is:

$$f_{wc,k} = K \cdot f_{bc}^{0.65} \cdot f_{mc}^{0.25} \tag{3.3}$$

where K depends on the selected type of blocks, f_{bc} is the mean compressive strength of the blocks, and f_{mc} is the mean compressive strength of the mortar, calculated according to the tensile strength of mortar $f_{mt,k}$.

Tassios-Chronopoulos [15]

$$f_{wc} = \xi \left[\left(\frac{2}{3} \sqrt{f_{bc}} - \alpha \right) + \beta f_{mc} \right] \tag{3.4}$$

$$f_{wt} = \frac{2}{3} f_{mt} \tag{3.5}$$

These formulae are combining all parameters affecting the value of f_w. These parameters are, "α" a factor describing the influence of the blocks' shape and the type of the construction, "β" a factor describing the kind of the masonry (stone or brick), "ξ" a factor describing the bed-joint width and the volume of the included mortar, "f_{bc}"

the compressive strength of stones, and "f_{mc}/f_{mt}" the compressive/tensile strength of mortar.

Values of f_{bc}, have to be estimated experimentally. The values of mortar's strengths f_{mc} (compressive) and f_{mt} (tensile), also have to be estimated experimentally by the specially adopted method of fragments' and scratch-width tests. The modulus of elasticity can be estimated as described later in paragraph 3.8.2.

- Broker [10] $$f_{wc} = 0.7 \cdot \sqrt{f_{bc}} \cdot \sqrt[3]{f_{mc}} \quad (3.6)$$
- Brenner [11] $$f_{wc} = 1.4 \cdot \sqrt{2.5 f_{bc}} \cdot \sqrt[4]{f_{mc}} \quad (3.7)$$
- Kirtchig [12] $$f_{wc} = 1.59 \cdot 0.205 \cdot f_{mc} + 0.189 \cdot f_{bc} \quad (3.8)$$
- Hendry [5] $$f_{wc} = \sqrt{f_{bc}} \quad (3.9)$$
$$f_{wc} = \sqrt[3]{f_{mc}} \quad \text{or} \quad f_{wc} = \sqrt[4]{f_{mc}} \quad (3.10)$$
- Mann [13] $$f_{wc} = 0.83 \cdot f_{bc}^{0.66} \cdot f_{mc}^{0.18} \quad (3.11)$$
- Tassios [14] $$f_{wc} = \frac{f_{bc}}{6} + \frac{\sqrt{f_{bc} \cdot f_{wc}}}{4} - \frac{f_{mc}}{20} + 1.4 \quad (3.12)$$
- Tassios [15] $$f_{wc} = \left(1 - 0.8 \cdot \sqrt[3]{\alpha}\right) \cdot f_{bc} \quad \text{if } f_{bc} < f_{mc} \quad (3.13)$$
$$f_{wc} = \left(1 - 0.8 \cdot \sqrt[3]{\alpha}\right) \cdot f_{mc} + 0.4 \cdot (f_{bc} - f_{mc}) \quad \text{if } f_{bc} > f_{mc} \quad (3.14)$$

where the factor α is a shape factor.

3.7.5 Failure criterion

Taking into account all the above data concerning the materials' characteristics, a failure criterion is established. On the basis of the FEM analysis results, this criterion is used for the definition of the failed regions of the structure.

A "classical" experimental investigation on the behaviour of masonry in the case of failure is presented here. This experimental work (biaxial compressive strength of brick masonry) has been elaborated by Page [16].

To define fully the failure characteristics of brick masonry under biaxial stress, a large number of biaxial tests on half scale brick masonry panels were conducted. A biaxial stress state was induced by loading the specimen in orthogonal directions by means of hydraulic jacks. All panels were tested under monotonically increasing load with a constant principal stress ratio. A part of the results of the above experimental investigation is depicted in Figure 3.7.

As can be seen from Figure 3.7, masonry has a strong anisotropic nature depending on the joints which act as planes of weakness.

Experimental and theoretical studies (strength of brick masonry under biaxial tension - compression) have also shown that the failure criteria for brickwork under biaxial compression-tension stress condition is as shown on Figure 3.8 [17]. On this diagram it can be observed that due to the anisotropic strength properties of brickwork it is necessary to consider not only the magnitude of the two principal stresses but also their inclination relative to the bed joint of the masonry.

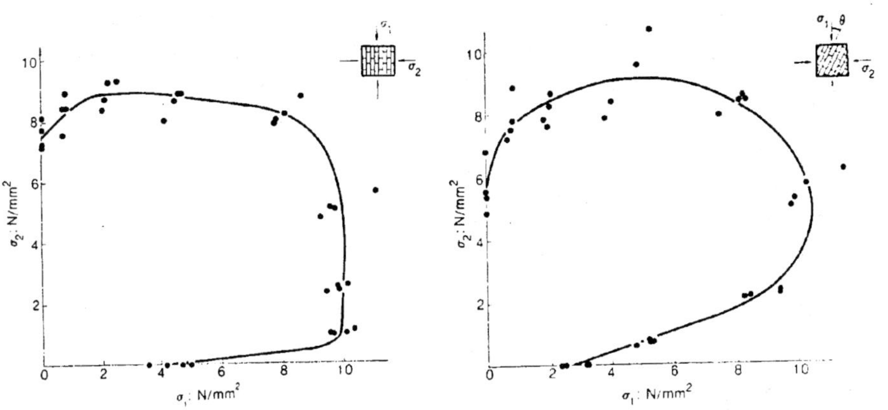

Figure 3.7: Masonry failure curve under biaxial compression for $\theta = 0^o$ and $\theta = 22.5^o$.

To apply the failure criterion defined by the surface shown in Figure 3.8, for example to a shear wall, it is necessary to calculate the principal stresses at a sufficient number of points in the wall and to compare the result at each point with the failure surface. An accurate calculation is complicated if there is cracking at the base of the wall but, if this does not arise, a standard elastic analysis is sufficient and the critical area is readily identified. Strictly speaking, this procedure reveals the point and the load level at which the first crack appears.

Taking into account assumptions and calculations mentioned above from experimental works and Eurocode 6 recommendations, it is possible to develop a semi empirical failure criterion for masonry.

This failure surface is based on a Von Mises modified failure criterion proposed by C.A. Syrmakezis et al., and consists of the four surfaces [18, 19]:

S1: Compression - Compression,

S2: Tension - Compression,

S3: Tension - Tension,

S4: Compression - Tension.

These surfaces are developed making the assumption that the masonry wall is an orthotropic material, and that the calculation of $f_{wc}^{\perp}$, $f_{wc}^{//}$, $f_{wt}^{\perp}$, $f_{wt}^{//}$ are as detailed in Figure 3.9.

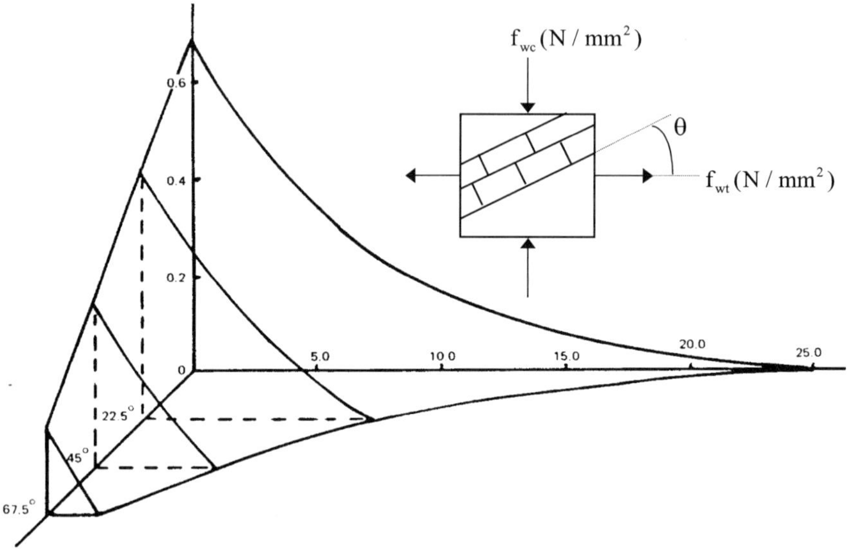

Figure 3.8: Idealised failure surface for brickwork under biaxial tension-compression. The effect of relative inclination of bed joint can be observed.

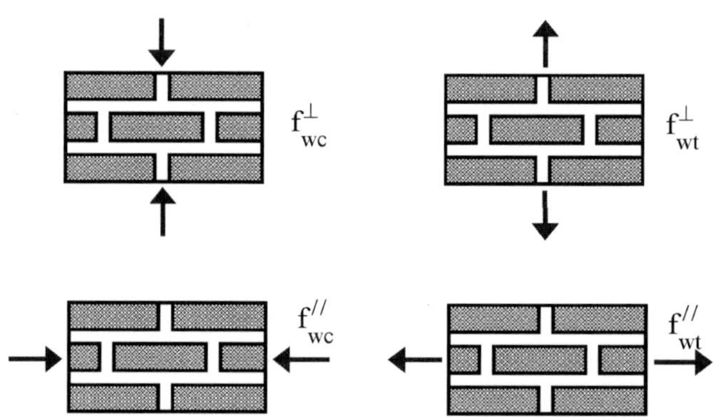

Figure 3.9: Compressive and tensile strength of masonry, normal and parallel to the bed face.

By defining the characteristic and design values of the tensile and compressive strengths of the masonry, the above mentioned Von Mises modified masonry failure criterion is established. This criterion is presented in Figure 3.10.

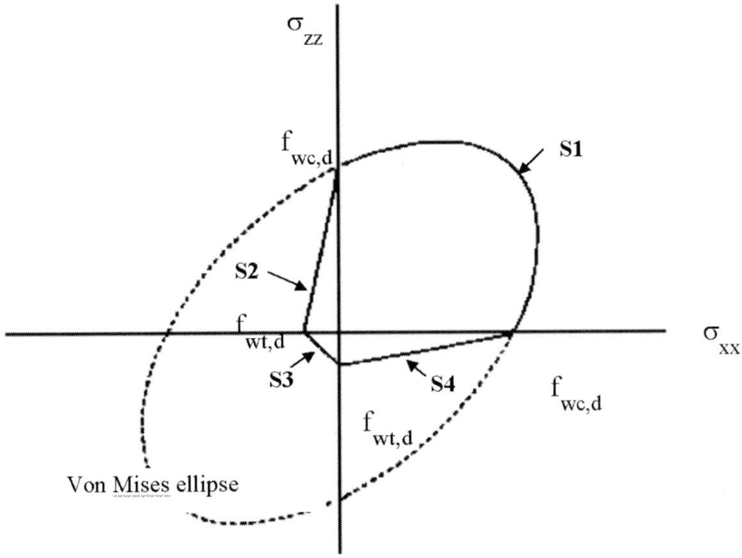

Figure 3.10: Von Mises modified masonry failure criterion in $(\sigma_{xx}, \sigma_{zz})$ terms.

The four surfaces can be expressed by the following equations:

S1: $\sigma xx^2 + \sigma_{yy}^2 - \sigma xx \cdot \sigma_{yy} + 3\tau^2 - f_{wc}^2 = 0$, c (Von Mises ellipse),

S2: $\sigma_{yy} + \left(1 - \frac{\sigma_{xx}}{a}\right) \cdot \sqrt{f_{wc}^2 - 3\tau^2} = 0$, σ_{xx} and $\sigma_{yy} = 0$,

S3: $\sigma_{xx} + \sigma_{yy} - a = 0$, σ_{xx} and $\sigma_{yy} = 0$,

S4: symmetrical of S2 (Figure 3.10), where $a = \frac{f_{wt}}{f_{wc}} \cdot \sqrt{f_{wc}^2 - 3\tau^2}$.

3.8 Applications

In this paragraph the investigation of two characteristic case studies of historical masonry structures are presented. Namely the analysis and redesign procedure for an old church in Greece (Figure 3.11) and a lighthouse in Cyprus (Figure 3.13) are presented in parallel.

Through the presentation of these case studies the methodology developed by the authors, at the Institute of Structural Analysis and Aseismic Research of the National Technical University of Athens, for the analysis - redesign - strengthening of a masonry structure is followed.

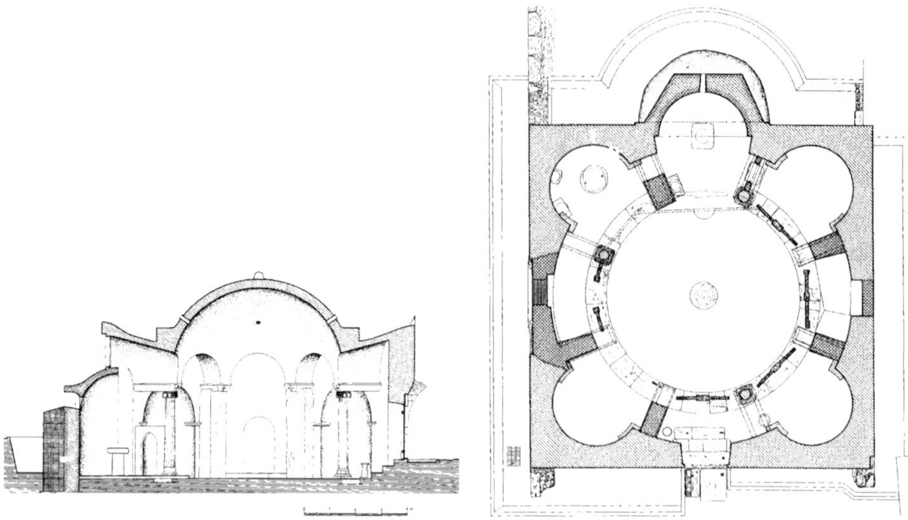

Figure 3.11: Church: section and plan.

3.8.1 Description of the structures

The first case refers to the structure of the Saint Ioannis holy church which is an almost square 12 by 12 metre structure, located on the island of Kos in Greece. The whole structure consists of masonry walls on the perimeter, and a roof shell with a diameter of 7m, supported on eight internal vertical elements. Three of these vertical elements are marble columns and the remaining five are masonry piers. Eight arches connect the perimeter walls with the internal vertical elements supporting the roof. Figure 3.11 shows a section and the plan of the structure.

The second structure is an old masonry lighthouse in Cyprus. This structure suffered a strong earthquake of 6.8R in 1996. Considerable damage was observed and strengthening has been decided upon. The structure consists of a 12m high masonry shell and a 4.25m high steel structure above the masonry. The masonry shell has a cylindrical part up to the level of 2.56m and a variable conical shaped section with a diameter changing from 4.15m to 3.30m for the remaining part. Two sections through the structure are shown in Figure 3.12.

3.8.2 Structural simulation of the structures

For the simulation of the structural characteristics of the structures, 3-D finite element models have been developed. Shell elements, taking into account in and out of plane bending, have been used for the masonry simulation, and for the frame elements of the steel and wooden members.

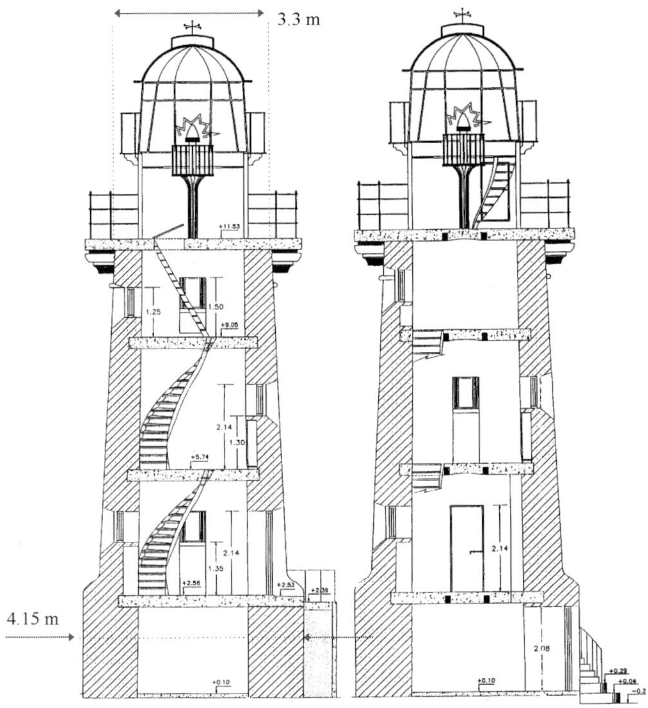

Figure 3.12: Sections of the lighthouse.

In both models all structural elements that add to the stiffness and hence to the response of the structure have been taken into account. Figure 3.13 shows the developed models for the two cases. For the case of the old church 2485 nodes and 1969 elements, while for the case of lighthouse 2259 nodes and 2189 elements have been used.

3.8.3 Simulation of actions

Throughout the analyses, loadings foreseen by the codes have been taken into consideration.

Combinations of dead loads, live loads and earthquake action, base seismic acceleration as well as force distribution along the structure's height have been taken into account following the Eurocodes' requirements, properly modified according to the requirements for each country.

Taking into account the high seismic risk in the areas where these structures are located, as well as the fact that they were almost damaged due to previous destructive earthquakes, seismic loading has been considered along all unfavourable directions of the structures.

The base seismic acceleration values accepted for the analysis were 0.660g and 0.375g for the church and lighthouse respectively. These values are determined by taking into account the design spectra values for each area (Figure 3.14) as well as the dynamic characteristics of each structure.

- for the lighthouse, taking into account its natural period $T_1 = 0.4$s:

$$R_a(T) = \bar{R}_{as} \cdot I \cdot \frac{1}{K} \cdot A_{max} = \frac{2.5 \cdot 1.0 \cdot 0.15g}{1.0} = 0.375g \tag{3.15}$$

- for the church, taking into account its natural period $T_1 = 0.15$s:

$$R_d(T) = A \cdot \gamma_I \cdot \theta \cdot \frac{\beta_0}{q} = 0.24g \cdot 1.20 \cdot 1.0 \cdot \frac{2.5}{1.05} = 0.660g \tag{3.16}$$

3.8.4 Material simulation

After an extensive in-situ experimental investigation for the different types of stones and mortars, and after analytical elaboration of the results, the masonry material data shown in Table 3.1 was adopted for the structures under consideration.

	Compressive strength f_{wc}(MPa)	Tensile strength f_{wt}(MPa)	Modulus of elasticity E(MPa)	Specific weight γ(kN/m^3)
Lighthouse	6.00	0.20	4000	22.0
Church	1.60	0.21	1600	22.1

Table 3.1: Masonry material data.

3.8.5 Analysis results

Using the structural models, loading data and material data described above, FEM linear elastic analysis has been performed and stresses (normal-shear) and displacements at the joints of the mesh have been calculated. In Figure 3.15 the graphical output of these results is presented indicatively.

3.8.6 Failure analysis results

Taking into account the materials' characteristics, a failure criterion was established as described in paragraph 3.7.5. On the basis of the FEM analysis results, this criterion was used for the definition of the failed regions of the structure.

A special computer programme called "FAILURE" has been developed in the Institute of Structural Analysis and Aseismic Research of NTUA for the failure analysis

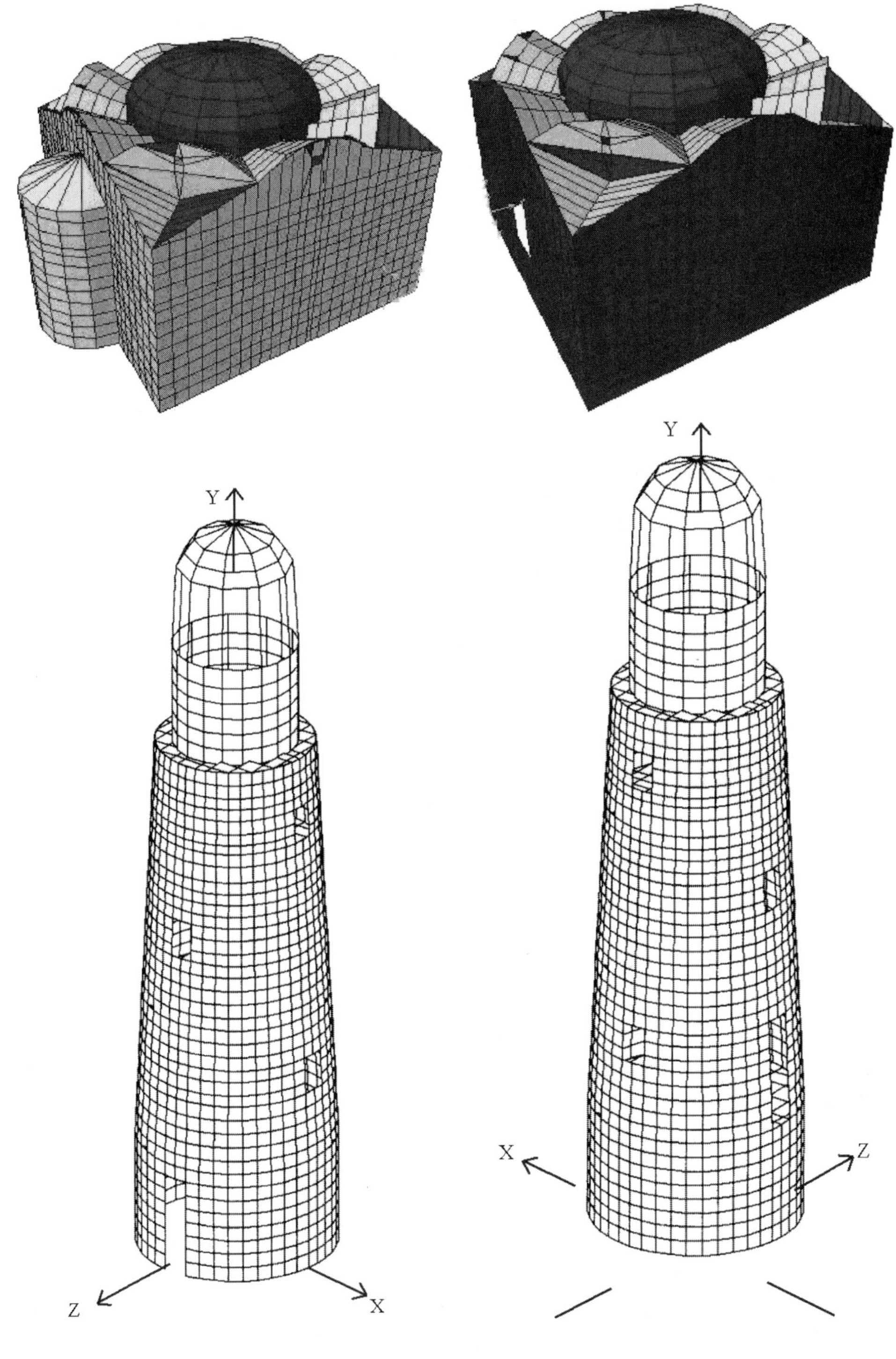

Figure 3.13: The 3-D model of the structures

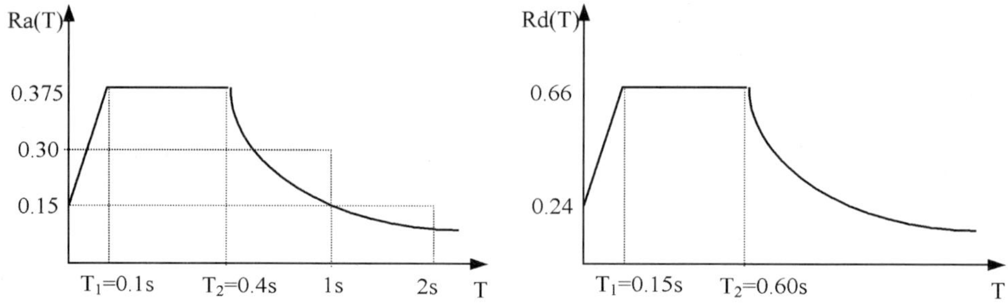

Figure 3.14: The design spectra.

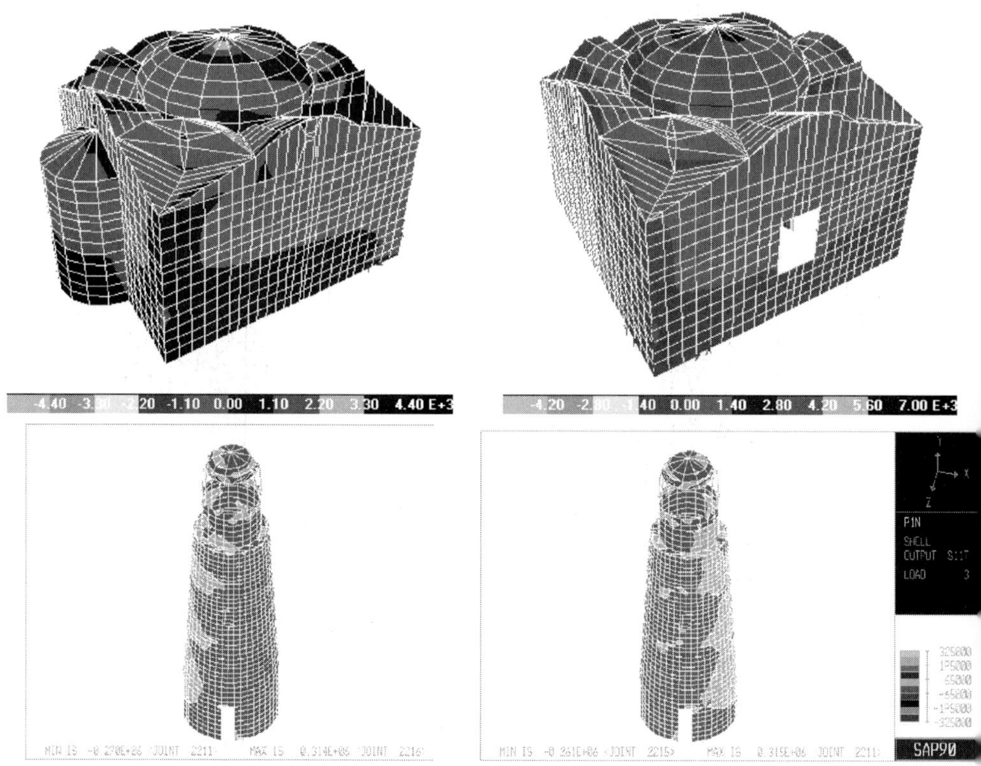

Figure 3.15: Analysis results.

of the structure. The program uses the FEM analysis results and the mechanical characteristics of the materials for the determination of the failure regions of the structure. The failure regions can be automatically plotted directly on the shape of the corresponding wall as well as the type of failure (e.g. BC, BCT, BTC, BT). This information provides a general view for the probable damage level and the main types of damage of the structure. Figures 3.16 shows such views of the structures under consideration.

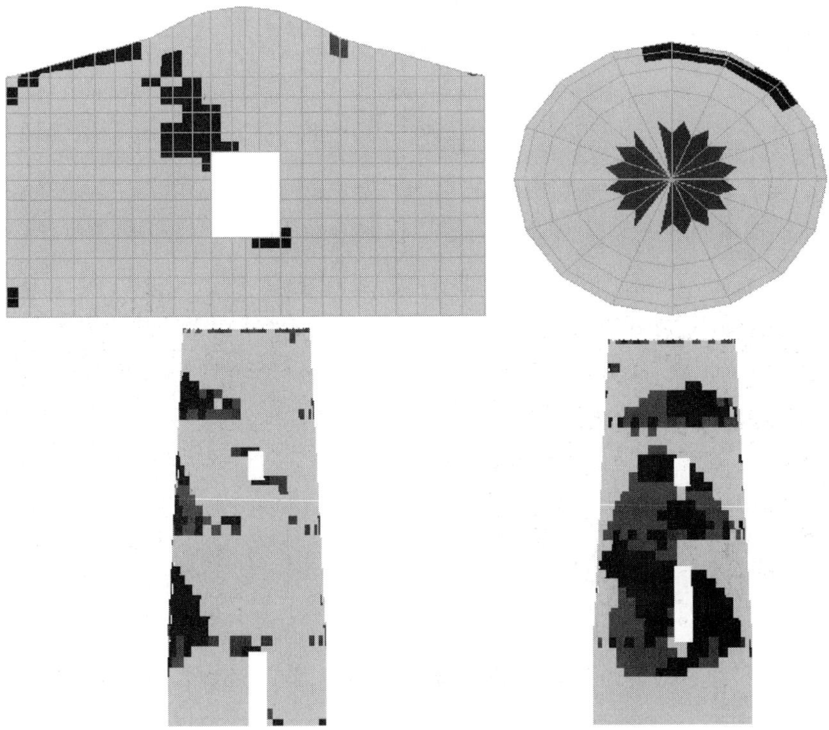

- Failure under biaxial Tension/Tension
- Failure under biaxial Tension/Compression
- Failure under biaxial Compression/Tension
- Failure under biaxial Compression/Compression

Figure 3.16: Failure analysis results.

3.8.7 Repairing and/or strengthening decisions

According to the failure analysis results, decisions have to be taken concerning repair and/or strengthening of the structures. The methods to be used, the extent of the interventions, the type of the materials, etc., are directly related to these failure results.

In order to achieve an appropriate safe/effective/economical repair scheme many scenarios with different repairs have been tried [9]. Finally the following repairs have been decided for each case:

- For the church: deep rejointing, cement-mortar injections, tie-rod confinement elements applied around the base of the masonry shell-roof of the church, foundation strengthening by application of a peripheral reinforced concrete beam, and steel anchorages between masonry walls, vertical columns and piers. All the above measures are shown in Figure 3.17.

- For the lighthouse: deep rejointing, cement-mortar injections, steel anchorages between masonry and reinforced concrete slabs (which were demolished and reconstructed), an internal reinforced concrete jacket at the base (first level) and an application of a steel plate ring at the top (last level) internally. All the above measures are shown in Figure 3.18.

For the estimation of the final mechanical characteristics of masonry after the application of the repair and/or strengthening measures semi-empirical expressions are used. The final improved values for the material mechanical characteristics are as shown in Table 3.2.

	Compressive strength f_{wc}(MPa)	Tensile strength f_{wt}(MPa)	Modulus of elasticity E(MPa)	Specific weight γ(kN/m^3)
Lighthouse	6.30	0.30	4200	23.0
Church	4.00	0.50	4000	23.5

Table 3.2: Improved masonry material data.

3.8.8 Reanalysis

Within the frame of a final redesign, a new structural analysis has been performed, using all new material, loading and structural data. Results of the analysis have subsequently to be used in the process for the final approval (or rejection) of the decisions already taken, concerning repair and/or strengthening of the existing structure.

Taking into account the expected improved response of the structures after the application of the measures concerning repair and strengthening of them, seismic loading has been changed. Mainly behaviour factor q has been changed in order to express the effectiveness of the proposed measures concerning the post-elastic response and mainly the increased energy absorption capacity of the structure. As a result, new design acceleration values have been adopted $R_d(T) = 0.46$g (q=1.50) for the church and $R_a(T) = 0.25$g (K=1.50) for the lighthouse.

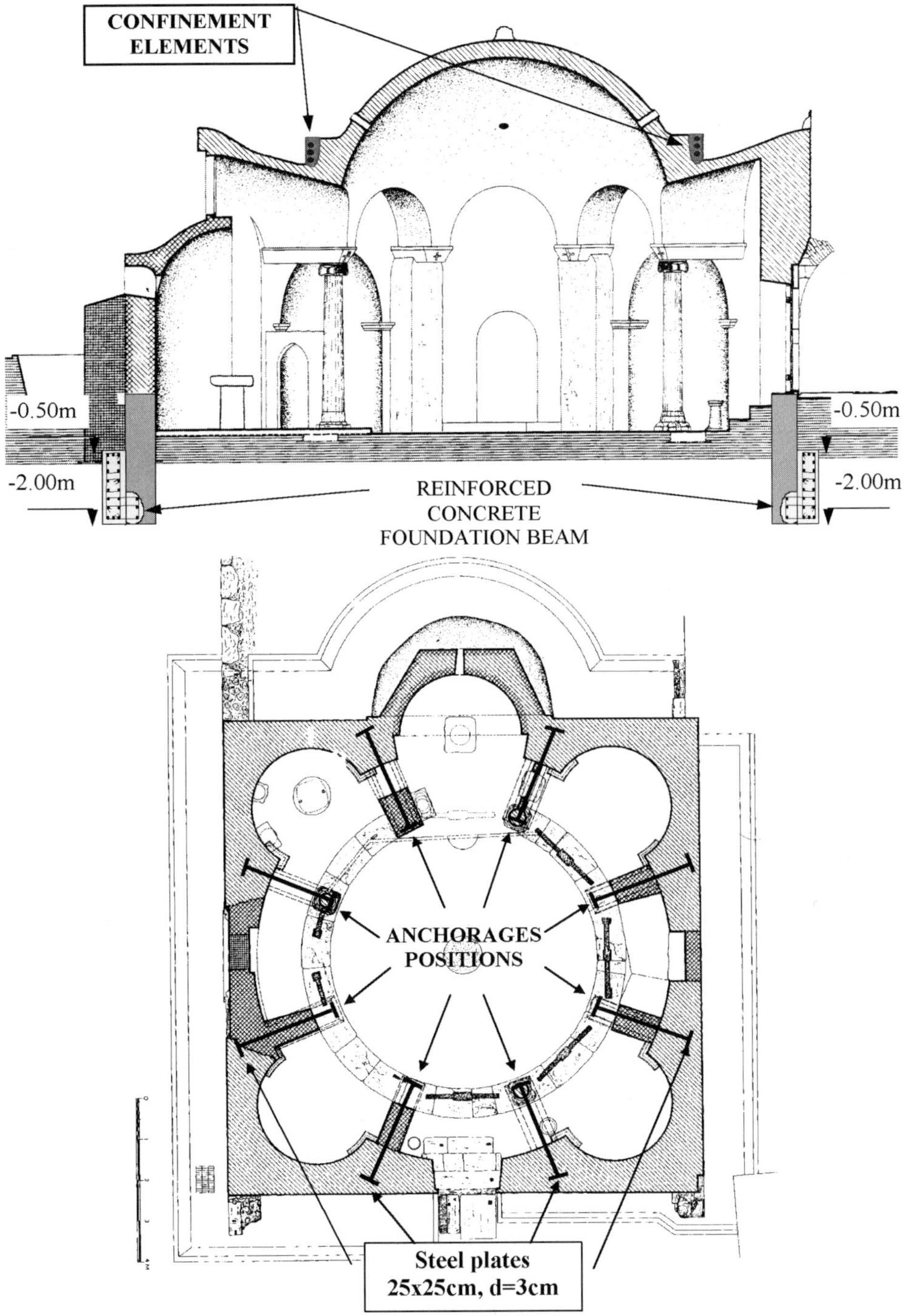

Figure 3.17: Repair/strengthening measures for the church.

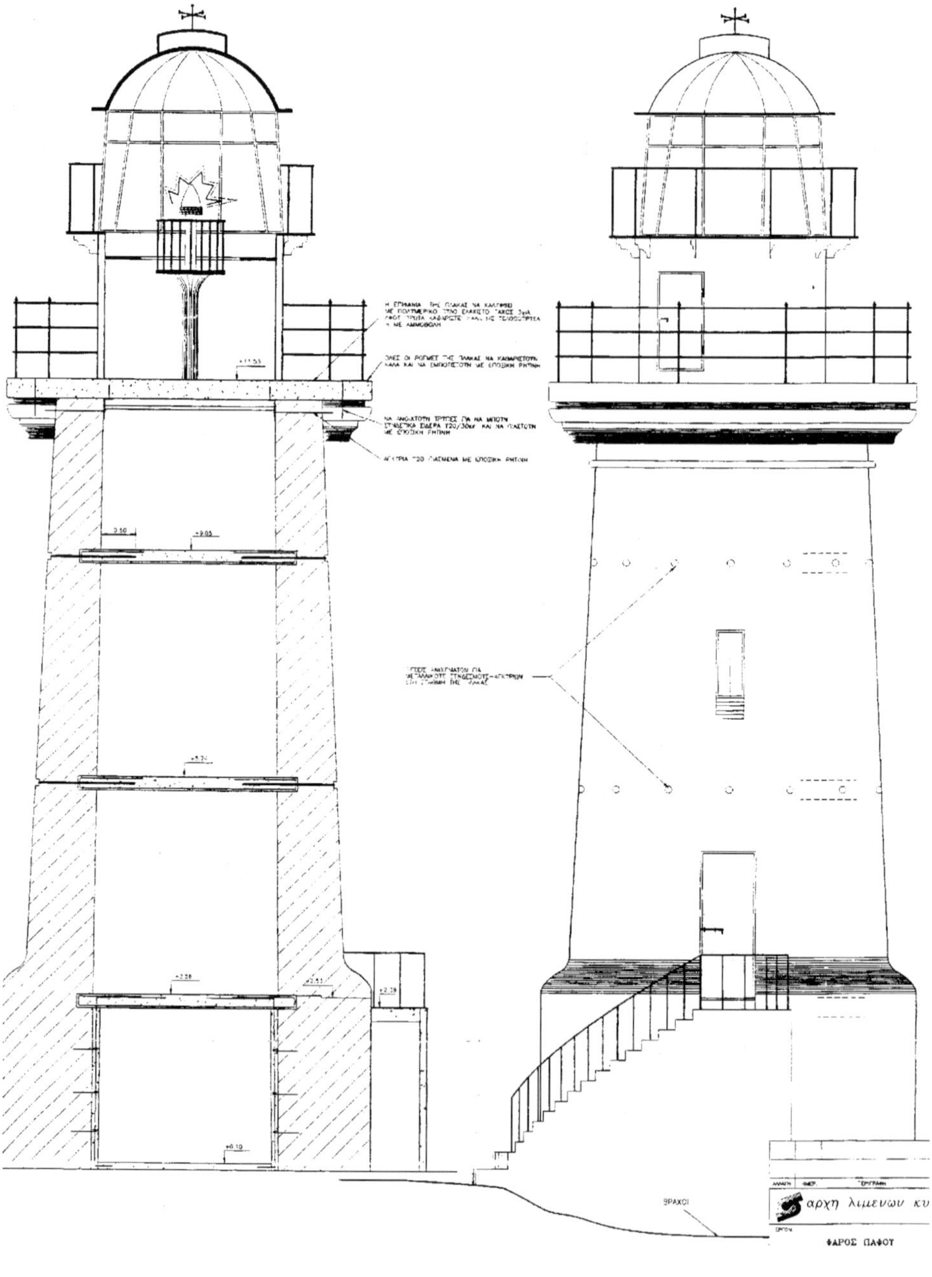

Figure 3.18: Repair/strengthening measures for the lighthouse.

3.8.9 Final Failure Analysis

Using the new analysis results the failure criterion has been applied again. In Figures 3.19, graphical output of the new failure analysis is depicted for both structures. Comparison of these Figures shows the significant decrease in the damage level proving the significant improvement of each structure's structural behaviour and consequently the suitability of the measures proposed.

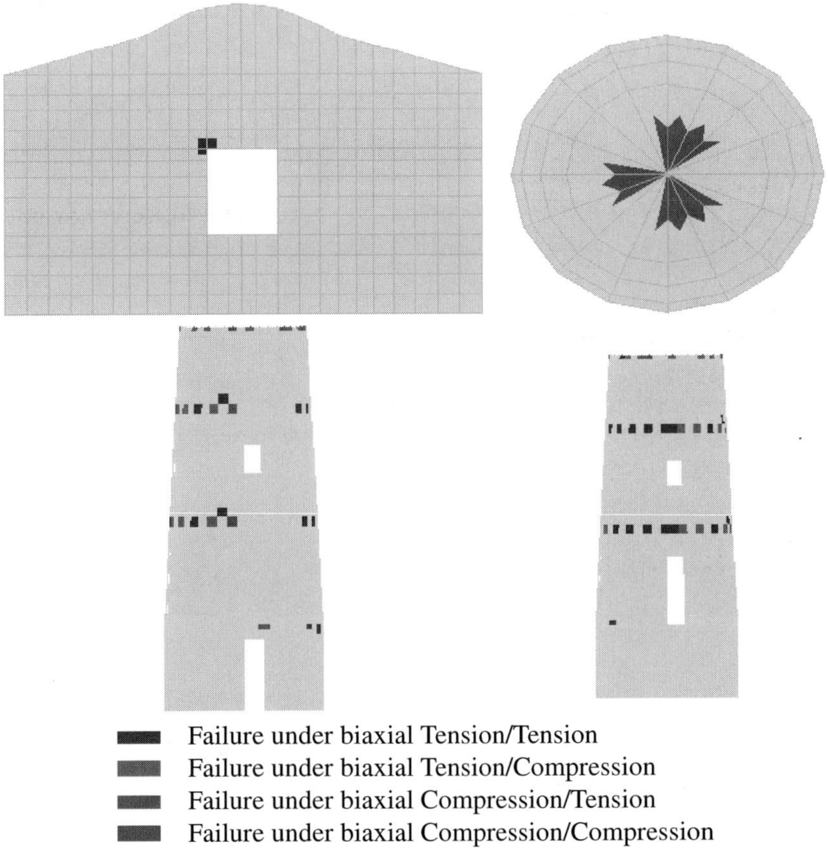

Figure 3.19: Failure analysis results.

3.9 Conclusions

The chapter refers to the problem of the computational modelling of masonry structures due to earthquake and vibration effects, with emphasis on masonry structures. A methodology for the effective structural, material and action modelling is proposed.

The methodology combines theoretical and experimental data, and exploits all modern computer techniques, thus it could become a powerful tool for the designer of all types of masonry structures. The application of the methodology to two characteristic case studies illustrates the fundamental role of each step of the analysis procedure for the design or redesign of masonry structures, and especially of historical masonry structures.

References

[1] Tomazevic M., "Masonry structures in seismic areas: a state-of-the-art report", Proc. of the 9th European Conference on Earthquake Engineering, Moscow, Sept., 1990.

[2] Syrmakezis C.A., Giuffre A., Wenzel F. and Yorulmaz M., "Seismic aspects of preservation of historical monuments", General Report of Working Group 7 of the European Association on Earthquake Engineering, 303-346, Proceedings of the 9th European Conference on Earthquake Engineering, Moscow, Sept., 1990.

[3] Syrmakezis C.A., Giuffre A. and Yorulmaz M., "Seismic aspects of preservation of historical monuments", General Report of Working Group 7 of the European Association on Earthquake Engineering, 3099-3128, Proceedings of the 10th European Conference on Earthquake Engineering, Vienna, Sept., 1994.

[4] Eurocode No 6, "Common unified rules for masonry structures", Commission of the European Communities, Report EUR 9888 EN, 1988.

[5] Hendry A.W., "Structural Brickwork", The Macmilan Press, 1981.

[6] CIB, "International Recommendations for Design and Erection of Masonry Buildings", 1958.

[7] "Mauerwerk-Kalender", 1982.

[8] Chuxian, S., "Analisys of the strength for compressive members of brick masonry under eccentric loads", CIB Third International Symbosium on Wall Structures, Warsaw, June, 1984.

[9] Tassios T.P. and Chronopoulos M.P., "Aseismic dimensioning of interventions (repairs/strengthening) on low-strength masonry building", Middle East and Mediterranean Regional Conference on Earthen and low-strength masonry buildings in seismic areas, Ankara, Aug.-Sept., 1986.

[10] Brocher O., "Steinfestigkeit und Wandfestigkeit Betonstein Zeitung 27", Heft 3, 1961.

[11] Hendry W., "Structural Masonry", MacMillan Education LTD, London, 1990.

[12] Kirtschigg K., "On the failure mechanism of masonry subject of compression", Seven International Brick Masonry Conference, Melbourne, 1985.

[13] Mann W., "Statistical Evaluation of Tests on Masonry by Potential Factors", Proceedings of the Sixth International Brick Masonry Conference, Rome, 1982.

[14] Tassios T.P., "Physical and mathematical models for redesign of damaged structures", Introductory Report, IABSE Symposium, Venice, 1983.

[15] Tassios T.P., Chronopoulos M.P.: "Aseismic dimensioning of interventions

(repairs/strengthening) on low-strength masonry building", Middle East and Mediterranean Regional Conference on Earthen and low-strength masonry buildings in seismic areas, Ankara, Aug.-Sept. 1985.

[16] Page A.W., "The Biaxial Compressive Strength of Brick Masonry", Proceedings of Institution of Civil Engineers, September, 1981.

[17] Page A.W., "The Strength of Brick Masonry under Biaxial Tension-Compression", International Journal of Masonry Construction, Vol.3, No1, 1983.

[18] Syrmakezis C.A., Chronopoulos M.P., Sophocleous A.A. and Asteris P.G., "Structural analysis methodology for historical buildings", Vol. 1, 373-382, Proceedings of the Fourth International Conference on Structural Studies of Historical Buildings, STREMA 95, 1995.

[19] Syrmakezis C.A., Sophocleous A., Asteris P. and Liolios A., "Earthquake resistant design of masonry structural systems", Vol. 2, 717-726, Proceedings of the First International Conference on Earthquake Resistant Engineering Structures, ERES 96, 1996.

©2001, Saxe-Coburg Publications, Stirling, Scotland
J.W. Bull, (Editor), Computational Modelling of Masonry, Brickwork and Blockwork Structures,
Saxe-Coburg Publications, Stirling, Scotland 79-108.

Chapter 4

The Dynamics of Masonry Bell Towers

A.R. Selby and J.M. Wilson

School of Engineering, University of Durham, United Kingdom

Abstract

The English system of bell-ringing is to swing the bell through 360° from the mouth-up position. The bell acts as a compound pendulum, and applies significant transient horizontal forces on to the bell tower.

The dynamic responses of 19 church towers in the North-East of England were recorded using low frequency accelerometers, during bell-ringing. The bell forces were estimated from *in-situ* measurements.

Finite element methods were used to estimate natural frequencies, mode shapes and transient response of the bell towers. Effective stiffness and mass of the masonry sandwich walls were deduced for each tower.

Serviceability and ultimate limit states are discussed.

4.1 Introduction

As we enter the new millenium it is particularly timely to consider the stability and dynamic behaviour of church bell towers. At noon on January 1st 2000, thousands of church bells rang out across the United Kingdom [1]. Over 100 projects for enhancing existing rings of bells, and of refurbishing defunct towers, were funded from a £3 million grant from the Millenium Commission with a further £3 million needing to be met from local fund-raising and donations.

Traditional church bell towers are normally part of the overall structure of the church building [2]. The tower may be at one end of the nave, usually the west, it may be in the NW or SW corner, or it may rise above the intersection of the nave and transepts. A small church tower may be only 15m high, while a cathedral tower may rise to 60m or more. Towers may be square-topped or they may be surmounted by a pointed spire. They are typically square in plan. Foundations may be spread wall footings at a depth so as to bear on to stiff soil strata. The walls are of sandwich construction, with an outer skin of ashlar stone and an inner skin of regular or random stone [3]. The cavity between the skins is normally filled by loose rubble, which may be lightly cemented. The bells are hung on a frame built into the tower some 5m or more below the top.

Church bells are common throughout the world. They are generally cast in bell metal, a copper tin alloy of bronze, which gives a pure resonant note when struck by a clapper. Bell-ringing in the United Kingdom is unique (or nearly so) in that the bell is rung by swinging in a complete circle of 360° starting from the mouth-up position [4]. A clapper hangs inside the bell, and is carried round the circular path with the bell until the bell slows towards the top of the circle; the clapper continues onward and strikes the opposite rim of the bell. A rope running around a wheel attached to the bell axle is used to initiate and control the bell swing. This technique is in contrast to the common continental practice of hitting the rim of a static bell in mouth down position by a mechanical hammer, and to the Austrian style of swinging bells through some ± 60° from mouth down.

An English church tower may house from three to twelve bells of different sizes [5]. They will be tuned and matched to provide several notes within a musical major scale. The smallest, treble, bell provides the highest note, and the largest, tenor, bell gives the deepest note. While a single bell may be tolled at a steady pace, change ringing or rounds may be performed by ringing all the bells in progressive patterns. There is mathematical order to the rounds, and many schemes have been developed over hundreds of years [4].

The full-circle swing of a heavy bell about its axle imparts to the tower a significant time-varying force with both horizontal and vertical components. This force causes the tower to respond dynamically, primarily in lateral sway [6]. The tower may be considered as a vertical stocky cantilever, which responds laterally with three components of movement. The three components are rocking on the foundations, shearing and bending [7]. The shear contribution may be considerable because the cantilever is short and deep, and because of the compliance of the mortar joints. Other forms of response, including torsion and vertical bounce are generally very small.

The tolling of a heavy bell in a masonry tower offers a rare opportunity to the structural engineer to make detailed measurements of its structural response to a defined dynamic force. With care, it is possible to measure accurately the fundamental frequency of sway, damping, the mode shape and the peak response. Detailed measurements of the transient behaviour of 19 towers were undertaken by Lund [8], and others, and will be used here for comparisons with computational solutions.

It is a straightforward task to build a finite element (FE) computational model of the tower for small strains, but there are several parameters which are difficult to quantify. An FE model can be calibrated against the measured response data, which allows deduction of effective stiffnesses of tower and foundation and effective mass of the walls. These parameters can be estimated for a number of church towers of varying configuration, to give typical and extreme values. This procedure has been followed for a number of church towers in the north-east of England.

None of the 19 towers visited showed evidence of structural degradation due to bell-ringing, but in one tower the bells had been re-hung lower down the tower because of fears by the architect of excessive motion. In this and two other towers the bell-ringers considered the tower to be very 'lively' and close to the limit at which it was feasible to ring rounds. From these observations, some deductions are made regarding effective serviceability limit states. Simple calculations of tensile or shear failure and of overturning show very large factors of safety.

If a tower has sustained damage either from bell-ringing or from other causes such as differential settlement, repairs may be needed and modifications to the bells or bell frame may be necessary. Neiderwanger [9] reported some cases of damage and repair. Modifications may include counter-weights or extra tuck-up of bells, frame stiffening or ringing restrictions.

4.2 Tower construction, bell frames and bells

The construction of a typical bell tower is similar to that shown in Figure 4.1, which refers to the tower of St Brandon's, Brancepeth [10]. The tower is built above an area of the west end of the nave and shares part of the church west end wall. The walls are of sandwich construction, and are thicker towards the base. The outer leaf is of random coursed ashlar blocks while the inner leaf is of regular coursed blocks. Only the overall thickness of wall can be measured easily and without penetrative methods, although Lilley & March [11] have suggested probing radar techniques may be helpful. Experience suggests that the leaf thicknesses vary considerably, from 20% each up to solid masonry. The cavity between leafs may be empty, or partly or totally filled with rubble which may be lightly cemented. The leafs may be connected at intervals with key stones and at the returns around openings, giving some structural integrity.

In St Brandon's tower there is a large cut-out arch in the lower level of the east tower wall. There are a number of smaller openings, to provide natural light to the several chambers. More significantly at the level of the bells there are large openings on all four sides of the tower with timber louvres to allow the sound of the bells to be heard within a distance of a kilometre or more. These latter openings make little change to the stiffness or mass because of their arched form and stiffened returns. However, major archways at nave level do affect both effective area and second area moment, and should be incorporated in analysis. Conversely, massive columns may be present in framing to arches, which enhance the area of active material. Many towers are stiffened over their lower levels by external buttresses at the corners.

Traditional towers in the north-east of England are constructed of sandstone masonry with lime mortar joints. The sandstone is initially a honey-coloured cemented quartzitic or siliceous stone, which is easily worked and is open-quarried from many outcrops. Its stiffness and density have been measured and reported [12] although

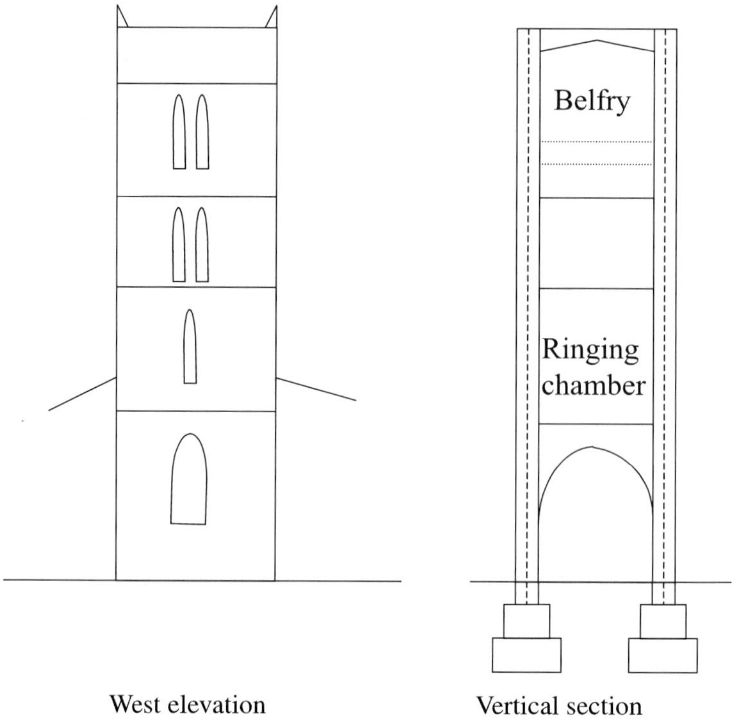

Figure 4.1: The bell tower of St Brandon's, Brancepeth.

its durability does not match that of the limestones used in churches in central and southern England [13]. Typically, the stiffness of the intact sandstone in uniaxial compression at small strain is 17 GPa, and its density is 2200 kg/m^3. Little is known of the mortar properties.

There are several floors at different levels up the tower, and a flat roof where the tower has no spire. The floors are generally of timber beams and planking, and the roof will be clad in lead. These floors give negligible stiffness or mass to the tower. The bells are supported on a bell frame, which is a trussed timber or steel system spanning both east-west and north-south.

The foundations of a tower will be spread stone footings or a timber raft at such a depth as to bear on to stiff stable soils or rock. Again, a non-intrusive survey cannot determine their extent, nor can the stiffness of the soil or rock be estimated readily, for an appropriate level of strain.

To summarise, the overall dimensions of the tower can be obtained by simple standard surveying techniques, but the stiffness and mass of the walls and foundation are difficult to quantify because of several unknown parameters. These include proportions of the sandwich walls, cavity fill, properties of the mortar, partial attachment to the church nave, and the type and size of the foundations.

From surveys of 19 church and cathedral towers in the north-east of England, primary dimensions of the towers and the mass of the bells tolled in the survey are summarised in Tables 4.1 to 4.4. The churches are grouped into four main types: those with a spire, those with a traditional square topped tower, those with a squat, square topped tower and the two Cathedrals. The division between the second and third types is arbitrarily taken for a height to width ratio of greater than or less than 3.

	Height to spire tip m	Height to bells m	Width m	Wall thickness m	Bells, number and dir.	Bell mass kg
St Mary's Shincliffe	27.1	11.3	5.32	1.05	6 N/S	210
St Cuthbert's Chester-le-street	43.5	13.3	7.24	0.92	8 E/W	966
St Nicholas Durham	43.5	18.4	5.18	0.7	6 N/S 1 E/W	516 203
St Cuthbert's Benfieldside	33.8	10.0	5.01	1.26	6 N/S 3 E/W	642 263

Table 4.1: Primary dimensions of towers with spires.

	Height of tower m	Height of bells m	Width m	Wall thickness m	Bells, number and dir.	Bell mass kg
St Oswald's Durham	24.8	16.9	7.2	1.16	8 E/W 7 N/S	650 447
All Saints Lanchester	19.7	14.9	6.03	0.885	6 E/W	432
St Matthew's Newcastle	28.0	17.6 (ex 22)	9.16 7.19	1.04	8 N/S 7 E/W	1588 1103
St Andrews Bishop Auckland	23.2	20.0	7.2	1.11	8 N/S 6 E/W	609 356
St Edmund Sedgefield	25.4	20.1	7.29	0.97	5 N/S 4 E/W	533 469
St Brandon Brancepeth	20.0	16.0	5.6	1.0	8 E/W 7 N/S	693 498
St Mary's Richmond	24.0	18.9	5.91	1.13	8 E/W 7 N/S	408 559

Table 4.2: Dimensions of traditional square topped towers.

	Height of tower m	Height of bells m	Width m	Wall thickness m	Bells, number and dir.	Bell mass kg
Christchurch Consett	16.25	14.0	8.55	0.995	8 N/S 6 E/W	822 434
St Margaret's Tanfield	16.5	11.5	6.2	0.75	8 E/W 7 N/S	660 432
St John's Shildon	18.5	13.5	6.86	0.9	8 E/W 7 N/S	829 589
St Michael's Heighington	16.2	14.1	6.04	0.8	6 E/W 4 N/S	782 432
St Michael's Houghton	21.4	16.5	7.48	0.84	8 N/S 7 E/W	610 464
St Stephen's, Willington	14.2	12.1	5.55	1.00	2 N/S 3 E/W	330* 330*

* estimated from bell dimensions

Table 4.3: Dimensions of 'squat' towers.

	Height of tower m	Height of bells m	Width m	Wall thickness m	Bells, number and dir.	Bell mass kg
Durham Cathedral	66.0	57.0	13.5	1.7	10 N/S 9 E/W	1425 1096
Newcastle Cathedral	32.8 (main tower)	18.8	11.0	1.17	12 N/S	1913

Table 4.4: Dimensions of the Cathedral towers.

4.3 Forces from a swinging bell

Each bell within a peal is supported by a headstock, which rests on a pair of trunnion bearings fixed to the upper chord of the bell-frame truss. A large lightweight timber rope-wheel is fixed to the headstock and to one side of the bell, and the rope passes round the wheel and then runs down the tower to the ringing chamber. A slider prevents the bell from swinging much beyond mouth-up, Figure 4.2.

The ringer controls the bell by judicious pulling of the rope, see Figure 4.3. During tolling, the bell starts from mouth-up and a small pull on the rope causes it to swing though 360° (or a few degrees more); the bell pauses at this point and then swings back through 360° to its start position. The complete cycle of forward and return swing causes very characteristic dynamic horizontal and vertical forces which are imparted through the bell frame into the tower.

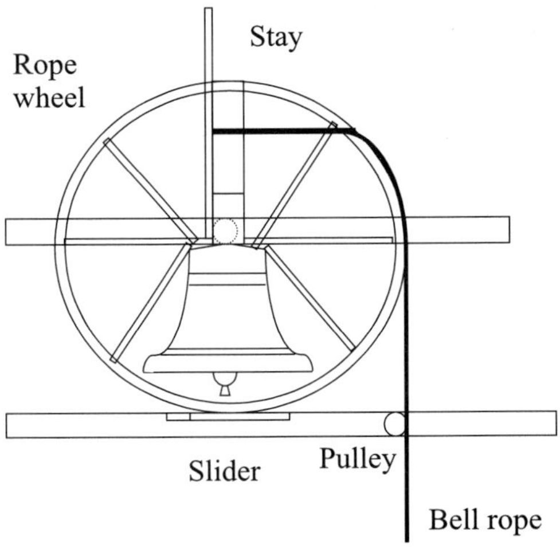

Figure 4.2: Side elevation of bell, rope wheel and support frame.

Essentially, the bell acts as a large compound pendulum, Figure 4.4, which generates dynamic forces because it is constrained to accelerate around a circular path centred on a fixed point O. Additionally the bell has its own moment of inertia $J_g(= mk^2)$ requiring a moment to accelerate rotation about its own axis. Lewis [14] originally set up the equations of motion of a bell as a compound pendulum.

The moment of inertia of the bell about the point O, J_o, at distance h from the centroid is

$$J_0 = J_g + mh^2 = m(k^2 + h^2) \tag{4.1}$$

For small oscillations, the frequency of oscillation, w is given by

$$w^2 = gh/(h^2 + k^2) \tag{4.2}$$

corresponding to a period, τ, of

$$\tau = 2\pi\sqrt{(h^2 + k^2)/gh} \tag{4.3}$$

Use of an inertial form factor c introduced by Heyman & Thelfall [15], of

$$c = h^2/(h^2 + k^2) \tag{4.4}$$

and introducing p, a parameter slightly in excess of unity to accommodate initial velocity and dwell position, where

$$p = \cos\alpha + w_\alpha^2\tau^2/8\pi^2 \tag{4.5}$$

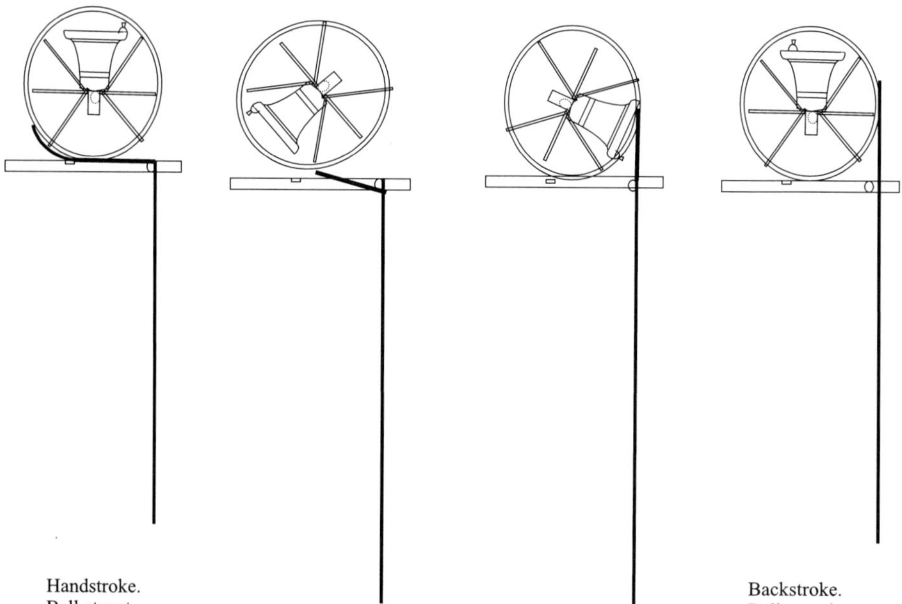

Figure 4.3: Bell positions during one half of the complete ringing cycle.

leads to expressions for horizontal force H and vertical force V in terms of the angular position of the bell, θ, of :

$$\begin{aligned} H &= mgc\sin\theta(3cos\theta - 2p) \\ V &= mgc(1/c - 1 + 3\cos^2\theta - 2p\cos\theta) \end{aligned} \tag{4.6}$$

The maximum value of the horizontal force occurs twice during a single swing but in opposite sense, when θ has an angle of 124° or 236°; the maxima occur again during the reverse swing. If the initial velocity is sufficiently small for p to approach unity then the maximum horizontal force, H_{max}, is

$$H_{max} = 3.05mgc \tag{4.7}$$

The vertical force has a maximum of V_{max} as it swings through the mouth down position, i.e.when $\theta = 180°$, which acts in the downward direction and hence is a stabilising action. For the condition of $p \approx 1.0$,

$$V_{max} = mg(1 + 4c) \tag{4.8}$$

Typical values for the parameter c range from 0.4 to 0.7, depending upon the degree of "tuck-up" of the bell into its headstock.

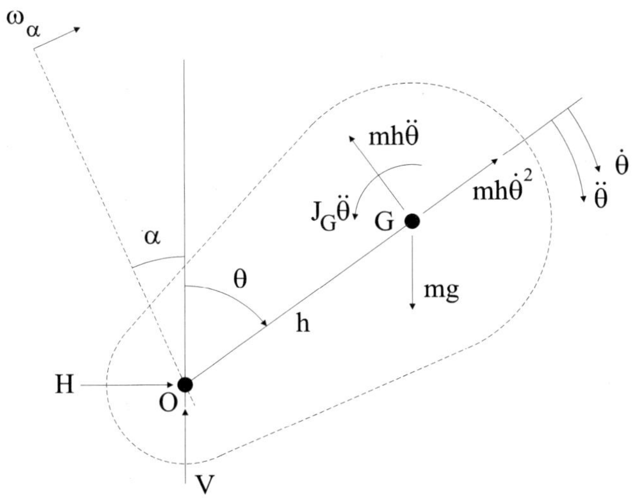

Figure 4.4: Dynamic equilibrium of a compound pendulum.

The definition of bell force as a function of *position* is not what is required in a time-stepping computation of tower response. Rather, the horizontal force is required as a function of *time*. This requires that the bell is moving with a small finite velocity at the top dead centre position. This can be achieved by recognising that the bell moves a small angle α beyond the vertical, where it rests briefly in a dwell position. The solution for this condition is by an elliptic integral of the first kind, (Birlisch [16]) which depends upon α, θ and w_α This procedure requires a computational approach. Alternatively, the position of the bell after a certain elapsed time may be calculated by the use of Jacobian functions.

For certain simple methods of ringing, including tolling and ringing rounds, the forces on the tower are periodic, and can be represented by Fourier series as described by Wilson [17]. For any one bell the horizontal force has a very characteristic shape, e.g. Figure 4.5.

Before such computations can be made for a given bell, it is necessary to quantify the three parameters of bell mass, m, the distance of the centroid from the axis of rotation, h, and the radius of gyration of the bell, k. The mass of the bell can be found from records held in the bell-tower or at the foundry (if it is still in existence), and this may be enhanced by 20% to account for the mass of the headstock. Following the procedure of Heyman and Threlfall [15], the quantities k and h may be estimated from some simple measurements which can be made in the belfry. The first process is to hang incremented weights on the bell rope, and observe movement of the rope-wheel. By moment equilibrium, as shown in Figure 4.6,

$$Wr = mgh \sin \beta, \quad \text{and} \quad d = r\beta \tag{4.9}$$

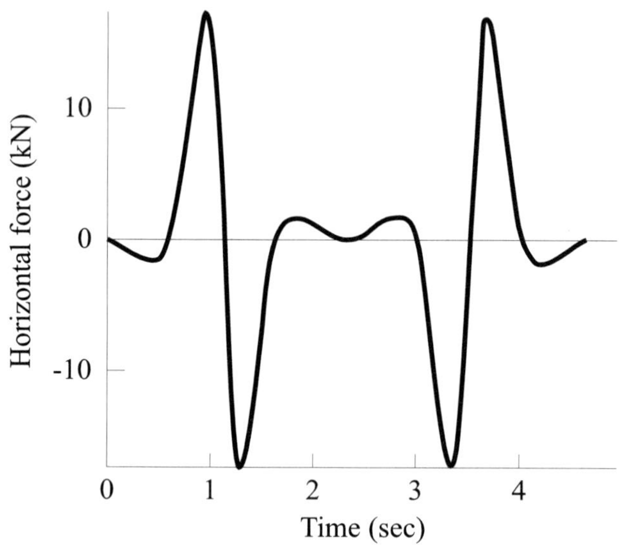

Figure 4.5: Horizontal forces from a swinging bell as a function of time. (Bell 10 of Durham Cathedral).

so for small angles of β,

$$W = dmgh/r^2, \tag{4.10}$$

and a plot of W against d will yield a value for h. It is perhaps worth mentioning the obvious, that the measurement of d should be taken at the rope-wheel to avoid the effects of rope stretch.

The second measurement is the frequency of small amplitude oscillation, w, of the bell about its mouth down position, which is defined by

$$w^2 = gh/(h^2 + k^2) \tag{4.11}$$

This is estimated by recording the time required for say 20 or 50 small oscillations, which can easily be initiated by manual displacement of the bell.

Calculations of k, H_{max} and V_{max} follow directly. Tables 4.5 to 4.8 give values of the maximum bell forces for the heavier bells in the several towers investigated in the study.

A laboratory study to measure directly the vertical and horizontal forces during full circle swinging of a compound pendulum was undertaken by Kipling [18]. Traces recorded during this work are reproduced in Figure 4.7, verifying the preceding theory.

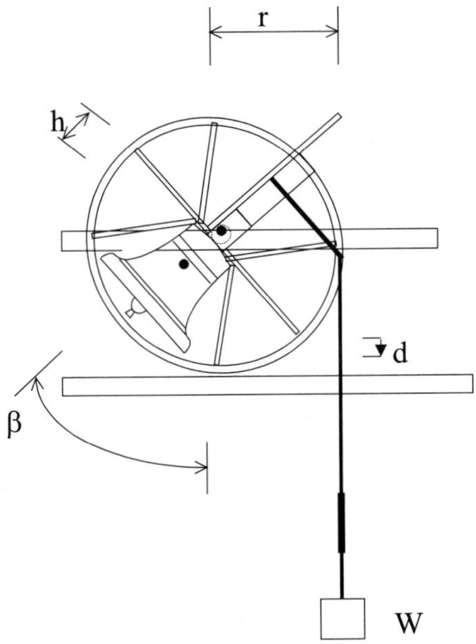

Figure 4.6: Measurements to determine the position of the bell centroid.

4.4 Measured Tower Response

To date, a total of 19 towers have been surveyed, [19], for their structural geometry, for the dynamic loads from the heavier bells, and for their transient motion. Three churches and a cathedral are shown in Figure 4.8, to indicate the range of size and architectural styles within the survey sample.

Tower dimensions were measured by taping, both in plan and vertically, and by theodolite. Overall wall thickness could be found from inner and outer dimensions and also at window openings. However, there was no non-intrusive method by which to estimate leaf thicknesses or cavity width. Connection to the main structure of the church was difficult to assess, and in some instances vertical cracks indicated some separation. Minor window openings were noted but were generally ignored in later analyses on the "give-and-take" principle of extra stiffness provided by the returns. Larger archway accesses were recorded for the analyses, as were internal columns and stairways and external buttresses. Foundations again could not be investigated.

Detailed measurement of transient tower motion is best done using high sensitivity low frequency accelerometers with integral charge amplifiers to allow signal transmission over distances of 20m or more without attenuation. Signals are then amplified and fed to a two-pen chart recorder for instant checks and reliable long signal storage. B and K type 8318 accelerometers with line-drive supply unit 2813 were found to be

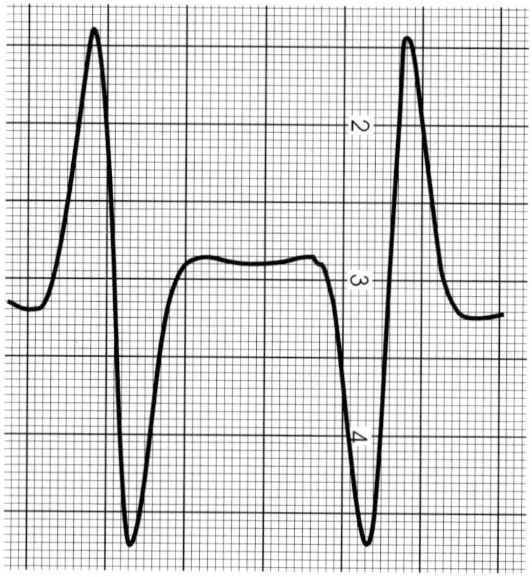

Figure 4.7: Measured horizontal forces from a laboratory model of a bell.

ideal for this work, although very low frequency geophones (velocity transducers) are an acceptable alternative.

The procedure followed was to set one accelerometer at ringing chamber level, orientated horizontally and parallel to the swing of the chosen bell (either north-south or east-west). The second, wandering, accelerometer was then placed at roof level. The bell was then tolled for some ten cycles, and the two traces were recorded during and after tolling. The wandering accelerometer was then moved and the bell was tolled again. The process was repeated with the wandering accelerometer positioned in many locations around the tower, in both horizontal and vertical mode.

A typical pair of traces of acceleration against time is shown in Figure 4.9. The form of response is characteristic of all the towers in the survey, in that on one swing of the bell the tower responds and then during the dwell period, the tower oscillates at its fundamental frequency but with diminishing amplitude due to material damping; the bell then swings again and the tower responds with a repeated pattern; when tolling ceases the tower continues to oscillate towards rest with clearly displayed lightly damped behaviour. Measurements from the free-oscillation decaying signal gave the fundamental frequency and the damping factor.

By referencing all measurements to the signal from the first, static, accelerometer details of both vertical and horizontal mode shapes could be assembled, even if the ringing procedure varied slightly. Vertical signals at ground level indicated the degree of rocking movement of the tower on its foundations. Critical evaluation of the tower slope and curvature gave indications of relative shear and of bending of the tower

Figure 4.8: Views of Durham Cathedral, St. Oswald's Durham, St. Cuthbert's Chester-le-Street, and Christchurch Consett.

above ground [8]. While the components of base-rocking, shear and bending were present in all towers studied, there was negligible display of vertical *bounce*, torsion or of higher modes beyond the fundamental.

Tables 4.5 to 4.8 summarise the measurements from the towers. Where accelerom-

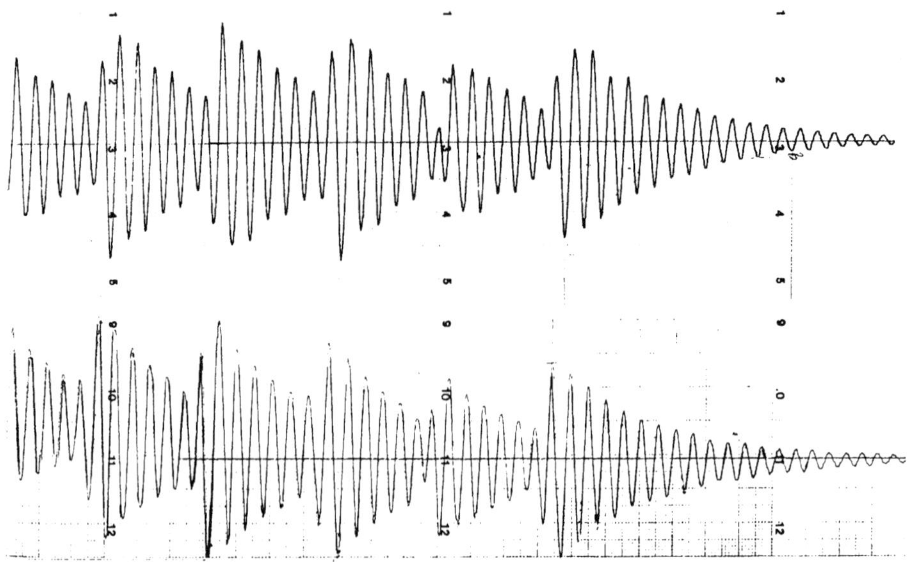

Figure 4.9: A typical pair of acceleration/time traces.

eters were used, then values are quoted in terms of m/s^2, and maximum displacement is by deduction using the assumption that $u'' = (2\pi f)^2 u$, as is the case for a pure sine wave. In the few cases where velocity was measured, conversion was again based on the pure sine wave assumption. It should be noted that in only one case did peak amplitude exceed 1mm, despite the typical subjective observation that *the tower must be swaying several inches*.

The results of Tables 4.5 to 4.8 show a number of general features. The fundamental frequencies range from 1.28Hz in Durham Cathedral, which is the tallest at 66m, to 4.21Hz in St Stephen's, Willington, which is a mere 14.2m tall. The frequency for each tower was often slightly different in the north-south and the east-west directions. This is attributable to the main cut-outs at ground level and to the degree of build-in to the main church structure, either to the nave or to the end cross wall; only St Matthew's in Newcastle is significantly rectangular in plan. The largest horizontal transient movement was 1.3mm measured near the top of the spire of St Cuthbert's in Chester-le-Street.

The proportions of base rocking, shearing and bending varied considerably, with that for rocking being as much as 40%, shearing being up to 70% and bending being up to 70% for the tall slender towers. Damping factors ranged between 0.014 and 0.078, although the majority of towers showed values between 0.02 and 0.03. The high value of 0.078 in Christchurch indicates some structural cracking or sliding mortar joints, which were not identified visually.

	Bell, number and dir.	Max. horiz. bell force N	Nat. freq. Hz	Dam-ping factor	Max. accn. mm/s^2	Max. displ. mm	R/S/B %
St Mary's Shincliffe	6 N/S	3479	2.36	0.018	11*	0.05	20/20/60
St Cuthbert's Chester-le-street	8 E/W	12620	1.59	0.028	130**	1.27	20/36/44
St Nicholas Durham	6 E/W	3787	1.82	0.014	32*	0.6	6/34/60
St Cuthbert's Benfieldside	6 N/S 3 E/W	7178 3846	2.25 2.27	0.027	24* 17*	0.12 0.08	n/a

* Measured at base of spire **Measured at 5m from point of spire

Table 4.5: Response characteristics of towers with spires.

	Bell, number and dir.	Max. horiz. bell force N	Nat. freq. Hz	Dam-ping factor	Max. accn. mm/s^2	Max. displ. mm	R/S/B %
St Oswald's Durham	8 E/W 7 N/S	7012 4793	2.01 2.05	0.026 0.026	60 45	0.38 0.33	22/35/43 30/23/47
All Saints Lanchester	6 E/W	8081	2.56	0.026	85	0.33	16/72/12
St Matthew's Newcastle	8 N/S 7 E/W	9987 7314	1.92 1.49	0.026 0.028	58 74	0.40 0.84	22/60/18 9/60/31
St Andrews Bishop Auckland	8 N/S 6 E/W	5827 3721	2.51 2.20	0.065 0.051	29 33	0.12 0.17	26/44/30 27/45/28
St Edmund Sedgefield	5 N/S 4 E/W	3844 4163	2.28 2.20	0.018 0.017	33 41	0.16 0.21	40/47/13 35/45/20
St Brandon Brancepeth	8 E/W 7 N/S	9385 6397	2.59 2.84	0.02 0.03	60 71	0.23 0.22	n/a
St Mary's Richmond	8 E/W 7 N/S	6110 4781	2.55 3.04	0.038 0.029	33 17	0.13 0.05	25/59/16 29/32/39

Table 4.6: Response characteristics of traditional square topped towers.

4.5 Computational modelling

Finite element computation is now used universally for the analysis of structures, and was the automatic choice for this work on the transient sway of masonry towers. How-

	Bell, number and dir.	Max. horiz. bell force N	Nat. freq. Hz	Dam-ping factor	Max. accn. mm/s^2	Max. displ. mm	R/S/B %
Christchurch	8 N/S	7477	2.48	0.078	97	0.40	6/37/57
Consett	6 E/W	4358	3.38	0.031	48	0.11	14/62/24
St Margaret's	8 E/W	5143	3.13	0.045	38	0.10	14/32/54
Tanfield	7 N/S	5513	3.88	0.027	10	0.02	18/11/71
St John's	8 E/W	9499	3.49	0.046	19	0.04	16/40/44
Shildon	7 N/S	6812	3.68	0.028	16	0.03	20/26/54
St Michael's	6 E/W	6439	3.16	0.023	35	0.09	9/64/27
Heighington	4 N/S	8547	3.63	0.016	20	0.04	6/64/30
St Michael's	8 N/S	6776	2.98	0.049	36	0.10	19/43/38
Houghton	7 E/W	4716	2.71	0.031	39	0.13	14/38/48
St Stephen's	2 N/S	5516	4.21	0.037	15	0.02	n.a.
Willington	3 E/W	5463	3.92	0.033	16	0.03	

Table 4.7: Response characteristics of 'squat' towers.

	Bell, number and dir.	Max. horiz. bell force N	Nat. freq. Hz	Dam-ping factor	Max. accn. mm/s^2	Max. displ. mm	R/S/B %
Durham	10 N/S	17109	1.28	0.016	23	0.35	n/a
Cathedral	9 E/W	13877	1.31	0.016	11	0.16	
Newcastle Cathedral	12 N/S	15861	2.04	0.014	36	0.22	6/58/36

Table 4.8: Response characteristics of the Cathedrals.

ever, because of the several unknown material and section parameters, the procedure now followed is to construct an FE model which is then calibrated against the *measured* frequency, mode shape and transient acceleration. The derived characteristic values can then be viewed over a number of towers for typical and extreme values. Initially, a number of decisions are necessary for a successful analysis of firstly an eigenvalue solution for the fundamental sway frequency and mode shape, and secondly for peak response in a time-stepping transient analysis.

The first decision relates to the type of analysis, which for masonry might be linear elastic, elastic-pure plastic, elastic with cracked joints, or biased towards shear response because of the soft mortar joints. Because the behaviour of a tower during bell ringing is essentially at small strain levels (typically less than 10×10^{-6}), plasticity can be ignored safely. The opening of horizontal tensile cracks due to bending can also be discounted for two reasons; firstly the self weight of the masonry above the bell level is sufficient to avoid tensile stress during ringing; additionally, at the instant in time when the horizontal force from a swinging bell is at maximum, the vertical force from the bell is of similar magnitude in the downward direction which has a stabilising effect.

The existence of any vertical cracking in the tower is much more significant in reducing the section resistance to bending and to shear; when such cracks are clearly in evidence then their effects should be incorporated into the analysis. In the 19 churches studied, no such cracks were identified. Enhancement of shear response may be achieved for beam elements by the use of Timoshenko beams with shear deformation factors based on the section area; however, enhancement is less easy in 3D brick elements where it is perhaps more appropriate to take a balanced choice of elastic modulus.

A tower acts as an open box-section vertical stocky beam resting on a non-rigid foundation. For the tallest, most slender towers it may be sufficient to model the structure as a Timoshenko beam of variable cross section, with estimation of effective area a and of second moment of area I made on the assumption of *solid* masonry walls, and ignoring minor openings [20]. Further, any structural connection to the main church walls is ignored. Finally, the flexibility of the foundation in rocking mode is represented by a rotational spring *to ground* while lateral and vertical ground movements are restrained.

4.5.1 Timoshenko beam models

Consider for example the tower of St Nicholas in Durham City centre. The tower is very slender, being 43.5m tall to the top of the spire and only 5.18m wide; the height to width ratio is 8.4, the greatest in the survey. Unusually, the tower is built outside of the body of the church and part way along the south wall of the nave, so that it shares part of its inner wall with the nave wall. The stiffness contribution from the nave is difficult to estimate. The dimensions of the tower are shown in Figure 4.10, together with the very simple (and economical) vertical cantilever *stick* model. Areas and second moments of area of the walls were determined taking account of arched doorways and windows and the stiffening from some external buttressing at the lower levels.

The tolling of bell number 6, the tenor bell, imposed a horizontal dynamic force of characteristic form with a peak of ± 3787N at bell chamber level in the E/W direction. The calibrated analytical model found to match effectively the recorded natural frequency of 1.82Hz and the peak acceleration at roof level of 0.032m/s^2, was based upon the stiffness and mass properties shown in Table 4.9. Finally, the mode shapes from measurement and computation are compared in Figure 4.10, and close correlation can be observed. This back-calibration procedure identified critical values for the effective elastic modulus and effective density of the sandwich walls of 0.21 and 0.55.

The church towers of St Cuthberts, Benfieldside, St Cuthbert's in Chester-le-Street

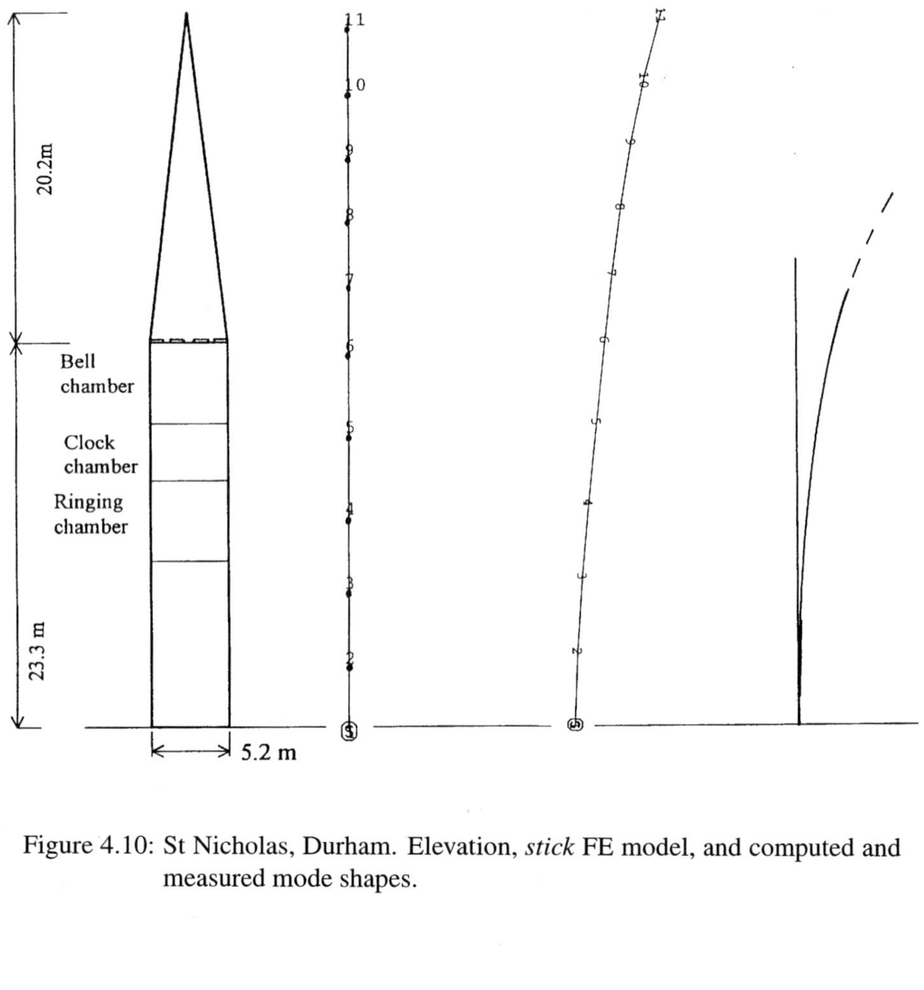

Figure 4.10: St Nicholas, Durham. Elevation, *stick* FE model, and computed and measured mode shapes.

and St Mary's in Shincliffe are each capped with a spire, and have overall height to width ratios of 6.7, 6.0 and 5.7 respectively. They were each back-analysed using vertical cantilever Timoshenko beam elements, with results as summarised in Table 4.9.

The tower and spire of St Cuthbert's, Chester-le-Street are of particular interest because of the rather unusual architectural form. The tower is rectangular in plan from ground level to 12.3m high (housing the base chamber and the ringing chamber), with an octagonal tower from 12.3m to 18.5m for the bell chamber, and a circular section spire rising to 43.5m. In addition, the survey revealed some complexity to the form of free oscillation, with an overall but slightly unstable frequency of 1.59Hz, with some tendency for the lower tower to vibrate at 2.7Hz. The reason for this behaviour became apparent during the calibration runs for the tower, when the fundamental mode shape *toggled* between a single monotonic curve and a mode shape of very similar curve except near the top 5m of the spire, which bent in the opposite direction. The

	Effective modulus as a proportion of 17×10^9Pa	Effective density as a proportion of 2200kg/m^3	Shear factor k_y	Foundation stiffness, K_y (Nm/rad)
St Mary, Bell 6 Shincliffe	0.118	0.45	0.5	100×10^9
St Cuthbert, Bell 8 Chester-le-Street	0.082	0.68	0.3	60×10^9
St Nicholas, Bell 6 Durham,	0.21	0.55	0.4	200×10^9
St Cuthbert, Bell 6	0.07	0.41	0.4	rigid
Benfieldside Bell 3	0.07	0.37	0.4	rigid

Table 4.9: Parameters of towers with spires, analysed using Timoshenko beams.

computed mode was an unstable outcome, which was dependent upon the defined second area moment of the upper portion of the finely tapered spire. In addition, the base tower was a very stocky short element, which was not well represented by Timoshenko beam elements, so that the mode shape was not entirely coincident with the measured deformation.

To summarise, the tower of St Nicholas, being very tall and slender, is adequately represented by a vertical cantilever of Timoshenko beam elements and a rotational spring for the foundation. Effective modulus and density are as tabulated and are noticeably higher than for the other spires in the group. It may be that the very tall slender tower was constructed with solid or near-to-solid walls. The towers in Shincliffe and Benfieldside are less slender than St Nicholas, but again are modelled satisfactorily by the beam elements. As discussed, the behaviour of St Cuthbert in Chester-le-Street is less satisfactorily defined despite its height of 43.5m.

4.5.2 3-D finite element modelling

For the majority of the towers investigated in the survey, and especially those of short and wide dimensions, a Timoshenko beam fails to describe correctly the shear deformation in both the side and cross walls. Only a full three-dimensional analysis in which the total wall thickness is represented by one or more brick elements, is adequate. Major cut-outs, columns and buttresses can be defined by the element mesh as appropriate. The foundation is included effectively by a layer of 3-D elements immediately below ground surface, sufficient to allow some tower base rotation but restricting vertical and transverse movement to negligible levels. A half-symmetric mesh is normally sufficient to incorporate the several structural response contributions including side-wall in plane shear, shear lag across the end-walls (with respect to direction of tower sway), whole-section bending and structural non-uniformities. The preferred rectangular block element is the 20-noded linear strain, quadratic displacement type, with a single element in the total wall thickness, and a reduction in effective stiffness and effective mass, to take account of the sandwich construction. A typical mesh might contain a minimum of 13 elements in the half-plan and 11 in height. It is apparent that a transient dynamic analysis by time-stepping by the Newmark β method becomes a computation requiring considerable computer power.

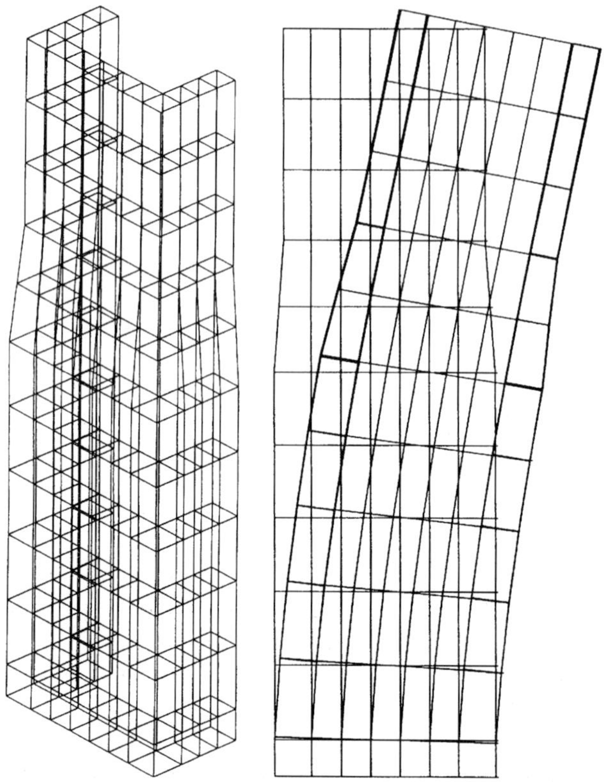

Figure 4.11: 3D half model of All Saints, Lanchester.

Whilst a number of these square topped towers have been analysed adequately using Timoshenko beam elements, [21,22], it is preferable to use a full 3-D computation where possible. A typical mesh for the tower of All Saints, Lanchester is shown in Figure 4.11, which satisfies the several requirements as listed in the previous paragraph. Also shown is a view of the fundamental mode shape, which includes base response, shearing and bending. 3-D transient analyses have been carried out on the 15 square-topped towers in the survey, and their essential derived equivalent properties are summarised in Tables 4.10, 4.11 and 4.12.

From the calibrated analyses of towers with height to width ratio in excess of 3, the deduced effective modulus for the sandwich walls was found to lie typically between

	Bells, number and dir.	Effective modulus as a proportion of 17×10^9Pa	Effective density as a proportion of 2200kg/m^3	Foundation modulus for a 1m slab. Pa
St Oswald's Durham	8 E/W 7 N/S	0.06 0.07	0.32 0.46	ignored
All Saints Lanchester	6 E/W	0.053	0.47	0.5×10^9
St Matthew's Newcastle	8 N/S	0.124	0.64	1×10^9
St Andrews Bishop Auckland	8 N/S 6 E/W	0.18 0.10	0.68 0.45	0.6×10^9 0.3×10^9
St Edmund Sedgefield	5 N/S	0.053	0.41	0.45×10^9
St Brandon Brancepeth	8 E/W 7 N/S	0.10 0.08	0.64 0.47	ignored
St Mary's Richmond	8 E/W	0.047	0.25	2×10^9

Table 4.10: Effective material properties derived from 3-D analyses of traditional towers.

0.05 and 0.12 as a proportion of the pure sandstone modulus of some 17×10^9 Pa. This low value is attributable to the soft mortar joints, to the very weak or unbonded rubble fill in the sandwich construction and to the several openings which were not included in the FE analysis. The effective density of the sandwich construction was found to be within the range of 0.3 to 0.7.

The foundation stiffness was less easily quantified by a single parameter, but when a 1m thick layer of elements was used in the analysis, then an effective modulus for this stratum was found to be of the order of 1×10^9Pa.

Variations among the deduced parameters can be attributed to differences in sandwich proportions and to different levels of restraint from composite action with the church nave. The towers of St Edmund at Sedgefield, and St Mary in Richmond were rather different from the rest of the group in that they had substantial buttresses at each corner of the tower, aligned at 45° to the plane of the side walls. These features had to be incorporated into the numerical models, although the exact dimensions and constitution were indeterminate. The large cut-outs in St Mary's tower, which were not included in the model, resulted in very low values of mass and stiffness.

The response of the squat towers was generally dominated by shear, Table 4.7. Typical values for the effective material properties of the sandwich walls were found to be 0.03 to 0.09, which implies a rather greater loss of stiffness due to mortar bed shear than in the more slender towers.

St Michael's in Houghton comprised a short stocky tower centred above the inter-

	Bells, number and dir.	Effective modulus as a proportion of 17×10^9Pa	Effective density as a proportion of $2200kg/m^3$	Foundation modulus for a 1m slab. Pa
Christchurch Consett	8 N/S 6 E/W	0.032	0.55	65×10^6
St Margaret's Tanfield	8 E/W 7 N/S	0.060	0.50	1×10^9
St John's Shildon	8 E/W 7 N/S	0.088	0.52	1×10^9
St Michael's Heighington	6 E/W 4 N/S	0.053	0.41	0.8×10^9
St Michael's Houghton	8 N/S 7 E/W	0.126	0.5	2×10^9
St Stephen's Willington	2 N/S 3 E/W	0.07 0.10	0.48 0.71	- -

Table 4.11: Effective material properties derived from 3-D analyses of squat towers.

section of the nave and transepts, and supported by four slender columns some 1m $\times$ 1m square; lateral support was obtained by arched walls in either the nave and choir or in the transepts. Structural response was dominated by its significant restraint from the main church structure, and so the deduced parameter values are specific to this church. A realistic match was obtained for the parameters as tabulated, but with lateral stiffness from the arched walls modelled by springs at the level of the ringing chamber floor.

4.5.3 Durham Cathedral and Newcastle Cathedral

The two Cathedrals in the survey deserve special mention, and results are summarised in Table 4.8. Durham Cathedral is a magnificent example of Norman architecture, which celebrated its 900th anniversary in 1993. The tower was probably completed by the year 1490, a smaller tower having been in place since 1262. The whole Cathedral, except for the extreme east end is founded directly onto the sandstone of the Durham *peninsula*. The main central tower rests on four crenelated columns which are free-standing at floor level, within the boundary formed by the nave, transepts and choir. At clerestory level arches link the pillars into the upper walls, until the tower rises above the nave construction to reach a full height of 66m. A fine peal of ten bells hangs at a height of 57m, and their voices can be heard over distances of a kilometre or more when ringing takes place on a still day. A study of measured tower response to bell tolling, and of finite element modelling is reported by Wilson & Selby [23]. In that study it was noted that tolling of the tenor bell, number 10, in a north-south plane, produced a response approximately twice as large as that during tolling of bell number 9 in the east-west direction. This was observed despite the similarity of natural frequencies of sway in the two directions, and the small difference in bell masses.

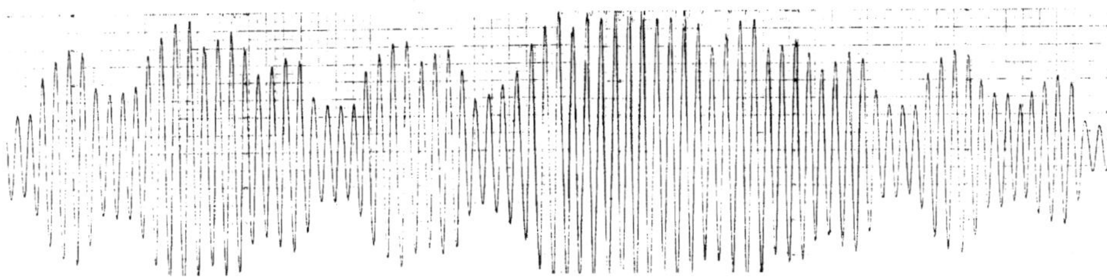

Figure 4.12: *Beating* response of Durham Cathedral.

Further study of the original measured traces in the N-S direction showed that the response to bell 10 was not of the characteristic pulse and decay repetition. Instead, a beating phenomenon occurred such that occasionally the response built up to a large value, while between times a period of small oscillations occurred, see Figure 4.12. By slightly changing the *dwell* period of the bell between swings, a similar effect was computed, showing the same uncharacteristic pattern. Taking this behaviour into account, and matching both frequency and transient peak response, the values for effective modulus and density were found to be 0.34 and 1.0, as in Table 4.12. These values imply an unusually solid *sandwich* wall construction.

	Bells, number and dir.	Effective modulus as a proportion of 17×10^9Pa	Effective density as a proportion of 2200kg/m^3	Foundation modulus for a 1m slab. Pa
Durham Cathedral	10 N/S	0.34*	1.0	rigid
	9 E/W	0.34	1.0	rigid
Newcastle Cathedral	12 N/S	0.108	0.45	1.8×10^9

*deduced after compensation for the beating behaviour.

Table 4.12: Deduced properties of the Cathedral Towers.

Newcastle Cathedral is some 400 years younger than Durham, and is resplendent in its late Gothic architecture. Externally the nave is of a plainer style. Unusually, the bell tower, built within the west end of the nave is surmounted by pinnacles and a lantern which is supported by four semi-arches each springing from the corners of the square topped tower. The structure of the tower is not easily defined (or even surveyed) since it includes major cut-outs in the belfry walls, louvred for sound direction, and many slab and corner buttresses, while at ground level the tower rests on a length of stiffened west wall and two major columns. The bells are a peal of 12, the tenor bell having a mass of 1913kg, the heaviest encountered during the survey. Tolling of the tenor bell resulted in an observed natural frequency of 2.04Hz, and a peak transient acceleration at roof level of 36mm/s^2. A 3-D model of half of the tower was used to derive a match for both fundamental frequency and transient response, Table 4.12. The primary mode shape showed significant deformation of the discrete elements within the belfry, Figure 4.13.

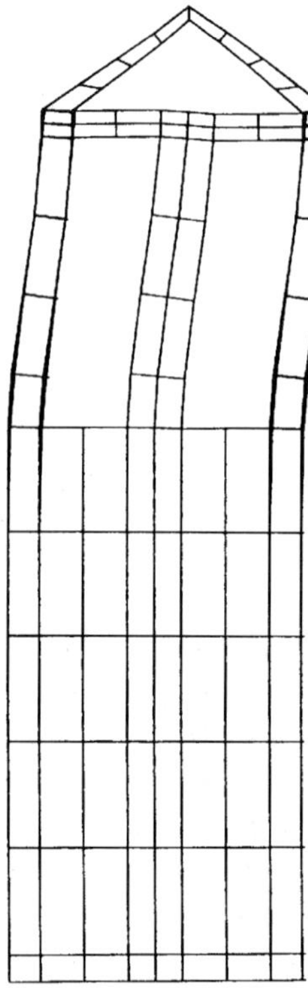

Figure 4.13: Response of arched lantern of Newcastle Cathedral.

4.5.4 Summary of FE Analyses

In the preceding sections, it has been shown that construction of a FE mesh of a bell tower, using either Timoshenko beam elements with bending and shear, or 3-D *brick* elements, is straightforward. However, accurate modelling of the tower dynamic behaviour is less easy and requires a number of considerations and assumptions. Firstly, the masonry walls of traditional church bell towers are of sandwich construction, which is less stiff than the solid isotropic sandstone; the overall density of the wall, too, is lower than intact sandstone. While the overall tower geometry may be well defined, stiffness changes from buttresses and from arched cut-outs may be difficult to represent. Compliance of the foundations should usually be incorporated, unless the tower is founded on bedrock. Finally, there is often some stiffening from integral walls and arch elements of the nave.

A practical solution to these various factors has been found by the adopted procedure of building an FE model of the tower and foundation as if the walls were solid masonry, but then tuning the values of elastic modulus and of density so as to achieve a representative model which matches the natural frequency of sway, the mode shape, and the peak transient response to tolling of a heavy bell of known characteristics. Application of this procedure to a range of towers in the north-east of England has resulted in realistic ranges of *effective* elastic modulus and density.

4.6 Serviceability and ultimate limit conditions

4.6.1 Serviceability

Conditions which imply a reduction or termination of use through a loss of serviceability include initiation and propagation of cracks in the masonry, fretting of the bell-frame in its sockets in the tower, increased bell-frame flexibility, difficulties for the ringers in ringing the bells under full control, and unease felt by people present in the ringing chamber during ringing.

Superficial surveys of the masonry of the towers revealed a surprising absence of cracking, the exception being a crack of some 2mm width running vertically between the tower and the nave walls in St Cuthbert's, Benfieldside. Apart from this single observation, the masonry of the many towers appeared to be in very good condition. Niederwanger [9] reported that a number of bell towers in Austria contain serious cracking in the form of vertical open lines running from ground level or arch points upwards. The greater prevalence of cracking may be a result of heavier bells being swung through only some $\pm$ 90° about the mouth down position. He describes methods of repair by prestressed tendons, which reduced transient displacements from 1.1mm to 0.05mm.

In Durham Cathedral a timber bell frame of considerable age was replaced in 1980 having shrunk and warped over many years. The new steel frame is much stiffer and is well fixed into the masonry of the tower. Christchurch in Consett is one of the few remaining towers to have a timber bell frame, and this does show some compliance and racking during ringing. The tower has attracted some criticism for its response during ringing.

Within the survey, three towers emerged with a reputation as 'lively', although ringers were rarely affected by ringing problems. The three towers were St Matthew's in Newcastle, where the bells had been re-hung lower down the tower than their original position, St Cuthbert's in Chester-le-Street and Christchurch in Consett. The bells in all three towers are heavy, especially the tenor in St Matthew's which has a mass of 1588 kg.

A number of parameters for structural acceptability were considered by Selby & Wilson [6]. These included geometrical properties of ratios of tower height to width and of bell mass to tower mass. However, the non-dimensional parameter which described most effectively the lively towers was found to be the ratio of peak transient displacement to tower height. St Matthew's showed a value of 30×10^{-6}, with St Cuthbert's at 29×10^{-6} and Christchurch at 24×10^{-6}. Other towers in the survey were found to be around 10×10^{-6}, see Figure 4.14.

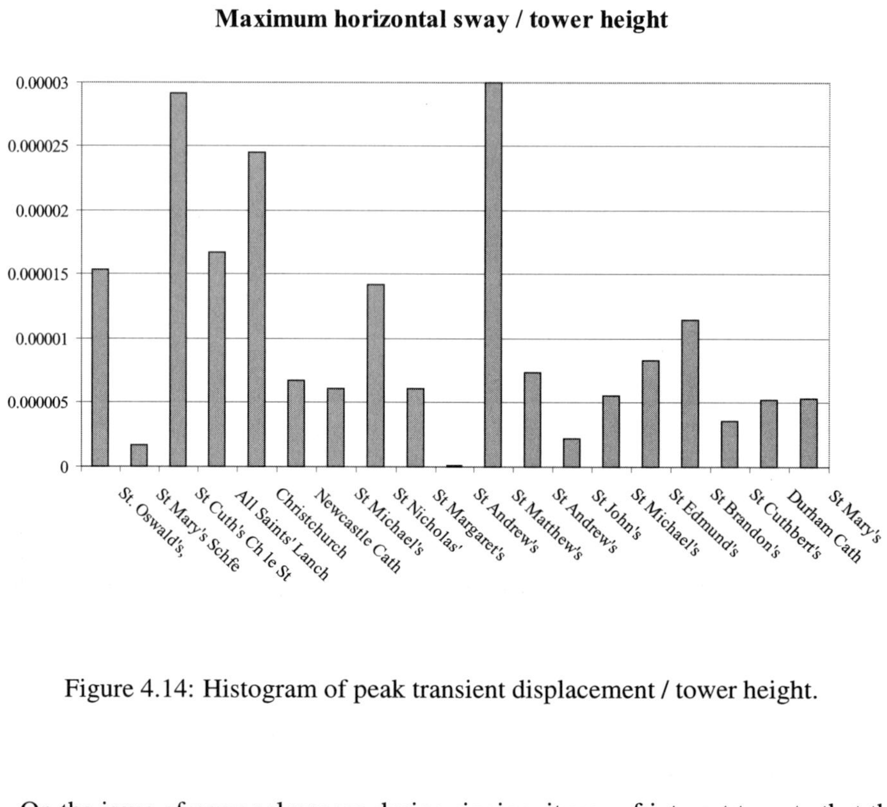

Figure 4.14: Histogram of peak transient displacement / tower height.

On the issue of personal unease during ringing, it was of interest to note that the ringing chamber of St Matthew's is located proportionately high up the tower. The peak acceleration recorded in the chamber was 43 mm/s^2. By coincidence the condition for low adverse comment to vibration as defined by BS 6472 [24] is 30mm/s^2 r.m.s or 42mm/s^2 peak. When the bells were hung higher up the tower, this condition would have been exceeded. Peak acceleration in the ringing chamber of Christchurch also reached 44mm/s^2, and in St Cuthbert's a level of 36mm/s^2 was recorded.

To summarise the significant elements relating to reaching a serviceability limit, none of the towers were showing structural degradation. Excessive tower/frame movement was causing some difficulty only in Christchurch, Consett, where the bell frame was in poor condition. When a ringing chamber is located high in a tower, the subjective estimate of vibration is high. Definition of a *lively* tower is described most effectively by the peak response to tower height ratio.

4.6.2 Ultimate limits and factors of safety

Consideration of the potential modes of material failure leading to rapid tower degradation or collapse suggests three most probable conditions:

1. Vertical stress due to bending and rocking.

   The issue of concern is whether the transient tensile stress can exceed the static self-weight stress causing opening of mortar joints, which would normally have very limited tensile strength. The 3-D analysis of St Matthew's tower was studied and the peak transient tensile stress was found to be 8.5kPa at ground level, while in-situ self weight stress at the same location was 200kPa. This implies a safety factor in excess of 20. At locations higher in the tower the ratio was even greater. These calculations ignored the simultaneous beneficial downward force from the swinging bell. An increase in compressive stress from the vertical force from a swinging bell is clearly insignificant.

2. Shear stress in the side wall.

   Computed shear stresses in the side wall were nowhere in excess of 1kPa. This may be compared to a limiting value for vertical stress of some 200kPa [25], and an enhanced value in the horizontal mortar joints due to compressive stress. There is very considerable strength reserve in this case. Shear failure as reported by Niederwanger [9] may have been initiated by a combination of load effects. Transient stresses from bell ringing alone seem to be an unlikely cause.

3. von Mises stresses.

   In a fully 3-D analysis a preferred estimate of shear effects is obtained from von Mises stress calculations. Referring again to the St Matthew's transient analysis, the peak von Mises stress was computed as 8kPa, which again offers a factor of safety of some 20.

In summary, structural calculations indicate a very high factor of safety for uncracked masonry under transient bell force loading. However, small imperfections in the masonry, and in the frame to tower connections, may initiate cracking. This may spread under the combined actions of ringing, wind forces or settlement. Fortunately, a simple test is available to indicate whether structural deterioration is occurring. Tower resonance should be measured annually by a single accelerometer and analyser during bell tolling. A reduction in natural frequency will indicate a loss of stiffness, presuming that tower mass is unchanged. If such indication is perceived then more detailed structural surveys may be required.

4.7 Conclusions

A study of the dynamic behaviour of 19 traditional masonry bell towers has been undertaken. Measurements of the transient response during full-circle ringing of a single heavy bell gave natural frequency in sway mode for each tower, mode shape and damping, and peak response. The transient bell forces were deduced from simple measurements and definition of compound pendulum mechanics.

Finite element analyses of the towers for frequency and mode shape and for transient response correlate with the observed behaviour provided that effective modulus of elasticity and of density are used for the sandwich walls, and that foundation compliance is included.

Limiting values of tower response for human acceptance can be defined. Calculations suggest very large factors of safety against damage to sound masonry from bell ringing.

4.8 Acknowledgements

Assistance in the measurements of tower movements was undertaken by a number of collaborators of whom John Lund deserves particular mention. Thanks go also to many bell-ringers who generously *rang to order*. Thanks are due to the many parish vicars who gave permission for access to the towers.

References

[1] Dymond C., "Clanger that could muffle celebrations", The Times, 19, 25th Oct 1997.

[2] Pevsner N., "The buildings of England, Co. Durham", 2nd ed., Penguin, 1983.

[3] Towers and Belfries Committee, "The Towers and Bells Handbook", Central Council of Church Bell Ringers, Guildford, 1973.

[4] Wilson W.G., "The Art and Science of Change Ringing", Faber & Faber, London, 1965.

[5] Heywood A.P., "Bell towers and bell hanging - an appeal to architects", Central Council of Church Bell Ringers, Guildford, 1914.

[6] Selby A.R. and Wilson J.M., "The dynamic response of a church tower to bell-ringing", Proc 2nd Int Conf. on Structural Repair and Maintenance of Historic Buildings, eds. Brebbia C.A., Dominguez J. and Escrig F., Comp. Mech. Pub., 3-16, 1997.

[7] Selby A.R. and Wilson J.M., "Computation and measurement of the dynamics of masonry bell towers", Proc 2nd Int Conf. Structural Dynamics Modelling, ed. Creechan A., NAFEMS Glasgow, 165-174, 1996.

[8] Lund J.M., "The dynamic response of church bell towers to bell ringing", Internal Report, Durham University, 1994.

[9] Niederwanger G., "Structural repair of damaged old bell towers based on dynamic measurements", Proc 5th Int Conf. on Structural Studies, Repair and Maintenance of Historic Buildings, eds. Sanchez-Beitia S. and Brebbia C.A., Comp. Mech. Pub, 447-455, 1997.

[10] Evans N.C., "The Response of Bell Towers to Bell Ringing", Internal Report, Durham University, 1990.

[11] Lilley D.M and March A.V., "Problems in rubble-filled random masonry walls", Proc 5th Int Conf. on Structural Studies, Repair and Maintenance of Historic Buildings, eds. Sanchez-Beitia S. and Brebbia C.A., Comp. Mech. Pub., 417-426, 1997.

[12] Attewell P.B., "Deterioration of the exterior stone at Durham Cathedral" in Engineering a Cathedral, ed. Jackson M.J. and Telford T., 120-145, 1993.

[13] Bulson P.S., "Cathedrals and Quarries", Proc 2nd Int Conf. on Structural Repair and Maintenance of Historic Buildings, eds. Brebbia C.A., Dominguez J. and Escrig F., Comp. Mech. Pub., 97-109, 1991.

[14] Lewis E.H., "Calculation of the forces acting upon a bell tower", Central Council of Church Bell Ringers, Guildford, 1914.

[15] Heyman J. and Threlfall B.D., "Inertia Forces Due to Bell Ringing", Int. J. of Mech. Sci., 18, 161-164, 1976.

[16] Birlisch R., "Numerical calculations of elliptic integrals and elliptic functions", Numerische Mathematik, 7, 78-90, 1965.

[17] Wilson J.M., "Periodic Forces on Bell Towers Arising from Bell Ringing", Internal report, Durham University, 1988.

[18] Kipling S.M., "A model bell tower", Internal Report, Durham University, 1997.

[19] Lund J., Selby A.R. and Wilson J.M., "Dynamics of Bell-Towers - a Survey in Northeast England", Proc 4th Int Conf. on Structural Repair and Maintenance of Historic Buildings, eds. Brebbia C.A. and Leftheris B., Comp. Mech. Pub., 45-52, 1995.

[20] Cowper G.R., "The shear coefficients in Timoshenko's beam theory", J. App. Mech, 335-339, 1966.

[21] Selby A.R. and Wilson J.M., "Dynamic Behaviour of Masonry Church Bell Towers", ASCE Structures Congress 96, Stability Analysis of Masonry Structures, Chicago, 1996.

[22] Wilson J.M., Selby A.R. and Ross S.E., "The Dynamic Behaviour of Some Bell Towers During Bell Ringing", Proc 3rd Int Conf. on Structural Repair and Maintenance of Historic Buildings, eds. Brebbia C.A. and Frewer R.J.B., Comp. Mech. Pub., 491-500, 1993.

[23] Wilson J.M. and Selby A.R., "Durham Cathedral Tower Vibrations During Bell-Ringing" in Engineering a Cathedral, ed. Jackson M.J., Telford T., 77-100, 1993.

[24] British Standard 6472, "BS6472 Guide to evaluation of human exposure to vibration in buildings (1Hz to 80Hz)", BSI London, 1992.

[25] Hendry A.W., "Structural Brickwork", Macmillan Press, 1981.

[26] Selby A.R. and Wilson J.M., "The structural safety and acceptability of bell towers", Proc 5th Int Conf. on Structural Studies, Repair and Maintenance of Historic Buildings, eds. Sanchez-Beitia S. and Brebbia C.A., Comp. Mech. Pub, 321-330, 1997.

©2001, Saxe-Coburg Publications, Stirling, Scotland
J.W. Bull, (Editor), Computational Modelling of Masonry, Brickwork and Blockwork Structures,
Saxe-Coburg Publications, Stirling, Scotland 109-142.

Chapter 5

Settlement Induced Damage to Masonry Buildings

C. Augarde

School of Engineering, University of Durham, United Kingdom

Abstract

Existing masonry structures are often affected by adjacent construction activities. An example is the construction of a new tunnel beneath an existing building to provide a new infrastructure link. In soft ground, the main effects of tunnelling likely to cause damage are the vertical and horizontal movements produced at the surface. While the magnitude of these movements is usually quite small, their differential nature along a structure can lead to damage from cracking. There is a need to improve the methods by which this damage is predicted. Current methods used in practice usually ignore interaction between the building and the ground on which it is founded. This interaction will change as the building is damaged and its stiffness changes. Models are necessary, therefore, that include simulation of tunnelling in the presence of a building. This chapter describes the development and use of a complex three-dimensional finite element model to study this problem undertaken by the author, and others at the Department of Engineering Science, Oxford University. Some results are presented for simple arrangements of masonry façades as well as more complex buildings for which field data are available. While the complexity of the models presented currently confine this work to the research field, increases in computing power will make their use in practice feasible in the future.

5.1 Introduction

The effects of construction activities on adjacent existing structures have assumed greater importance in recent years due to the need to develop more congested urban sites. An area of current interest in the UK is the effect of tunnelling operations on masonry structures when the tunnelling is relatively shallow and is carried out in soft ground (i.e. not rock).

The construction of a tunnel in soft ground leads, almost inevitably, to movements in the ground above. The source of these disturbances is the excavation of a larger volume of ground than is required to house the final tunnel lining. At a minimum, there is a gap between the outside of the tunnelling shield and the final lining. Usually, however this *volume loss* is greater than this minimum value due to over-excavation and movement during construction.

The volume loss at the tunnel leads to vertical and horizontal movements at the surface with magnitudes that are seen to decrease rapidly away from the tunnel centreline. A building situated in the zone affected by tunnel construction will experience differential movements that can lead, in some cases, to distortion or cracking. Masonry structures, in particular, are prone to cracking induced by tunnelling movements. It is important, therefore, that methods of damage assessment are available to predict the likely extent and magnitude of this damage, during the planning of a new tunnel.

This chapter briefly describes the current procedure for predicting movements and damage to masonry structures resulting from tunnelling operations. The procedure usually applied in practice does not take into account any interaction between the soil and the structure and can therefore give over-conservative predictions of damage.

New methods are described here to model the behaviour of a masonry structure affected by tunnelling-induced movements using three-dimensional finite element techniques. The choice of constitutive model for masonry is an important aspect of the development of these methods. The chapter concludes with some examples of the use of these new techniques.

5.2 Current procedures used to assess settlement damage due to tunnelling

The assessment of the likely damage to a masonry structure due to nearby tunnelling is a three-stage process. Firstly, the settlements expected for the arrangement of tunnels at a greenfield site, i.e. one without surface structures, are calculated. This is done using a semi-empirical approach that assumes the transverse settlement profile to have an inverted Gaussian distribution. (This procedure is often termed the *semi-empirical approach* since its basis lies in assessment of field data). Longitudinally, (i.e. above the axis of tunnelling) the settlements are assumed to follow a cumulative probability curve. Figure 5.1 shows the settlement trough predicted by this approach. Horizontal movements also occur but are usually determined by assuming movement is directed towards the tunnel axis. Extensive field data are available that confirm this approach to be valid. It should be noted, however, that this can only be relied upon for greenfield sites.

The second stage of a damage assessment involves the application of the settlements determined in the first stage to a simple model of the building, usually a linear elastic deep beam (i.e. where shear deformations are significant) [1]. By assuming the building to exactly follow the deformed surface, the maximum tensile strains due to bending and shear can be calculated. The maximum tensile strains are then compared to tabulated values linked to degrees of damage severity (Table 5.1). The degrees of severity range from *negligible* ($< 0.05\%$ strain) to *severe* ($> 0.3\%$ strain) and are themselves linked to examples of damage e.g. visible cracking. These data, given in Reference [2], have been compiled from field measurements relating to building movements not generally associated with tunnelling.

The third and final stage of a damage assessment is the use of more complex approaches if the second stage results are unacceptable.

Limiting tensile strain range(%)	<0.05	0.05 - 0.075	0.075 - 0.15	0.15 - 0.3	>0.30
Norm of degree severity	Negligible	Very Slight	Slight	Moderate	Severe to Very Severe

Table 5.1: Damage severity indices.

5.2.1 Numerical models of the tunnelling settlement problem

The procedure outlined above, for the assessment of settlement damage, has a number of deficiencies when considering masonry structures. Principally, the procedure fails to account for the stiffness and weight of the structure. Additionally, the building is modelled as linear elastic when its behaviour is likely to be quite different, due to the brittle nature of masonry. The development of the semi- empirical approach to rectify these omissions has not been attempted since there are very few instances of field data that include building measurements. Any attempt would also be complicated by the need to cover the wide range of buildings likely to be of interest.

Instead, numerical modelling has proved to be the most suitable way to improve predictions of tunnelling-induced settlement damage. Studies using finite element techniques have been undertaken in recent years in the UK by Potts & Addenbrooke [3], and by the author and co-workers [1, 4–6]. Potts & Addenbrooke have extended the semi-empirical procedure using the results of two-dimensional finite element calculations. In these calculations, a linear elastic beam is included on the surface to model a building. The results of this work are modification factors that can be applied to the results of a semi-empirical analysis to take account of the presence of

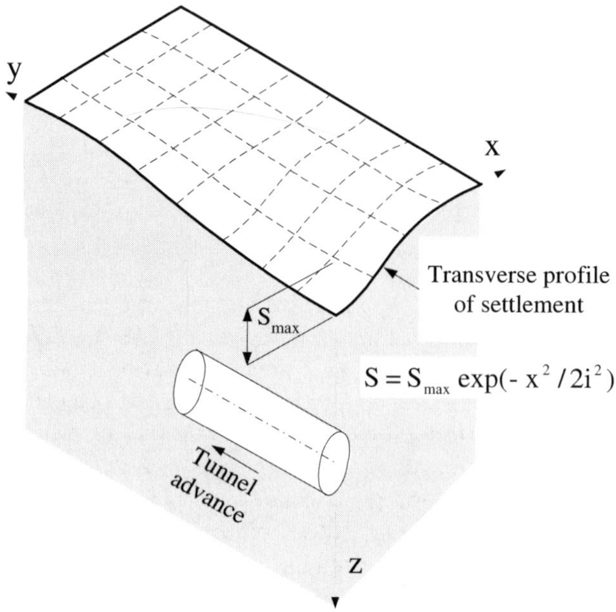

Figure 5.1: Settlement profile above a tunnel as predicted for a greenfield site by the semi-empirical approach.

the building. While this is undoubtedly an improvement on the original method it fails both to include the behaviour of masonry in the building, and to account for the three-dimensional nature of the problem.

The process of tunnel installation is incremental with the tunnel excavation and lining occurring progressively along the tunnel axis. The movements produced at the surface therefore also occur progressively, in a wave with the progress of the tunnel itself. This is one aspect of the three-dimensional nature of this problem. The other is the fact that building stiffness will be affected by the interconnection of components, located at differing angles to the tunnel axis. It can be seen that ideally, a three-dimensional model of this problem is necessary.

5.2.2 Modelling masonry

The most important components of a numerical model are usually the choices of constitutive laws for the materials involved in the analysis. Of prime importance, and interest here, is the need to simulate the behaviour of masonry in the model of the building. The obvious differences between linear elasticity and the brittle nature of masonry indicate the choice of masonry model to be critical for the success of any modelling of this problem.

The following Section describes the component parts of the model of tunnelling developed at Oxford which is three-dimensional and includes a building with a realistic masonry material formulation.

5.3 A three-dimensional finite element model

5.3.1 Simulation of tunnelling

The numerical model of tunnelling uses finite elements to represent the ground, a tunnel and a building at the surface as shown in Figure 5.2. For the analyses conducted to date, the ground has been assumed to be an overconsolidated clay deposit with increasing stiffness and strength with depth. The soil model is formulated to take account of the effect of past stress history on the current stiffness of the ground. The rapid decrease in stiffness of an overconsolidated clay deposit when subjected to changes in strain at low strain levels is thought to influence heavily the response of ground in problems such as this. The development of soil models that described this behaviour is currently an important area of research in soil mechanics [7]. At Oxford, a complex elasto-plastic constitutive model has been developed for this purpose. Its discussion is, however, beyond the scope of this work. Full details are given in Reference [8]).

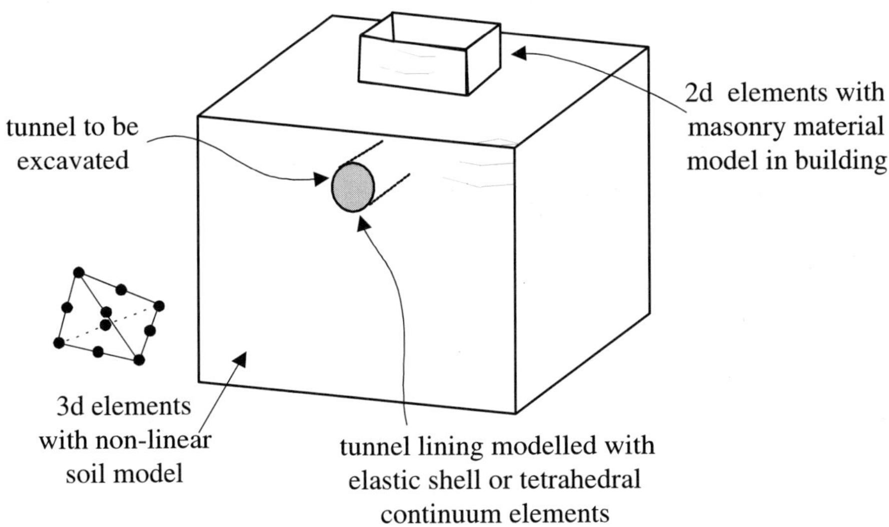

Figure 5.2: Components of the 3D finite element model of tunnelling.

The building is constructed from masonry, modelled as an elastic-no tension material; the full formulation is described in Section 5.4 below. Tunnel linings are linear elastic. Tunnel installation is simulated in the model in discrete stages, each of which consists of the simultaneous removal of soil and activation of shell elements to simulate the liner. Elements that are to be removed in the current stage are ignored when the structure stiffness is formed. Nodal loads are imposed on elements surrounding the excavated block to render the exposed faces free of surface tractions [9]. The calculation of these loads is described in Reference [4]. The volume loss that occurs at the tunnel, and which leads to the surface settlements, is simulated in the model by uniform shrinkage of the lining elements in a plane normal to the tunnelling centreline. This occurs at the same time as excavation and lining activation.

5.3.2 Modelling a building

While large masonry buildings are three-dimensional structures, their most significant feature, with respect to settlement damage, is the in-plane response of the main façades. The façades are represented in the model by meshes of six-noded triangular plane stress elements, each façade having its own local co-ordinate system. The façade meshes are joined to each other, and to the surface of the ground mesh, using a novel system of tie elements that implement the kinematic constraints linking each local two-dimensional co-ordinate system with the three-dimensional global system [10, 11].

The six-noded triangular elements used for the façade meshes are chosen to be compatible with the elements used to model the soil, to which the façades are attached. These are 10-noded tetrahedral elements, chosen for their suitability to model incompressible materials [12]. (The soil is assumed incompressible, in this model as the *undrained* response is of prime importance). It is also easier to generate unstructured meshes with triangular and tetrahedral elements, necessary to provide a reasonable representation of the complex geometries involved in the tunnelling problem.

5.3.3 Choice of masonry model

Masonry is a well-established structural material with a complex engineering behaviour. It can be characterised as non-homogeneous and anisotropic, having very high compressive strength and little tensile strength. Its brittle nature leads to the susceptibility of masonry buildings suffering differential settlement to crack. Two main approaches have been adopted for the development of numerical models of masonry structures. One approach is to model each brick and mortar joint separately [13, 14]. This would, clearly, be unsuitable for the majority of models of real structures. An alternative chosen by many researchers is to treat the masonry as homogeneous and to use a constitutive model that reflects the combined behaviour of masonry units and mortar. This approach has the advantage that the layout of a finite element mesh used to model a masonry façade does not have to depend on the location of individual blocks.

Examples of this approach can be found in many places in the literature [15]. For example, Reference [16] describes a failure surface for a homogeneous masonry material comprising a Von Mises type surface in the compressive region with tensile cut-offs elsewhere. A simpler model is described in Reference [17] in which tensile capacity is ignored and compressive strength is assumed infinite. This is the approach used here and its detailed description follows in Section 5.4.

A balance has to be drawn in developing complex finite element models of this nature in the levels of complexity of individual parts. Those with expertise in numerical modelling of masonry may be concerned that the model adopted here is too simplistic. It is, however, necessary to avoid over-concentrating effort in one area of the model if input from other areas is at a lower level of complexity. This is the reason for choosing a relatively simple formulation which nevertheless models the most important aspect of the behaviour of masonry with respect to this problem, i.e. changes in stiffness due to cracking.

5.3.4 Hardware & software

While the eventual purpose of the model is to provide design rules for planners of tunnelling schemes, it sits firmly in the research field at present. This is a result of the need for high-specification hardware. This situation is likely to change, however, as computing power becomes cheaper and the model is developed.

The various components of the model described above have been implemented in the finite element program, OXFEM, which has been developed over the past 15 years at Oxford University. OXFEM is written in Fortran 90 and has latterly been devoted to geotechnical problems involving complex constitutive models for soil behaviour. Non-linearity is accommodated in an incremental scheme, the Modified Euler method, where loads are applied in steps, followed by an equilibrium check to determine the out-of-balance force at each nodal degree-of-freedom. This out-of-balance force is then corrected in the next step. Incremental techniques such as this require the user to specify the numbers of steps, a decision that requires some experience of the non-linearities involved.

Calculations using the model are generally run on high-performance Unix workstations. Considerable use has been made of the Oxford Supercomputer, OSCAR, an 86-processor shared memory parallel machine. Its architecture is such that existing serial code can easily be moved to parallel with considerable reductions in running time. (Reference [18] describes in detail the transfer of the program OXFEM to run in parallel).

5.4 An elastic no-tension material model for masonry

5.4.1 The basic formulation

An important component of the model described in Section 5.3 is the formulation for the masonry material in the building façades. This is now described in detail. The material is assumed to have zero tensile strength and infinite compressive strength. When the stresses are compressive, the material is linear elastic. When the major principal stress becomes positive (i.e. goes tensile) the model "cracks" normal to the direction of this stress. The simplicity of this model brings with it various problems in implementation. These are discussed following a description of its numerical formulation.

The model is formulated within the standard displacement finite element method which is covered in many texts [19]. The modelling of cracking consists of changes to the material stiffness matrix D at integration points within elements deemed to have cracked. The element stiffness contribution is the matrix K where

$$[K] = \int_V [B]^{\mathrm{T}}[D][B]\,\mathrm{d}V \tag{5.1}$$

$[B]$ is the matrix linking strains to displacements and the integration is carried out over the volume of the element.

The material model is implemented in plane stress so the stress and strain vectors are respectively

$$\{\sigma\} = \{\sigma_x, \sigma_y, \tau_{xy}\} \tag{5.2}$$

$$\{\varepsilon\} = \{\varepsilon_x, \varepsilon_y, \gamma_{xy}\} \tag{5.3}$$

and

$$\{\sigma\} = [D]\{\varepsilon\} \tag{5.4}$$

When both principal stresses are compressive the material stiffness matrix is elastic and is given by

$$[D] = \frac{E}{1-\nu^2}\begin{pmatrix} 1 & \nu & 0 \\ \nu & 1 & 0 \\ 0 & 0 & \frac{1-\nu}{2} \end{pmatrix} \tag{5.5}$$

where ν is Poisson's ratio and E is Young's modulus. If the strains in the material cause the major principal stress to rise (from being negative) to zero then cracks will form, as shown in Figure 5.3b, where θ is the inclination of the minor principal stress to the x-axis.

In this case, the stiffness of the material in a direction normal to the crack field is set to zero.

The new material stiffness matrix, $[D]$ is obtained by transforming the material stiffness matrix in a co-ordinate frame rotated anti-clockwise by angle θ to the (x,y) frame, $[D^*]$ where

$$[D^*] = E\begin{pmatrix} 1 & 0 & 0 \\ 0 & 0 & 0 \\ 0 & 0 & 0 \end{pmatrix} \tag{5.6}$$

The material stiffness matrix referred to the (x,y) frame is

$$[D] = [T]^{\mathrm{T}}[D^*][T] \tag{5.7}$$

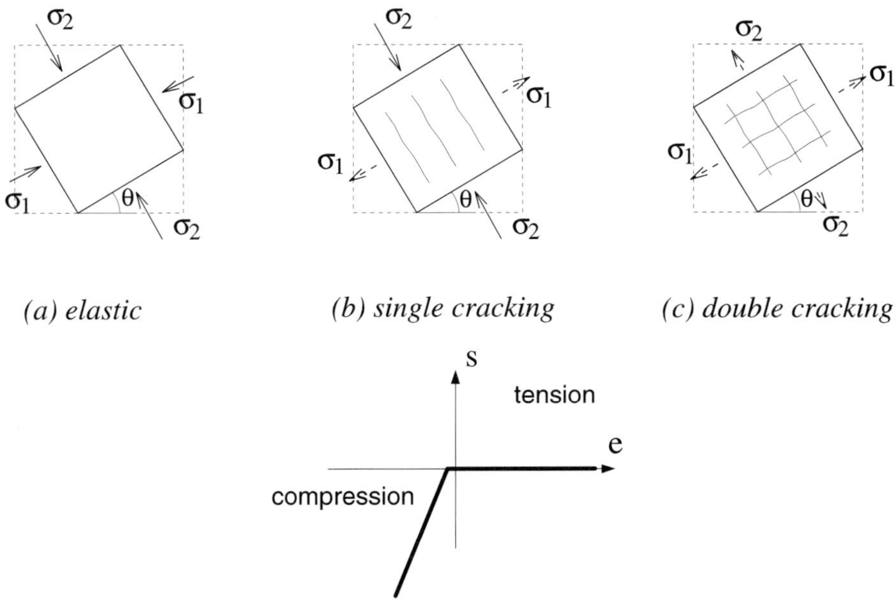

Figure 5.3: Stress states and constitutive model for elastic no-tension model.

where

$$[T] = \begin{pmatrix} \cos^2\theta & \sin^2\theta & \sin\theta\cos\theta \\ \sin^2\theta & \cos^2\theta & -\sin\theta\cos\theta \\ \sin\theta\cos\theta & -\sin\theta\cos\theta & \cos^2\theta - \sin^2\theta \end{pmatrix} \tag{5.8}$$

If both principal stresses fall to zero, then cracks form in two perpendicular directions and $[D] = 0$ (Figure 5.3c).

5.4.2 Validation of the masonry formulation

Once implemented into a finite element code, a new material formulation requires some verification. With the masonry model used here, validation is not made with data from test on masonry specimens. Instead, two problems are used that check the onset of zero stiffness following the change of the major principal stress to tensile.

5.4.2.1 Cavity expansion

The cavity expansion problem consists of a thick-walled cylinder initially subject to internal and external pressures p_0. As the internal pressure is increased to p_0 + Δp_0, the compressive hoop stress on its inner surface of the cylinder reduces. At a critical point, the hoop stress at this location becomes zero and then tensile. With increasing internal excess pressure, a region develops between the inner face and a radius r_c where hoop stress is tensile. Clearly, if the cylinder is made of the elastic-no tension material described above, the tensile region will crack and the hoop stress will be zero everywhere in that region.

The radius of cracking r_c can be found from the following quadratic equation [10]

$$(p_0 + \Delta p)\, a r_c^2 - 2p_0 b^2 r_c + (p_0 + \Delta p)\, a b^2 = 0 \tag{5.9}$$

where a and b are the inner and outer radii of the cylinder. Within the uncracked area, the elastic solution will be valid and the radial displacement on the cracked radius u_{rc} can be calculated as

$$u_{rc} = -\frac{1-\nu}{E} r_c p_0 + \frac{r_c}{E}\left(\frac{b^2 + r_c^2}{b^2 - r_c^2} + \nu\right)\Delta p \tag{5.10}$$

The radial stress in the cracked area σ_r, can be calculated by considering equilibrium and the fact that there is no hoop stress due to cracking

$$\sigma_r = -\frac{(p_0 + \Delta p)\, a}{r} \tag{5.11}$$

The radial displacement of the inner surface u_a can then be determined by integrating the radial strain

$$u_a = u_{rc} + \frac{a\,(p_0 + \Delta p)}{E} \ln \frac{r_c}{a} \tag{5.12}$$

from which a relation between the change in radial displacement due to the imposition of Δp can be found as

$$\Delta u_a = u_{rc} + \frac{a\,(p_0 + \Delta p)}{E} \ln \frac{r_c}{a} + \frac{1-\nu}{E} a p_0 \tag{5.13}$$

To test the finite element formulation, a problem where $a/b = 0.6$ is solved using two-dimensional finite elements with the elastic no-tension material. The results are compared to the analytical solution outlined above. The finite element mesh used for this problem is shown in Figure 5.4a. (Only a quarter of the cylinder is modelled due to symmetry).

Figure 5.4b shows a dimensionless plot of the variation in radial displacement at the inner face with increments of excess internal pressure. At $\Delta p/p_0 = 0.67$, cracks form through the entire thickness of the cylinder wall. At this point the finite element results show a slight overshoot, but this is corrected quickly and generally there is excellent agreement between numerical and analytical results.

5.4.2.2 Prestressed beam subject to end rotation

This problem consists of a simply-supported beam which is initially subjected to uniform compressive axial prestress, σ_0. Rotations θ are then applied to each end of the

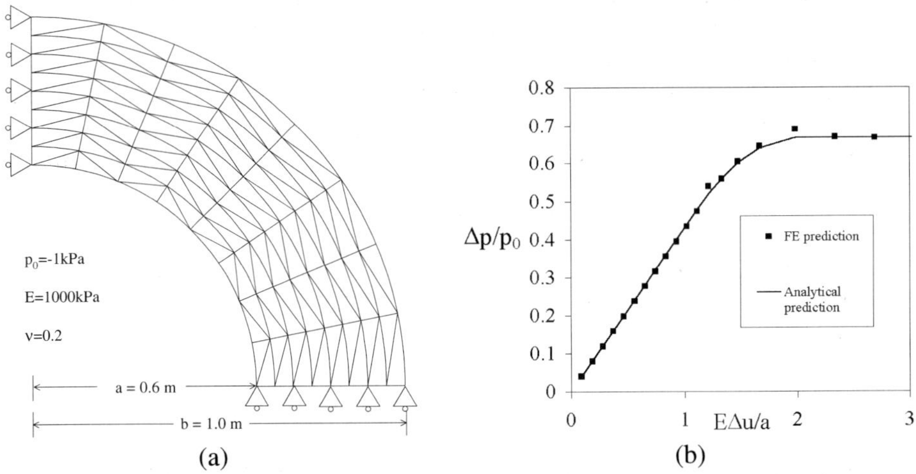

Figure 5.4: (a) Finite element mesh for cavity expansion problem, (b) comparison of analytical and FE solution of cavity expansion problem.

beam to reduce progressively the compressive prestress. Once the stress in the top fibre goes tensile, cracks form. This reduces the depth over which the longitudinal stress can act and hence alters the moment per unit thickness about the bottom edge of the beam M_b. The depth d of the cracked area can be determined as [10]

$$d = h - \frac{\sigma_0}{2}\frac{L}{E}\frac{1}{\theta} \tag{5.14}$$

where h is the depth of the beam, L is its length and E is the Young's modulus of the material in compression.

The bending moment M_b is found by integration from the bottom fibre of the beam to the crack tip

$$M_b = \int_0^{h-d} \left(-\sigma_0 + \frac{2\theta}{L}\frac{E}{}y \right) y \, dy = -\frac{\sigma_0}{2}(h-d)^2 \frac{2\theta E}{3L}(h-d)^3 \tag{5.15}$$

where y is measured upwards from the beam soffit. Substituting for d from Equation (5.14) gives

$$M_b = -\frac{1}{24}\frac{\sigma_0^3 L^2}{E^2}\frac{1}{\theta^2} \tag{5.16}$$

and removing the part corresponding to the initial stress gives the change in the moment per unit width measured about the beam soffit

$$M_b = \frac{\sigma_0^2}{2}h^2 - \frac{1}{24}\frac{\sigma_0^3 L^2}{E^2}\frac{1}{\theta^2} \tag{5.17}$$

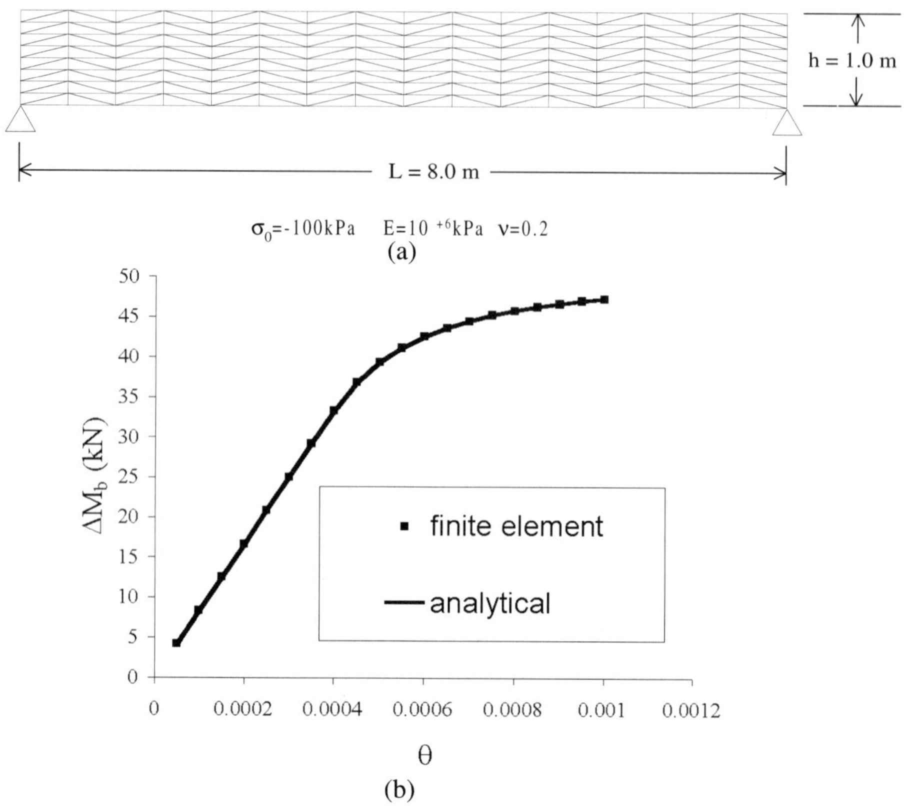

Figure 5.5: (a) Finite element mesh for beam bending problem, (b) comparison of analytical and FE solution of beam bending problem.

Equations (5.16) and (5.17) relate to the post-cracking behaviour. Figure 5.5a shows the finite element mesh used to analyse this problem where Young's modulus in compression was taken as 10^6 kN/m^2 and the beam length and height were 8m and 1m respectively. Figure 5.5b shows the close agreement between the response of the finite element model and the analytical solution outlined above. The section of the curve corresponding to Equation (5.17) is the non-linear section commencing at an angle of approximately 0.0005.

5.4.3 Implementation and numerical stability

Some problems are associated with the use of elastic-no tension material models and measures are incorporated in this formulation to deal with them. The problems stem from the sudden loss of stiffness when a major principal stress becomes tensile. In the conventional finite element scheme adopted here, stresses, strains and stiffnesses are calculated at discrete locations (integration or *Gauss* points) within elements. Numerical integration is necessary as the formulations adopted for the finite elements preclude closed form solutions. Ideally, once tension is registered at a Gauss point, all stiffness parallel to the minor principal strain direction must disappear. When an element is partially cracked, i.e. some points but not all are cracked, a highly non-linear variation in stiffness within the element results. Unsupported tension from a partially cracked element will be transferred, via nodal forces, to adjacent elements. If these are also cracked, the cracks there may, unrealistically, reclose.

In practice therefore, it is necessary to maintain a small residual stiffness across cracks, to prevent partial singularity in the element stiffness matrices. This residual Young's modulus is defined as

$$E_r = fE \tag{5.18}$$

where E is the Young's modulus in compression and f is a residual stiffness factor. Values of f between 0.01 and 0.001 have been tested in the validation process. Not surprisingly, larger residual stiffness is required for problems in which there are steep variations in stress distributions within the finite element mesh.

Over-integration is also used to accommodate the steep jumps in stiffness in elements affected by cracking. In common with many finite element codes, OXFEM uses Gauss-Legendre quadrature where integration of a function $\Phi(\xi)$ over a range $-1 \leq \xi \leq 1$ is approximated as

$$\int_{-1}^{+1} \Phi d\xi \approx \sum_{i=1}^{n} W_i \Phi\left(\xi_i\right) \tag{5.19}$$

and W_i is a weighting for integration point i. A polynomial of order $(2n - 1)$ can be exactly integrated with a Gauss-Legendre rule having n points. As the function of interest becomes more non-linear, the number of integration points increases. For the six-noded triangular element used for the masonry façades, the shape functions, linking nodal displacements to internal element displacements are complete quadratics. Calculation of an element stiffness matrix, under elastic conditions, therefore involves the integration of a quadratic since the strain displacement matrix $[B]$ will contain linear terms. Only three integration points are therefore necessary to obtain exact integration. Once cracking occurs, the stiffness calculation becomes highly non-linear and the success of a three-point rule requires the use of very small elements. Since relatively coarse meshes are necessary, to reduce solution times, a sixteen-point rule has been found to give a reasonable level of accuracy and stability.

To deal with the problem of unrealistic reclosure, due to redistribution of nodal loads following cracking in an adjacent element, a gradual, rather than abrupt, change of stiffness is incorporated. This can be visualised as a smoothing-out of the transition from stiffness in the compressive region to residual stiffness in the tensile (cracked) region.

Finally, a small residual tensile strength c is maintained at cracked integration points. The purpose of this is two-fold. Firstly, it avoids numerical difficulties involving uncertainty over material state at a zero-strength point. Secondly, zero tensile strength leads to the generation of large areas of unrealistic cracking, possibly hiding the important mechanisms arising from the differential movements applied. Studies have shown that a value for c of 10 kN/m^2 provides the right level of residual strength in cracked areas [10] and this is adopted in the analyses described later in this chapter.

5.4.4 Post-processing masonry data

The final stage of a finite element analysis is the post-processing of output data into a useful format. The large volume of data from most realistic models precludes any hand manipulation and the options are either in-house or commercially available software. In this research, use is made of both approaches with I-DEAS *Master Series* and an in-house program, 2CAN. The latter is used for generation of contours of surface settlements. 2CAN is also used to process and display the cracking of building façades. This is a feature not available in I-DEAS, and is one of the specific post-processing tools to be written for this project.

Output from the masonry model consists of the cracking strain ε^{cr} at integration points within elements together with the inclination θ of the cracks. This can be viewed as the plastic part of a total strain due to the applied settlement profiles. The exclusion of the elastic part of the total strain is reasonable when determining the level of damage to the masonry material. The cracking strain is resolved, not into principal components, but into three components that facilitate simple ways of displaying the damage to a façade. The three components are ε_1^{cr}, which represents the component normal to the first crack, ε_2^{cr} which represents the component normal to the second crack, if present, and γ_{12}^{cr} which represents the shear deformation across the crack. Generally, the most useful data are the normal first crack strains ε_1^{cr} as these can be compared with the strains obtained from the semi-empirical approach outlined in Section 5.2.

An improved idea of the intensity and inclination of cracks in the masonry can be obtained by plotting short lines at each cracked integration point in the mesh. These are displayed as collections of lines, at the given inclination, centred on the element integration point. The number of lines indicates the magnitude of the crack strain. This display gives an immediate indication of the damaged areas of the building façades.

A further refinement to the output display from the masonry model is obtained by averaging the cracking strains over regions in the mesh. The average cracking strain is defined as

$$\overline{\varepsilon}^{cr} = \frac{\sum_i \varepsilon^{cr} \Delta A_i}{A} \tag{5.20}$$

where the sum is over all the integration points which fall within a particular zone of area A and ΔA_i is the area associated with integration point i. The zones are specified by the user depending on the layout of the façade, perhaps due to the location of openings or other features modelled. Both this format and that without averaging can be converted to displays of the category of damage severity as used by the final stage of the semi-empirical approach.

All these forms of presentation of the damage data are used in the analyses that follow.

5.5 Two-dimensional studies of façades

As a pre-cursor to full use of the three-dimensional model described above it is first useful to look at two-dimensional models, concentrating on the behaviour of a single masonry façade.

Little is known of the mechanisms by which a masonry façade responds to the imposition of differential settlements or horizontal movements, especially those arising from tunnelling. This section shows the differences in damage predicted for a simple façade when assuming either elasticity or a more realistic masonry model.

5.5.1 Façade types analysed

Comparisons between elasticity and elastic-no tension material models are made using models of walls with and without openings. Both are typically found as components of the systems of façades in existing masonry structures. A wall of dimensions 20m by 8m with unit thickness is analysed here, representing a façade from a two-storey building. The arrangement of openings in the façade is shown in Figure 5.6. The finite element model of the façade includes lintels over the openings which are elastic throughout both analyses. This is necessary to prevent immediate failure above the openings.

The material properties used in these analyses are given in Table 5.2. Explanations of the parameters for the masonry model are given in Section 5.4.

5.5.2 Analysis procedure

An aspect of real masonry structures that has not been discussed up to now is the state of the structure before the application of the effects of tunnelling. During construction, parts of a masonry façade may become damaged for many different reasons. Following construction, consolidation settlement of the ground supporting the foundations of the structure may lead to further changes in the stresses in the façade. The structure may also be affected by changes in environmental conditions over a longer period of time.

Clearly, it is virtually impossible to determine the state of stress in a component of a masonry structure with any degree of certainty and therefore is not attempted in

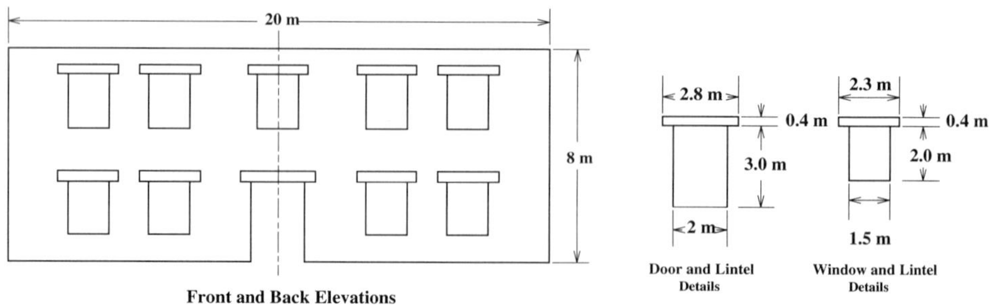

Figure 5.6: Layout of openings in two-dimensional façade analysis.

the models described here. Equally, it is obvious that the self-weight of the structure should be modelled. In all analyses using these models the self-weight is applied in a pre-analysis load step. The stresses generated in the structure due to this load are maintained, as are the changes in stiffness due to cracking. Strains and displacements are then reset to zero so that any subsequent changes are due to the effects of tunnelling alone.

Following application of self-weight, the input to these two-dimensional analyses is a settlement profile applied directly to the base of the façade. No account is taken of the interaction between the façade and the source of tunnelling settlements. For the reasons outlined in Section 5.2.1, this is clearly at odds with reality. These calculations are included to demonstrate the importance of the choice of material model rather than to accurately model the tunnelling settlement problem. All movement is assumed to take place within the plane of the façade.

	Youngs Modulus, E(kPa)	Poisson's Ratio ν	Unit Weight γ (kN/m^3)	f	c (kPa)	ε_0^{cr} (microstrain)
Elastic	1.0×10^7	0.2	20	-	-	-
Masonry				0.01	10	4

Table 5.2: Material property parameters for elastic and elastic-no tension masonry models.

Two widths of settlement trough are used, with trough width parameters $i = 5$m and 12.5m. In each case the maximum settlement S_{max} is assumed to be 25mm and to occur at the centreline of the façade. This means that only one half of a façade needs to be analysed, due to the symmetrical nature of the settlement trough. A vertical roller boundary is therefore modelled at the building centreline. Figure 5.7 shows the troughs in relation to the building base. Clearly, these troughs represent increasing amounts of ground loss and increasing tunnel depths.

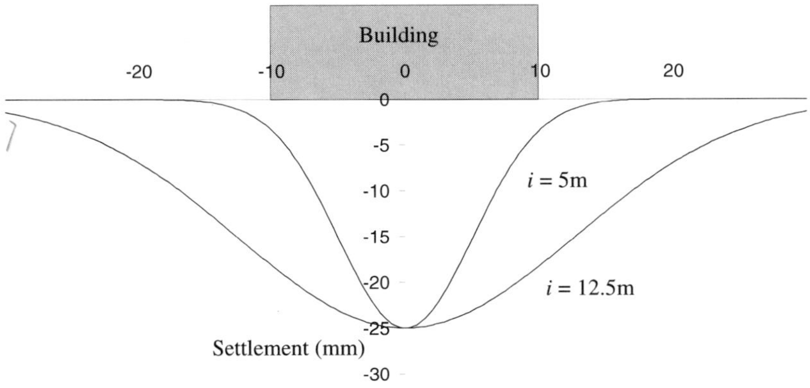

Figure 5.7: Settlement troughs applied to facades in 2D calculations.

For comparison with existing methods of damage assessment, the maximum tensile strains predicted can be calculated using the semi-empirical approach (outlined in Section 5.2). Given the location of the building, and considering sagging and hogging curvature, the maximum tensile strains are predicted to be $1430\mu\varepsilon$ and $521\mu\varepsilon$ for the troughs with $i = 5$m and 12.5m respectively. These correspond to damage severity categories of *slight* and *very slight* respectively. In both cases these maxima occur in the sagging region of the building. As the trough width increases so the damage predicted (measured in terms of maximum principal tensile strain) decreases, as one would expect due to the flattening of the trough.

5.5.3 Results for a plain façade

The results of the finite element modelling of the façades are presented here as contour plots of the damage severity category and as plots of crack positions. In the elastic façades, the major principal strain is used to determine the damage severity index. In each case, the façade centreline is the left-hand edge of the plot. (This is obvious for the façades with openings but not for the plain façades). The façades with openings are divided into 23 zones for crack strain averaging. The zones are located to follow the features in the façades, namely openings and lintels.

Crack strain averaging is not carried out for the plain façades. Cracks are plotted only where the normal crack strain is 0.05% or above.

Figure 5.8 shows the results for the plain façades. Figures 5.8a and 5.8b show the non-localised spread of strains across the façade with elastic material, no localisation being possible due to the nature of the material model. As the trough narrows the tensile strain levels increase. The results are roughly in line with the predictions of the

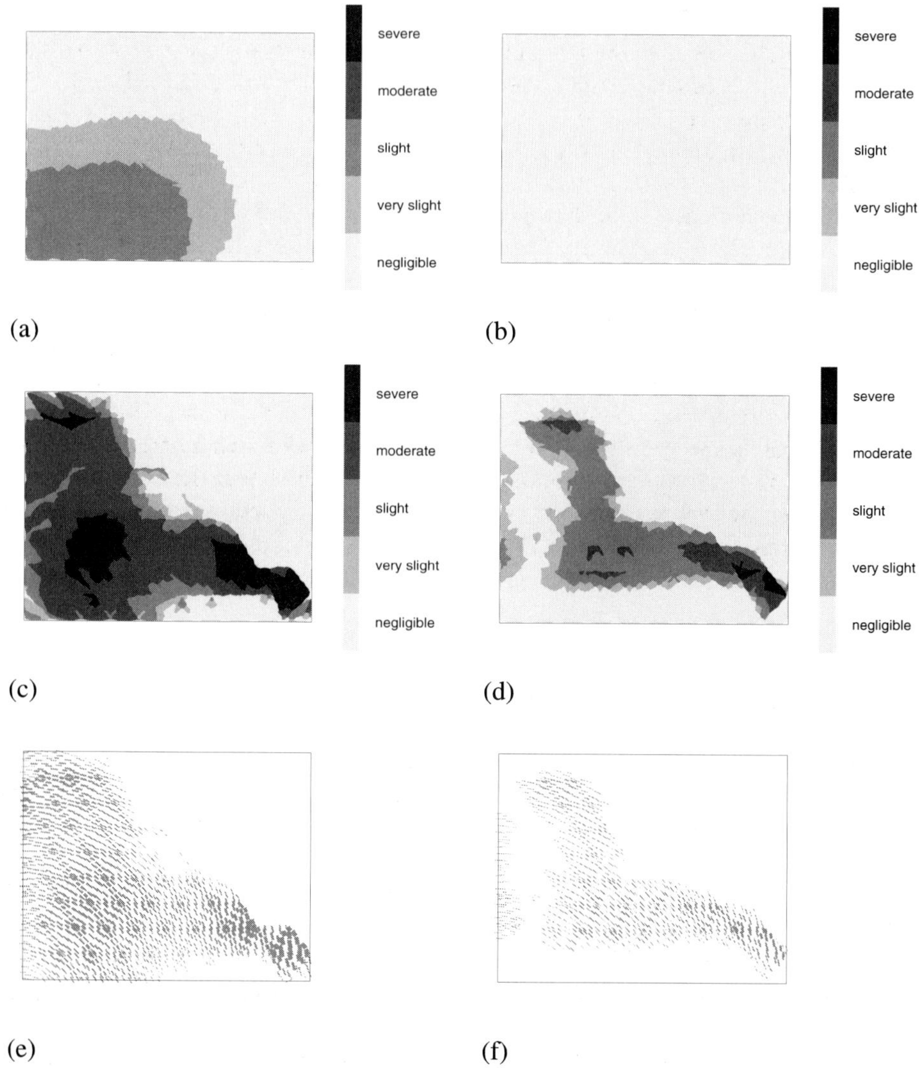

Figure 5.8: Results for analyses of plain façades. Damage levels for elastic material, (a) $i = 5$m, (b) $i = 12.5$m. Damage levels for elastic-no tension material, (c) $i = 5$m, (d) $i = 12.5$m. Cracking patterns for elastic-no tension material, (e) $i = 5$m, (f) $i = 12.5$m.

semi-empirical approach and slight damage is predicted for the case of $i = 5$m. The wider trough induces only negligible damage in the façade. Despite the inadequacy of elasticity to represent masonry, these results show the benefit of a numerical model just in providing greater information on the locations of maximum tensile strains.

Figures 5.8c and 5.8d are the damage results for the elastic-no tension façades. In addition, the predicted crack patterns are shown in Figures 5.8e and 5.8f. The immediate impression is one of localisation of damage. In the case of the narrow settlement trough, damage is concentrated close to the centre of the façade, resulting from shearing in the sagging area. There is also severe damage at the right-hand base corner. The general levels of damage predicted using the masonry model are much higher than predicted by the semi-empirical approach or by the elastic calculations. The wider settlement trough leads to considerable reductions in the area of the façade affected but the worst damage is the same as for the narrower trough.

The crack patterns indicate that the façade appears to form load-bearing arches that span across the areas of greatest differential settlement. The orientation of the cracks is horizontal over the centreline and becomes vertical at the right-hand base corner. The arches support the material lying at the top right-hand part of the façade which remains undamaged throughout both analyses. These analyses of plain façades illustrate the potential vulnerability of masonry façades if one assumes rapid loss of stiffness following cracking, as modelled by the elastic-no-tension formulation used here.

5.5.4 Results for a façade with openings

Figure 5.9 shows the results of the analyses where the same settlement profiles are applied to façades with openings. The elastic results (Figures 5.9a and 5.9b) show the effects of stress concentrations around the openings. The maximum severity indices are higher than for the plain façades for this reason. Once again, the wider settlement trough induces a lower level of damage although it is not *negligible*, due to the openings.

Figures 5.9c and 5.9d show the average levels of damage in each of the zones into which the façade has been split. The imposition of the narrow settlement trough (i = 5m) yields a *moderate* maximum damage level while it is only *slight* for the wider trough. The unaveraged results, which are not presented here, show areas of unrealistically high strain immediately next to stress concentrating features. In a real façade it is likely that there would be redistribution of stress as the material began to lose stiffness and therefore the averaging procedure would appear to produce a more realistic prediction of the true behaviour. The plots of cracking in Figures 5.9e and 5.9f again show the development of arches. For the narrow settlement trough there appears to be an arch formed so that the heaviest damage is largely confined to the ground floor of the façade.

Something similar appears to be happening with the wider trough. Columns between openings appear especially vulnerable. Comparison of these crack plots with those for the plain façades show the way that the openings force an alternative arching pattern.

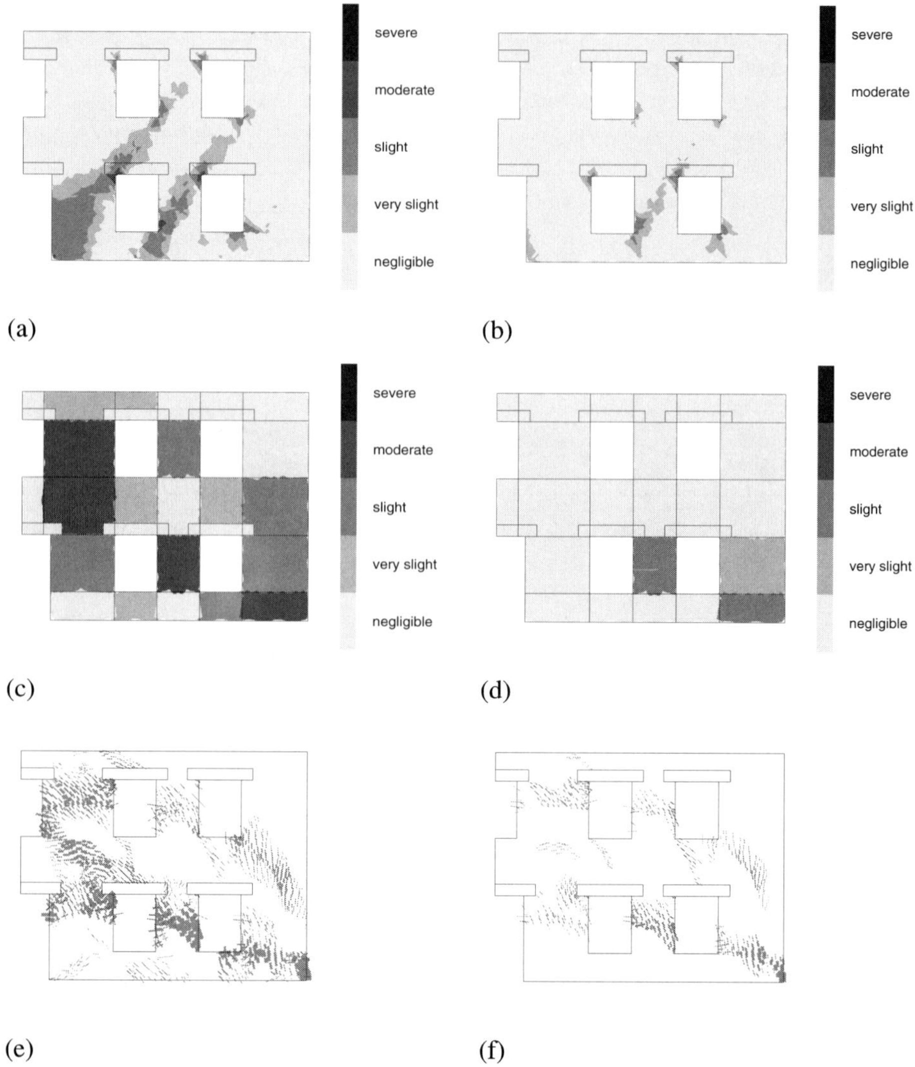

Figure 5.9: Results for analyses of façades with opening. Damage levels for elastic material, (a) $i = 5$m, (b) $i = 12.5$m. Averaged damage levels for elastic-no tension material, (c) $i = 5$m, (d) $i = 12.5$m. Cracking patterns for elastic-no tension material, (e) $i = 5$m, (f) $i = 12.5$m.

5.5.5 Discussion

These analyses highlight the differences in behaviour of façades dependent on the material model chosen to represent masonry. The use of elasticity leads to quite different patterns of major principal strains than the comparable normal crack strains in the elastic-no tension model. This indicates that the material models strongly influence the mode of failure of the façade as a whole. In the latter case, arching behaviour is evident. This observation alone could form the basis for a revised semi- empirical approach to settlement damage. In general, the use of a material model that incorporates cracking leads to higher predictions of damage.

5.6 The three-dimensional model of tunnelling

Having examined the behaviour of façades in two-dimensional analyses the next step is to use a fully three-dimensional model. The major difference between the analyses described in this section and those of the previous section is that the tunnelling settlements here are produced from the model rather than being superimposed by the modeller. This allows full interaction to develop between façade and ground; an aspect of the problem not modelled in the current damage assessment procedure.

Two examples of the three-dimensional model are given here. The first is an idealised model of a simple masonry structure affected by the construction of a shallow tunnel. The second example involves the modelling of shaft construction adjacent to an old church in London. Both illustrate the need for a three-dimensional simulation of this problem and show how this particular three-dimensional model is used. They also illustrate the practicalities in using these very large computational models.

5.6.1 Example analyses of a simple building

Figure 5.10b shows a building of plan dimensions 20m by 10m, comprising four façades: two identical plain walls and two with openings of identical patterns. (The pattern is the same as that used in the analyses described in Section 5.5). Neither the foundations, the internal walls nor the roof structure are included in the model, for simplicity, although they could be easily incorporated into the model if necessary. (The foundations would typically provide tensile resistance to the base of the building).

The effect of the construction of a straight tunnel of 5m diameter, with its axis at a depth of 10m beneath the ground surface is modelled in this analysis. The building is located at a 45° skew to the tunnelling direction as indicated in the figure so that the tunnel passes under the front façade and one side façade. A two-dimensional analysis of this particular problem could not model accurately the plan location of the building with respect to the tunnel. This is apart from the inability of such an analysis to model the incremental passage of the tunnel beneath the building. The masonry material properties in the façades are the same as for the two- dimensional analyses described in Section 5.5 and the façades are also assumed to be 1m thick. The soil model used is an elasto-plastic formulation with increasing stiffness and strength with depth, as mentioned earlier in the general description of the model developed at Oxford [8].

In this analysis, a total tunnel length of 60m is excavated in four increments, as indicated in Figure 5.10a. While tunnel excavation is a continuous process, the numerical model approximates by excavating in discrete increments. The greater the number of increments, the closer the approximation to the continuous process. The desire to use a large number of increments, however, has to be balanced with the need to limit the solution times for the calculation, which can be excessive as each increment is solved over a number of intermediate sub-steps.

The finite element model of this problem consists of a block of ground of plan dimensions 100m by 60m with a total depth of 60m. These dimensions are sufficiently large to avoid any boundary effects [1]. Boundary conditions consist of base fixity, vertical rollers on the sides of the block and full freedom on the ground surface. The ground is meshed with 2243 ten-noded tetrahedra and the building comprises 802 six-noded triangular elements. The tunnel lining is modelled using linear elastic shell elements of an unusual overlapping formulation (described in detail in references [1], [20]). There are 640 lining elements in total which are activated in blocks at each stage of tunnel excavation. While there may appear to be large numbers of elements in the mesh used in this analysis, the discretisation is relatively coarse. (The computer resources available to run this analysis were less powerful than possible in a typical research environment today).

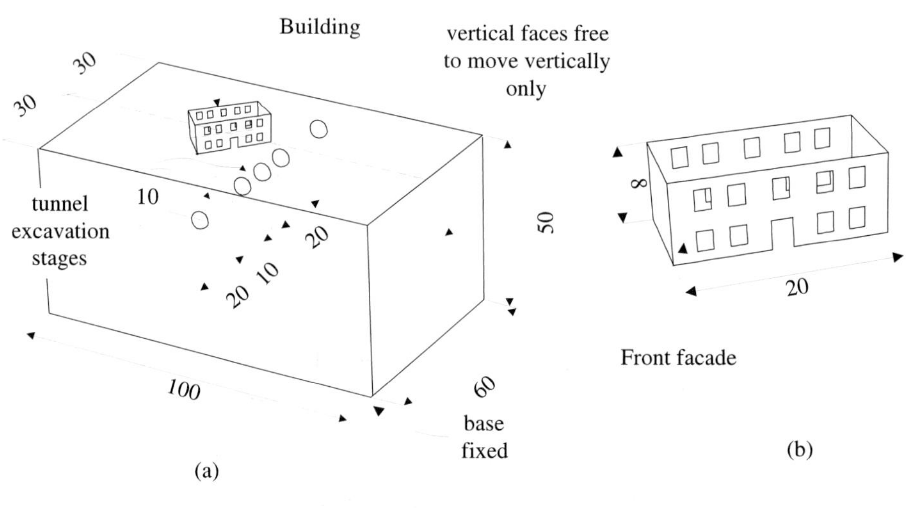

Figure 5.10: Example analysis of a simple building, (a) general layout (b) façade layout (all dims in metres).

The analysis was carried out in six stages. Stage 1 consisted of the application of the building self-weight, assuming a unit weight for masonry of 20 kN/m^3. Stage 2 involved no further loading but allowed the small residual out-of-balance forces in the building, present after Stage 1, to be relaxed. At the end of Stage 2, the strains and displacements throughout the soil and building mesh were reset to zero. The state of each integration point was, however, maintained to model the state of the ground and the masonry prior to tunnelling, (as discussed in Section 5.5.2).

Incremental excavation and lining were then carried out in Stages 3 to 6.

5.6.2 Results

As stated above, an important aspect of the three-dimensional model is that the settlement profiles are produced by the response of the finite elements modelling the ground, rather than by direct user input. Therefore, before considering the response of the building it is useful to compare the settlements predicted by the model with those from the semi-empirical approach. To do this, an additional analysis was carried out without a building but with the same simulation of tunnel construction.

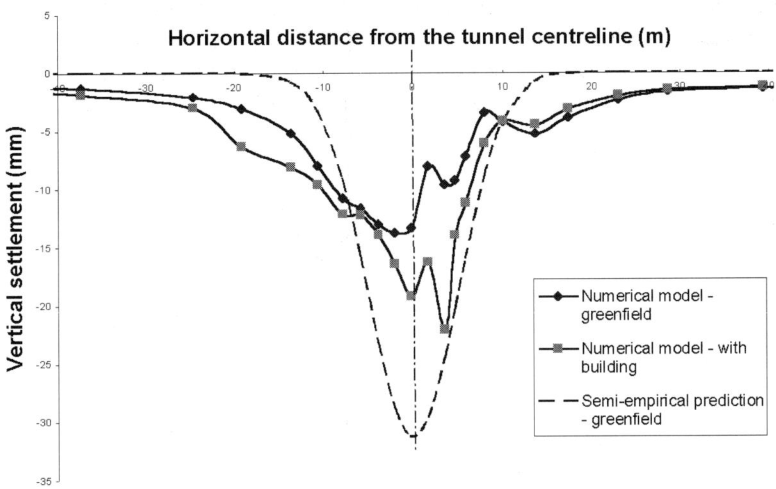

Figure 5.11: A comparison of the model predictions of greenfield settlement profiles with those of the semi-empirical approach.

Figure 5.11 shows the transverse settlement profiles at the end of the analysis at the centre of the mesh for the numerical models, with and without the building. The Figure also shows the semi-empirical prediction for the same tunnel and value of volume loss. The maximum settlement predicted for the numerical simulation of the greenfield case is approximately 14mm, as compared to 32mm from the semi-empirical prediction. This under-prediction of settlements with numerical models is similar to that found in previous work by other researchers investigating this problem [3,7]. It is thought to be a result of non-localisation of strains around the tunnel, leading to a wider, shallower trough. The maximum settlement over the tunnel axis when a building is present is around 21mm. The irregularity in profiles is probably due to the coarse discretisation used.

The effects of tunnelling on the building can be judged, as before, by examining the levels of damage in each façade in the analysis with a building present. In this analysis, however, there is progressive damage due to the passage of the tunnel beneath the building.

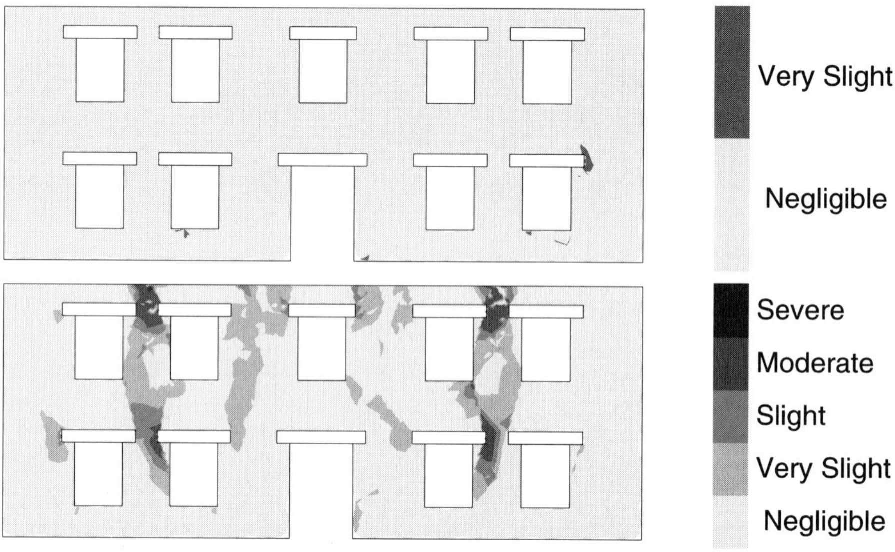

Figure 5.12: Damage categories in facades at the end of the analysis; (a) front (b) rear.

In some areas cracking occurs which then closes up due to the changing nature of the wave of settlement that passes. Unlike the two-dimensional analyses, the façades are interconnected and, while their mode of deformation is entirely in-plane, interaction between façades is evident. All these effects serve to move the model closer to the real situation.

A prediction of damage using the semi-empirical approach for this building is made difficult by its skewed position with respect to the tunnel axis. If the length of the front or rear façades is taken as its projected length onto the normal to the tunnel axis and a semi-empirical analysis undertaken, the maximum tensile strains in sagging and hogging regions are predicted to be $1390\mu\varepsilon$ and $350\mu\varepsilon$ respectively. These levels indicate worst damage to be either *slight* or *negligible*.

For comparison, damage categories are shown at the end of the analysis for the front and rear façades in Figure 5.12. The front façade results indicate lower damage levels than the semi-empirical predictions while the rear façade damage levels are generally higher than the traditional predictions with a maximum crack strain of between 3000 and $3500\mu\varepsilon$, although this is highly localised and averaging would probably obtain a more reasonable result.

Perhaps of greater, but qualitative interest are the **differences** in behaviour between front and rear façades. The crack patterns predicted for the excavation stages (3 - 6) of the analysis for the front and rear façades are shown in Figure 5.13. In the front façade cracking commences at stage 3 and at stage 4 is concentrated around the right-hand end, closest to the location of the tunnel face at that stage.

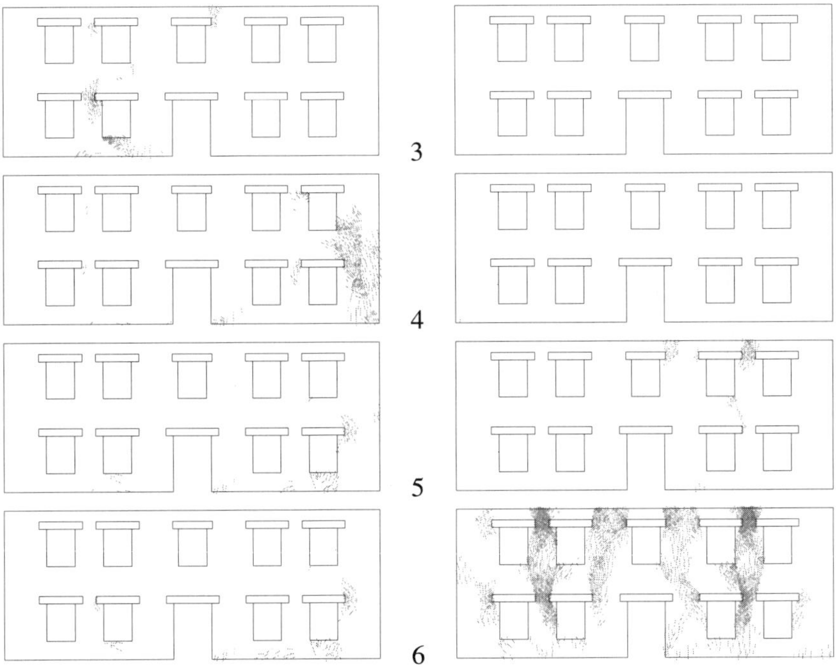

Figure 5.13: Cracking in front and rear facades after each stage of tunnel excavation. (The facades are viewed from outside the building).

At stage 6, the crack pattern in the front façade has changed, and generally reduced, with areas of cracking reclosing following passage of the tunnel face. Behaviour of the rear façade is quite different and is characterised by sudden widespread hogging bending failure. Stages 3 to 4 show little or no cracking, corresponding to the proximity of the tunnel face to this façade in these stages. At stage 6, major cracking is evident indicating overall hogging bending failure. Of interest is the suddenness of this major failure. Examination of the profile of the base settlements of this façade shows a noticeable increase in hogging curvature between stages 5 and 6 which explains this failure mode. The masonry model is clearly close to failure at stage 5 but widespread cracking, and hence loss of stiffness is only initiated by the effects of stage 6 excavation.

The cracking seen in the front and rear façades suggests that, at stage 3 of the analysis only the front façade is affected by tunnelling settlements. Stages 4 and 5 result in a stabilisation of the front façade crack pattern. At stage 6, the combined effects of the additional tunnelling settlements, as the tunnel face passes the rear façade, and weakening from earlier interactions through the side façades, leads to major cracking. An important conclusion, therefore, is that the location of a building façade with respect to the tunnelling direction is important if it is connected to other façades that are affected earlier. It appears that such façades are subjected to interaction effects that may lead to weakening followed by local movement which causes more damage.

5.6.3 Modelling the effects of shaft construction on an 18th century stone clad church, Maddox Street, London.

A second example of the use of the three-dimensional model of settlement damage concerns the construction of a shaft and tunnel adjacent to an historic church in Maddox Street, London. Monitoring of ground surface settlements, precision levelling around the structure and observations of damage were carried out during this scheme. These measurements can be directly compared with the predictions of the numerical model. Further details can be found in Reference [21]. Howard Humphreys Consulting Engineers designed and supervised the construction of a tunnel in 1991/92 to provide a route for a new electricity supply into the West End of London. The project required construction of a 2.5m tunnel, 15m to 25m below ground. Shafts were required along the route for access during construction. One such shaft was in Maddox Street, which is approximately 40m long and 10m wide between building frontages. It is bounded on the North side by a terrace of four storey brick Georgian houses, and on the South side along its entire length by an eighteenth century stone-clad brick church. The 4m diameter shaft was sunk to 15m depth in the middle of the street at the East end, 5m from the North East corner of the church. Figure 5.14 shows the site plan for the scheme adjacent to the church.

The church is the most important building likely to be affected by the construction works and is the only building modelled in the analysis. The terrace on the North side is on the opposite side of the anticipated settlement trough so it seems reasonable to assume that there is little interaction between it and the church. In addition, field measurements showed largest movements during the construction of the shaft so that it alone is simulated in this analysis.

A number of assumptions are required in the representation of the church. The first question is whether it is valid to model the church as a rectangular box of four façades, similar to the previous example. The church structure consists mainly of masonry walls. There are columns within the church but these support only the roof and balcony, which both being of wooden construction have little stiffness or self weight compared to the masonry walls themselves, which are 1m thick. The simple four façade model therefore seems appropriate.

There are windows at regular intervals around the church. All the openings in the façades could be modelled, but that would lead to unnecessarily complex meshes.

The main reason for modelling openings is to model crack growth from the corners realistically and to capture the reduction of stiffness due to the opening. Once again

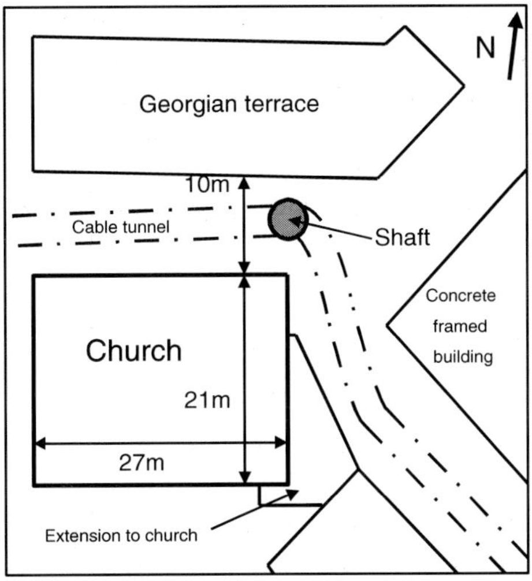

Figure 5.14: Schematic site location plan, Maddox Street, London.

a balance has to be drawn to limit the complexity of the analysis and hence provide a reasonable solution time.

A new approach is adopted here to account for the reduced stiffness and self-weight present where there are openings in a façade. In less critical façades of the building, an average stiffness for the whole façade is used (a "fully smeared" approach). In other areas, vertical zones or columns of reduced stiffness replace openings aligned vertically. This "semi-smeared" method is similar to the approach used in Reference [22] except that Simpson's "strata" of reduced stiffness represent rows of openings horizontally rather than columns. Horizontal zones may not, however, model so well the important formation of cracks from the top of the façade, propagating downwards in a hogging region such as those witnessed in the rear façade of the first three-dimesnional example.

A combination of discrete openings, "smeared" and "semi-smeared", are instead used in the model of the church. The openings closest to the shaft are modelled in detail, the next closest as columns of reduced stiffness. The two more distant façades are modelled using a fully smeared stiffness. The masonry properties are as the preceding examples.

While site investigations show an overconsolidated clay overlain by approximately 5 metres of gravel and made ground, the soil in the model is assumed to be a uniform overconsolidated deposit with similar properties to those used in the first three-dimensional example above. This approach is justified by the location of the foundations of the church, at 3m below ground. It is thought the response of the structure would likely be dominated by the behaviour of the clay in this instance. Four analyses were conducted to study the response of the building

1. A simulation of the shaft construction and the surrounding soil, i.e. a "greenfield" calculation without the building.

2. Analysis of the building alone, applying the displacements calculated from the above "greenfield" analysis directly to the base of the building.

3. Analysis of the building alone, applying the **observed** field displacements to the base of the building.

4. A fully coupled analysis including shaft, soil and building. A view of part of the finite element mesh for this run is given in Figure 5.15.

As for the first three-dimensional example above, this collection of analyses enable

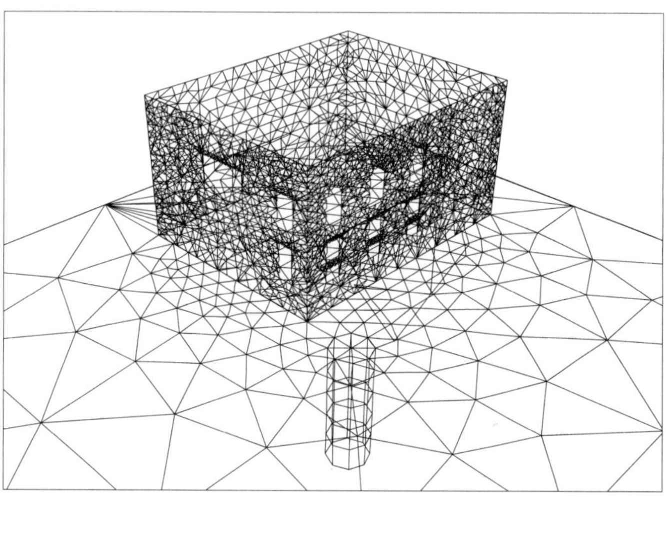

Figure 5.15: Detail from finite element mesh for analysis 4 at Maddox Street.

5.6.4 Results

A comparison of the predicted settlement profiles with those measured at the site can be made in Figure 5.16 where the settlements along the East and North façades are plotted. These façades are closest to the shaft. The South and West façades did not settle significantly, or experience any measurable damage in any of the analyses or in

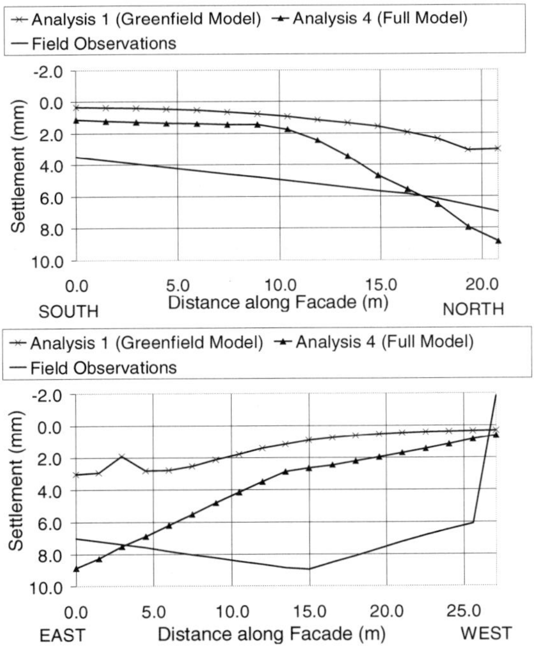

Figure 5.16: Observed and calculated settlements, (a) East facade, (b) North façade.

the field. The profiles for the observed site data indicate that the building tilted almost as a rigid body towards the North, with a fairly constant settlement of 68mm along the whole of the North façade. Some of this settlement may have resulted from the construction of the tunnel, which leads from the shaft along Maddox Street parallel to the North façade.

The results from the full analysis (4) of soil and building show the bases of both façades following half-troughs similar to the shape that would be predicted by a semi-empirical analysis but with a somewhat pronounced point of maximum hogging. The maximum settlement of 9mm calculated at the North-East corner is just 2mm greater than that observed. The calculated "greenfield" settlements without the building also follow Gaussian-type half-troughs, a bit smoother in general but with a reduced maximum settlement of only 3mm at the corner, and with signs of localised irregular behaviour (probably due to mesh discretisation) near the corner.

These concur with contoured plots of the surface settlement, which are not included here for brevity. The numerical predictions are reasonably close to the settlements measured on site. The inclusion of a model of the building (as in Analysis 4) improves the agreement.

Plots of damage category for the North and East façades of the building for Analy-

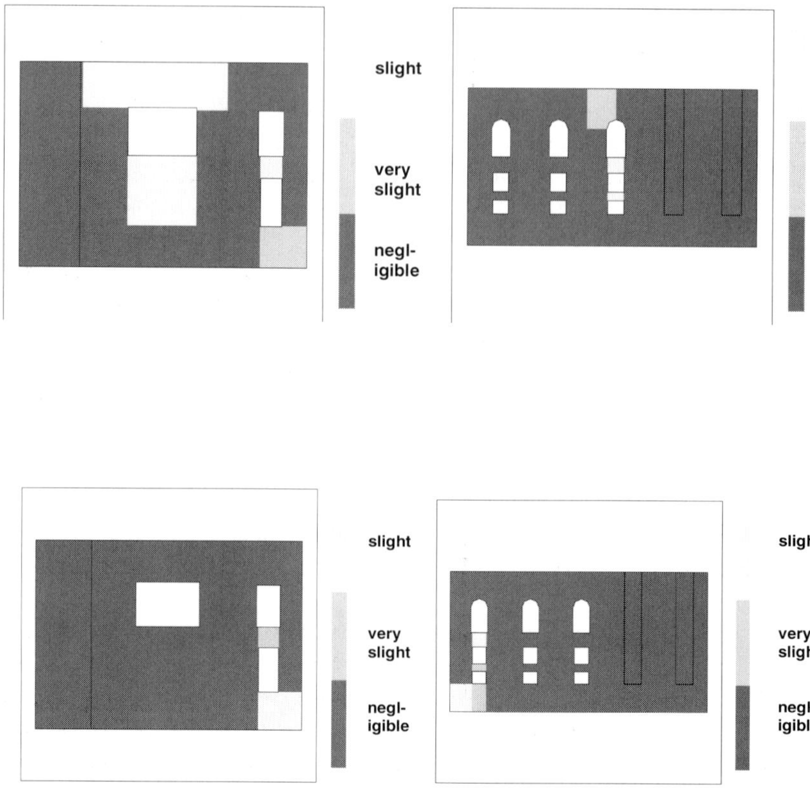

Figure 5.18: Damage categories from Analysis 2, application of greenfield profiles to East and North façades.

ses 2 and 4 are presented in Figures 5.17 & 5.18. (It should be noted that the greyscale is reversed in these plots, with **lightest** areas being most heavily damaged). All views are from outside the building. The façades have been divided into regions and the cracking strain averaged to give the damage category. As in the analysis of the two-dimensional façades, some judgement was required on the division of the façades into regions. These were chosen, in this case, to highlight regions adjacent to openings and areas of different stiffness. Two vertical semi-smeared zones are shown on the plots for the North façade. Reduced region sizes were used where more detail was required, *e.g.* in the North East corner of the building.

The distribution of the damage, in particular the position of the "slight" damage, is quite distinct for each run. The site investigations undertaken during and after shaft and tunnel construction failed to find any new cracking of this magnitude in the façades. The structure was, however, known to be cracked due to earlier nearby tunnel construction particularly in the East façade and at the basement level. The model predictions cannot be regarded as definitely proved accurate, but are entirely consistent with the observations in suggesting "slight" damage. Of prime importance is the fact that they do not under-predict the real damage.

Figure 5.17 shows the results for the full non-linear analysis. This predicts the largest amount of damage on the East façade, with "slight" damage around the larger opening, and further damage associated with the North East corner. The North façade experienced damage concentrated in one column of openings. It can be seen by comparison with the plotted settlements along the base of these façades (Figure 5.16) that in both cases these areas of damage occur in the hogging part of the settlement profile, which is quite pronounced for both façades. It appears that some interaction occurs.

The hogging region of the settlement bowl lines up with an opening or openings; this causes damage to that part of the façade, thus reducing its stiffness, thereby magnifying the hogging profile, producing a slight "hinge" and saving the remainder of the façade from damage. The position of such hinges would be hard to predict without a three-dimensional analysis.

The results for Analysis 2 (Figure 5.18) show damage in the *very slight* and *slight* categories concentrated at the corners of the façades nearest the shaft. This is consistent with the greenfield settlements applied, which were fairly small and slightly irregular at this corner.

The damage results from Analysis 3 (in which the observed settlements are imposed on the building) show no damage greater than *negligible* on the East façade, and on the North façade *slight* damage aligned with the second column of openings. This low damage is consistent with the building tilting as a rigid body towards the North, with almost straight settlement profiles along both façades. The fact that *slight* damage was still registered between openings shows the important effect of the presence of openings on the behaviour of masonry façades, causing stress concentrations at sections of reduced shear strength and stiffness.

The fact that the building tilted almost as a rigid body is thought to be due to the foundation contributing more to the stiffness of the whole system than had been taken into account in the analyses. This hypothesis has been tested with a further calculation in which a stiff foundation was added. The results of the numerical model indicated behaviour closer to this rigid body mode thus highlighting the importance of modelling the foundation stiffness.

5.6.5 Discussion

The two examples described above demonstrate early attempts at the use of the three-dimensional finite element model to assess the effect of nearby excavation works on masonry buildings. In the first example, the progressive damaging effect of tunnelling is modelled. The important conclusion is that façades in a masonry building may be affected in quite different ways by the passage of a single tunnel due to their locations with respect to the settlement trough. Interaction between façades appears also to be a significant feature that it is only possible to study with three-dimensional models. Improvements in ground response, to match greenfield predictions for greenfield sites, should come from further work on soil models and by using finer mesh discretisation. The latter will become more feasible with increases in computer power.

The second example shows that the representation of a building can be further simplified by "smearing" openings. Provided this is done away from the areas of main interest it appears to be an efficient and realistic way of accounting for stiffness variations. The results obtained from Maddox Street appear to be reasonable in terms of the damage category of *slight* predicted for the building, which is consistent with the observed result, perhaps a little conservative, but not overly so. There is, however, scope for refinement of the assumptions made in the modelling process, and improvements can be made in future analyses.

As in the first example, interaction phenomena between the building and the ground were observed in the Maddox Street model. Cracking of the building in hogging regions reduced the effective stiffness, thus causing the cracking to localise. The situation would have been quite different if the building had spanned across the whole settlement trough, so enabling arching action to take place. Thus the importance of the particular geometry of the problem is obvious.

5.7 Concluding remarks

The prediction of the effects of tunnelling, and other construction activities on sensitive masonry buildings is an important problem only partly solved by current procedures. The interaction between ground and building can only reasonably be accommodated in a numerical model. Accuracy depends on the material models chosen, especially for the masonry. The two- and three-dimensional models, whose development and use are described above, are examples of predictive tools for this problem. Development and improvement of these models is continuing to improve their capabilities, including further validation against field data.

It is hoped that with increasing computer power, the use of three-dimensional numerical techniques may become commonplace in design offices, leading to better predictions of the effects of tunnel construction and reduced accommodation measures on such schemes.

5.8 Acknowledgements

The work described here is the result of a research programme at Oxford University that began in 1993. Funding for this research has been received from a number of sources including the Engineering and Physical Sciences Research Council, Brown & Root Ltd. and the Royal Commission for the Exhibition of 1851. Considerable development work has been undertaken at the Oxford Supercomputing Centre. The research has been carried out by a number of existing and former members of the Civil Engineering Research group at Oxford led by Prof. Guy Houlsby and Dr Harvey Burd. In particular, the development of the masonry material model (Sections 5.4 and 5.5) is the work of Dr Liu Gang and the study of the church at Maddox Street, London (in Section 5.6.2) is the work of Alan Bloodworth.

References

[1] Augarde C.E., "Numerical Modelling of tunnelling processes for assessment of damage to structures", D.Phil. Thesis, University of Oxford, 1997.

[2] Mair R.J., Taylor R.N., Burland J.B.,"Prediction of ground movements and assessment of risk of building damage due to bored tunnelling" in Geotechnical aspects of underground construction in soft ground (eds Mair &Taylor), Balkema, 713-718, 1996.

[3] Potts D.M., Addenbrooke T.I., "A structure's influence on tunnelling-induced ground movements", Proc. ICE, Geotech. Engng, 125:109-125, 1997.

[4] Augarde C.E., Burd H.J., & Houlsby G.T., "A three-dimensional finite element model of tunnelling" in Numerical models in Geomechanics, Proc. NUMOG V, 457-462, 1995.

[5] Burd H.J., Houlsby G.T., Chow L., Augarde C.E., Liu G., "Analysis of settlement damage to masonry structures", Proc. 3rd European Conf. Numerical Methods in Geotechnical Engineering (ed. I.M. Smith), 203-208, 1994.

[6] Burd H.J., Houlsby G.T., Augarde C.E., & Liu G., "Modelling the effects on masonry buildings of tunnelling-induced settlement", Proc. ICE, Geotech. Engng, 143(1), 17-29, 2000.

[7] Stallebrass S.E., & Taylor R.N., "The development and evaluation of a constitutive model for the prediction of ground movements in overconsolidated clay", Geotechnique, 47(2), 235-253, 1997.

[8] Houlsby G.T., "A Model for the Variable Stiffness of Undrained Clay", Proc. Int. Symp, on Pre-Failure Deformation of Geomaterials, Torino, 26-29 September, 443-450, 1999.

[9] Brown P., Booker J., "Finite element analysis of excavation", Computers and Geotechnics,1(3), 207-220, 1984.

[10] Liu G., "Numerical modelling of damage to masonry buildings due to tunnelling", D.Phil. Thesis, University of Oxford, 1997.

[11] Liu G., Houlsby G.T., Augarde C.E., "A tying scheme for imposing displacement constraints in finite element analysis", Communications in Numerical Methods in Engineering, 16(10), 721-732, 2000.

[12] Bell R.W., Houlsby G.T., Burd H.J., "Suitability of two and three dimensional finite elements for modelling material incompressibility using exact integration", Communications in Numerical Methods in Engineering, 9(4), 313-329, 1993.

[13] Ali S., Page A., "Finite element model for masonry subjected to concentrated loads", ASCE Journal of Structural Engineering, 114(8), 1761-1784, 1988.

[14] Lotfi H., Shing P., "Interface model applied to fracture of masonry structures", ASCE Journal of Structural Engineering, 120(1), 63-80, 1994.

[15] Pande G.N., Liang J., Middleton J., "Equivalent elastic moduli for brick masonry", Computers & Geotechnics, 8(3), 243-265, 1989.

[16] Loo Y., Yang Y., "Cracking and failure analysis of masonry arch bridges", ASCE Journal of Structural Engineering, 117(6), 1641-1659, 1991.

[17] Di Pasquale S., "New trends in the analysis of masonry structures", Meccanica, 28, 173-184, 1992.

[18] Bloodworth A.G., Augarde C.E., Houlsby G.T., "Transferring a non-linear finite element code to the Oxford Supercomputer, Oscar", Proc. 7th Int. Conf. Civil & Struct. Eng. Comp., Oxford, Civil-Comp Press, Vol. B 85-94, 1999.

[19] Cook R.D., Malkus D.S., Plesha M.E., "Concepts and applications of finite element analysis", Third Edn, John Wiley & Sons, New York, 1991.

[20] Phaal R., Calladine C.R., "A simple class of finite elements for plate and shell problems. II: An element for thin shells, with only translational degrees of freedom", Int. J. Num. Meth. Eng., 35, 979-996, 1992.

[21] Bloodworth A.G., Houlsby G.T., "Three-dimensional analysis of building settlement caused by shaft construction". Proc. Int. Symp. Geotech. Aspects of Underground Construction in Soft Ground, Tokyo, 1999.

[22] Simpson B., "A Model of Interaction between tunnelling and a masonry structure", Proc. 3rd Eur. Conf. Num. Meth. in Geotech. Eng., Manchester, Balkema, 221-228, 1994.

©2001, Saxe-Coburg Publications, Stirling, Scotland
J.W. Bull, (Editor), Computational Modelling of Masonry,
Brickwork and Blockwork Structures,
Saxe-Coburg Publications, Stirling, Scotland 143-176.

Chapter 6

Modelling and Behaviour of Masonry Walls in Fire

M. O'Gara

Faculty of Engineering, University of Ulster, Newtonabbey, Northern Ireland, United Kingdom

Abstract

This chapter provides an insight into the thermo-structural behaviour of masonry walls at elevated temperatures, and the various parameters affecting their structural response. A detailed description of the mechanical material properties of various masonry materials at elevated temperatures is given. An involved high temperature concrete material mathematical model is presented and validated. The material model is implemented in a newly developed finite element model, capable of simulating the thermo-structural behaviour of masonry compartment walls subject to fire exposure on one side. The finite element model was used to produce numerical examples which are compared with experimental results.

6.1 Introduction

Preservation of a structure during the event of a fire is an ever increasing concern, and one which is foremost at the design stage. An adequately designed fire resisting construction will significantly reduce post fire repair costs, while also providing a safe building for the evacuation of any building occupants. Masonry materials have been successfully used for the construction of fire resisting elements since the need for fire protection became apparent. In particular masonry walls have been engineered to act as fire barriers and are aptly termed "fire walls". Fire walls are commonly employed in a vast array of constructional situations, ranging from domestic dwellings to large industrial applications. The term "fire resistance" in this instance is defined as the period of time a fire wall maintains three basic criteria during fire exposure. The first of these criteria is integrity, where the wall must not develop cracks or fissures that permit the passage of flame or hot gases from one side of the wall to the other. Secondly, the preservation of structural stability of a load bearing wall, to provide safe egress, safe access for fire fighters, and to mitigate against structural damage. Finally the third criterion is that of insulation, whereby the temperature of the unexposed face must not increase to a degree where it is capable of creating combustion of adjacent materials.

At present the design of fire walls is somewhat elementary, providing assessment by tabulated data, which are largely based on isolated standard fire wall test results. BS5628 Pt.3 [1] relates minimum wall thickness to periods of maximum fire resistance depending on material type. Eurocode Pt.6 [2] is somewhat similar to BS5628 Pt.3, but is expanded to incorporate magnitude of applied load. Although the codes provide a good basis for fire wall design they fail to incorporate several other factors such as slenderness ratio, boundary conditions and eccentricity of applied load.These additional parameters have proved to be influential in several experimental studies [3,4].

The knowledge of materials in high temperature environments has increased dramatically in the past two decades. In particular the application of concrete to high temperature nuclear reactor structures has necessitated further knowledge, leading to an extensive body of research in this area. Concrete has very good fire resistance properties in comparison to other structural materials, and can be employed effectively to protect other structural materials such as steel. The elevated temperature behaviour of concrete is complex and its physical properties are dependent on aggregate type, stress conditions, mix proportions, curing conditions and age. Concrete material properties do experience considerable degradation with temperature, however low thermal conductivity ensures this effect is largely localised to the fire exposed boundary. The investigation of the fire behaviour of other masonry materials, such as clay and calcium silicate, has been much less extensive, and a minimal amount of information exists.

The present chapter attempts to discuss the numerical modelling of masonry fire walls. Further understanding of these critical structural elements will do much to enhance their performance and efficiency. The attainment of further data by way of numerical analysis has long since been shown to be more economical and convenient than the experimental approach. Numerical modelling can also remove the limitations of small scale isolated fire wall testing, and extend studies to their behaviour with surrounding sub-frame assemblages.

6.2 Thermo-Structural Behaviour of Masonry Walls in Fire

6.2.1 Overview

A great number of factors determine the structural response of a masonry fire wall, and it is this which makes the prediction of structural behaviour so complex. Physical conditions such as wall geometry, boundary conditions, magnitude and location of applied loading all strongly influence results. High temperature mechanical material properties and thermal material properties also considerably alter response.

6.2.2 Thermal Bowing

In fire separating elements such as fire walls, heat is usually exposed to one face. This is particularly important in the case of masonry materials due to its generally low thermal conductivity, producing high thermal gradients over the cross section. If the wall is unrestrained this leads to differential thermal expansion of the material. With the hot face expanding more rapidly than the cool face, the unrestrained wall will tend to bow towards the fire. In cases of high slenderness ratio thermal displacements may be of a magnitude to cause instability buckling. Previous experimental investigations have revealed that this occurs when the lateral thermal displacement at the mid height of the wall reaches 90 per cent of the wall thickness [3, 5]. In tall cantilever fire walls thermal bowing may be of a magnitude to cause instability bucking under the self weight of the wall alone. The reduction of strength and modulus of the material on the face exposed to the fire can serve to reduce thermal bowing, depending on the magnitude of applied loading. The thermally degrading material on the fire exposed face will experience additional compressive strains when subject to axially applied loading, therefore counteracting thermal expansion. In some situations where, i) the magnitude of applied loading is high, ii) central lateral displacements are small, and iii) the material properties on the fire exposed face have significantly deteriorated, - the additional compressive strains may be sufficient to change the direction of the central displacement. This behaviour is known as reverse bowing, and eventual failure is normally due to crushing of the thermally degraded material on the fire exposed face.

The thermal conductivity of a masonry material together with the normally high rate of temperature increase ensures a predominantly non-linear distribution of temperature through the thickness of the wall. This in turn leads to a non-linear application of thermal strains, ε_{th}. Due to the shearing transfer in a continuum material the resulting strain response tends to a linear distribution, adhering to the numerical stipulation that plane sections remain plane (Figure 6.1a). This effectively means that the applied thermal strains are incompatible with the linear flexural strains, ε_{act}. Thus at any internal point the thermal strains cannot occur freely and are internally restrained producing thermal stresses σ_{ths}, as explained by Equation (6.1). Figure 6.1b shows the thermal stresses which may exist in a non-loaded section, due to the application of thermal strains as in Figure 6.1a. Compressive stresses are experienced in the outer sections, while tensile stresses exist in the central region of the wall. Thermal stresses are self equilibrating both axially and rotationally, as given by Equations (6.2,6.3). Tensile thermal stresses are often of a magnitude to cause cracking, and since thermal stresses are self equilibrating this effectively relieves compressive thermal stresses.

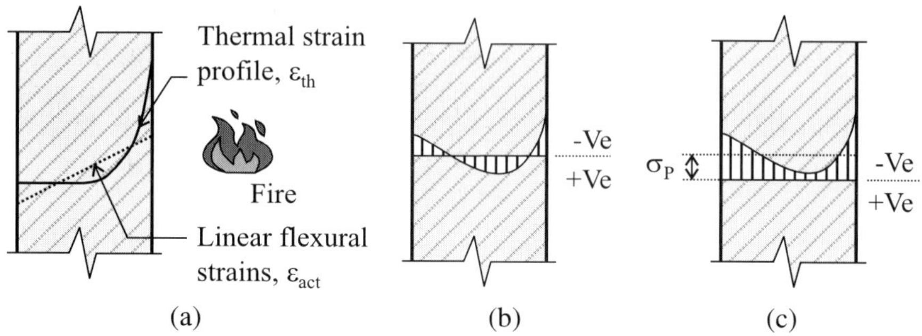

Figure 6.1: (a) The thermal strain and resulting linear flexural strain distributions, (b) Thermal stresses σ_{ths}, produced by an non-linear temperature profile, (c) Thermal stresses in a wall subject to an applied axial load producing compressive stress, σ_P.

The effects of thermal stresses are largely influenced by the external loading system. In the case of a wall which is subject to an axially applied load to produce uniform compressive stress σ_P, thermal stresses may never induce tensile stresses, but simply serve to reduce the compressive stresses in the central region of the wall thickness (Figure 6.1c).

$$\sigma_{ths} = D\left[\varepsilon_{act} - \varepsilon_{th}\right] \quad \text{where } D \text{ is the elasticity matrix} \tag{6.1}$$

$$P_{ths} = \int \sigma_{ths}\, \mathrm{d}x = 0 \tag{6.2}$$

$$M_{ths} = \int \sigma_{ths} x\, \mathrm{d}x = 0 \tag{6.3}$$

6.2.3 Masonry Material Properties

In a load bearing structural concrete wall the material compressive strength f_c, strongly influences the magnitude of thermal bowing. A stressed material will experience less resultant thermal expansion than an unstressed material [6]. The reduction in thermal expansion is proportional to the stress level, which is directly related to the compressive strength of the material. An increase in f_c results in a reduced stress level (σ/f_c) leading to an effective increase in thermal expansion. This behaviour is due to the additional temperature related transient strain component, or load induced thermal strain (LITS) and will be discussed in greater detail in Section 6.3.2. Therefore contrary to expectation an increase in strength can reduce fire resistance by increasing thermal deformations. It must be noted that this strain component is unique to concrete, and there is no evidence of its occurrence in other masonry materials. However, significant reduction of f_c can also impede fire resistance by causing earlier failure of a material, which is further reduced with increasing temperatures.

The thermal expansion coefficient is a thermal material property and one which significantly affects the extent of thermal deformations and the magnitude of thermal stresses. Thermal strain has been shown to be a non-linear function of temperature for most masonry materials and will be fully explored in Section 6.3.2. The consensus from experimental studies is that materials which exhibit lower thermal expansion, such as lightweight concretes or clay, provide longer periods of fire resistance in masonry walls.

Variation of tensile strength may slightly alter the advent of possible thermal cracking, however within the practical range of this parameter, behaviour is largely unaffected.

6.2.4 Wall Geometry

The occurrence of instability buckling in a laterally deforming wall due to thermal bowing, is largely determined by the slenderness ratio (height to breadth ratio of the wall cross section). Boundary conditions also influence the event of buckling, however discussion of this is reserved for Section 6.2.5.

Several experimental investigations have been conducted to date [3, 4], and considerable attention has been given to the effect of slenderness ratio (λ). Fire tests on walls constructed with clay masonry units conducted by Byrne have indicated that as λ increases the time to structural failure decreases [3]. At higher values of λ the effect of applied load was reported to have a minimal effect on the failure times, and was thought to be due to large thermal bowing ensuring failure was controlled by buckling. The same cannot be said for concrete masonry walls since the aforementioned influential transient strain component has a significant effect on the results, as fire wall tests performed by Lawrence show [4].

6.2.5 Boundary Conditions

Restrained boundary conditions can greatly increase fire resistance times. In the case of unrestrained thermal expansion, thermal bowing occurs freely. Where end fixities are present the thermal response is at least partially restrained causing induced loading and decreased thermal deformations.

The top and both vertical edges of a masonry wall may be fixed, pinned or free, resulting in various permutations and combination of fixities. The bottom boundary condition can be assumed to be fixed in all circumstances. Various researchers have recognised the inconsistency between the boundary conditions in a standard fire test and those which exist in real structural situations. While it is the intent of the standards to reflect real life end conditions, essentially the standard fire tests on which they are based generally employ either near full restraint or no restraint. It is also the case that in a real structural fire situation, restraint mechanisms may undergo transformation, and at present no information is available to this end.

Axial and rotational restraint of the top boundary condition will restrict vertical thermal expansion and the deformation of thermal bowing, imposing additional compressive stresses which may prolong fire resistance, or alternatively in cases of small λ and low compressive strength, fire resistance may be reduced by imposing failure of the masonry material. The former behaviour of improved resistance is more often the case in real structural situations.

Gnanakrishnan experimentally investigated the effect of various end fixities on both load bearing and non-load bearing walls [5]. The non-load bearing walls were constructed in a rigid steel frame and when tested exhibited deformation and cracking pattern of the form shown in Figure 6.2b. Results clearly indicated that as the end conditions of the specimen were relaxed, the magnitude and rate of displacements increased, thus significantly lowering the fire resistance level of the wall. Load bearing walls that were rotationally restrained top and bottom while the vertical edges were free, experienced cracking in both horizontal mortar joints due to the thermal bowing action as shown in Figure 6.2a. The opening of the joint on the unexposed face has the effect of redistributing the applied load towards the fire exposed face, serving to counteract thermal bow.

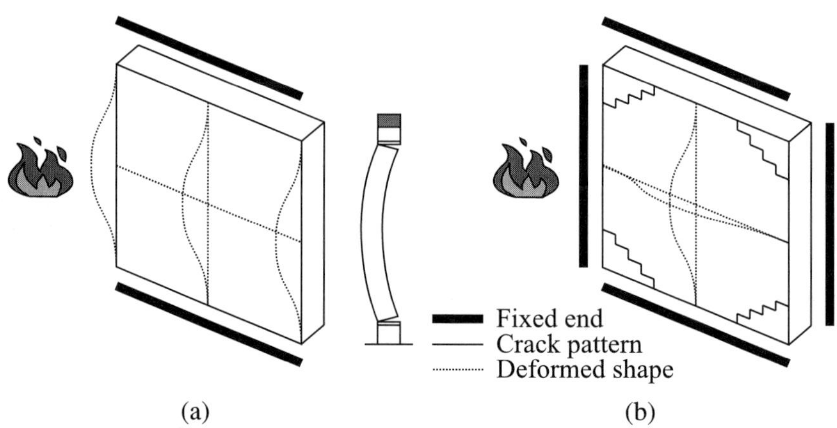

Figure 6.2: (a) Horizontal cracking arising in an axially loaded masonry wall, (b) deformation and crack patterns in a non-loaded masonry wall with all four sides restrained.

6.2.6 Applied loading

The effects of axially applied loading at elevated temperatures are very much dependent on the wall λ, material compressive strength and material type. As λ increases the effect of axial applied load on time to structural collapse is reduced. This has been reported for concrete, clay and calcium silicate materials. Of these three materials applied load has the greatest influence on concrete due to the transient strain component. In a concrete material an increased stress level will provide increased transient strain, hence effectively reducing thermal expansion. In many values of λ this is seen to increase time to structural collapse.

In the early stages of fire exposure when thermal bowing is relatively small, axially applied loading may have the effect of counteracting thermal bow. As temperature on the fire exposed face rises, the central horizontal displacements increase to a point where geometric non-linearity is of importance and the so called "$P - \Delta$" effects exacerbate central deflection to the critical point of structural collapse.

Applied loading has a more profound effect on walls with lower slenderness ratio. This is due to the fact that shorter walls experience less thermal bowing, therefore failure is less likely to be controlled by lateral instability. Increased applied loading has been shown to improve structural fire resistance times. In some cases of very stocky walls subject to high magnitudes of loading, reversed bowing has been experienced, however this situation is less likely to occur and is less relevant to common structural fire walls.

Eccentricities of load are commonly experienced in structural fire walls. Load eccentricity may have the effect of prolonging degrading fire resistance time; depending on weather the load is eccentric towards or away from the fire exposed face. Eccentricity of load will produce a distribution of stress through the thickness of the wall, possibly of the form shown in Figure 6.3, and is compared with the stress created by the same load applied concentrically. Eccentric loading of this nature, along with the event of a fire on side 1 of the wall, results in a reduction of fire resistance relative to the centrally loaded situation. The additional end moment $(P.e)$ imposed will contribute to central lateral displacements synonymous with the direction of thermal bowing. This behaviour is more pronounced in walls of higher slenderness ratio, as failure is largely controlled by lateral instability. Concrete materials are more affected by eccentric loading away from the fire side. As the axial load moves away from the fire exposed face the compressive stress on that side of the wall reduces from σ_c to σ_a, resulting in a reduction of the transient strain component of the material on the fire side. Consequently it produces additional effective thermal strains and the associated increase of thermal bowing displacements. On the other hand the occurrence of fire on side 2, leads to improved fire resistance. In this instance applied load counteracts thermal bowing, and as before in the concrete material the transient strain plays an important part due to the higher compressive stress on the fire exposed face. The transient strain is increased, therefore minimising thermal bowing, and in cases of large eccentricity and small slenderness ratio reversed bowing is more likely to occur.

Elevated temperature material property tests have also revealed that an increasing stress level tends to delay loss of strength and stiffness until higher temperatures are reached. This will be discussed further in Section 6.3.2.

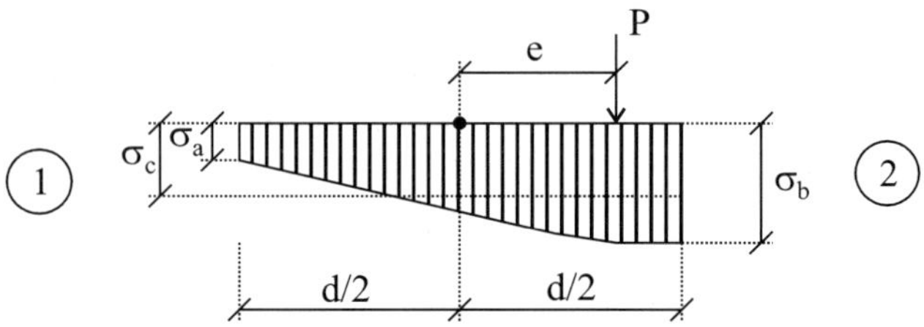

Figure 6.3: Resulting stress distribution over the thickness of a wall due to an eccentric load.

6.2.7 Moisture effects and material spalling

The presence of water can be both beneficial and yet detrimental. The gain in fire endurance is proportional to volumetric moisture content. In masonry, water used during construction acts as a heat sink absorbing latent heat required for evaporation and this has the effect of maintaining wall temperature at 100°C for a considerable time. Thermal conductivity depends largely on moisture content, varying linearly up to 100°C, after which it decreases. Thermal insulation is a function of thermal conductivity, thus an increase in moisture content enhances fire resistance.

In building elements, spalling of the surface on the material may occur due to the build up of pore water pressure across the drying region driven by a steep temperature gradient. The pore water pressure is generated from a moisture laden layer, and the phenomenon is termed "moisture clog". Moisture clog is dependent on porosity, and in less porous materials the moisture clog is unable to dissipate, therefore when pressure exceeds the yield strength of the material, explosive spalling occurs.

6.3 Mechanical Material Properties at Elevated Temperature

6.3.1 Overview

To begin to understand the complexities of elevated temperature structural behaviour it is necessary to achieve a knowledge of the temperature dependence of the material properties. Masonry material most commonly used can be categorised into three general groups namely concrete, clay, and calcium silicate. It is intended to discuss in depth all mechanical and thermal properties of each material that are seen to influence the thermo-structural behaviour.

Of the three material groups the most extensive body of research exists for concrete for reasons previously given. A vast number of publications exist regarding high temperature material properties of concrete, and many with conflicting and even contradictory findings. Due to obvious limitations only certain keynote prominent publications will be considered in the present work.

On the other hand, a very limited amount of information is available regarding the high temperature material properties of both clay and calcium silicate. However, some general conclusions can be arrived at based on what meagre research that exists.

6.3.2 Concrete material properties

Concrete undergoes changes in its chemical composition, physical structure and water content when subject to high temperature. The concrete material components of hardened cement paste and aggregate behave very differently at high temperatures, and it is partly this that contributes to the loss of concrete's mechanical properties. In short cement pastes continually contract and shrink between temperatures of approximately $300-850$°C, while most aggregates undergo thermal expansion up to around 1.0 to 1.4 percent which is complete at approximately 650°C. Thus within a temperature range of $20-1200$°C the incompatibility between the thermal expansion of each component creates tension in the cement paste causing internal micro-cracking, and a break up of the concrete matrix. Research has shown that many factors affect the precise nature of the above process, namely aggregate type, mix proportions, stress conditions, curing conditions and cement type. This brief insight will provide a clearer understanding of how individual properties are affected by increasing temperature.

6.3.2.1 Temperature dependence of individual concrete material properties

The effects of high temperature on the uniaxial $\sigma-\varepsilon$ relationship, uniaxial compressive strength, tensile strength, ultimate strain, creep strain and modulus of elasticity will be discussed herein. There is a considerable volume of high temperature material property information available, however a large percentage of results are incomparable as tests were performed under different conditions. Elevated temperature tests have either been conducted by "steady state" or "transient" methods. With the steady state test the specimen is heated to a specific temperature, then loading is applied by increment of stress or strain until failure occurs. The test is repeated over a range of temperatures. In the transient method, the unheated specimen is subject to a constant applied load, and then heated at a controlled rate until failure. This is repeated for a range of applied loads. The transient method is seen to be more analogous to conditions that occur during a fire.

Investigation of the compressive $\sigma-\varepsilon$ behaviour of concrete at high temperatures has predominately been conducted using the steady state test method, primarily due to the nature of the test yielding a direct $\sigma-\varepsilon$ relationship for a constant temperature. Figure 6.4 shows a series of $\sigma-\varepsilon$ curves for various temperatures obtained from steady state tests conducted by Schneider [7]. The strain rate controlled approach was used, enabling the descending branch of the curve to be recorded. The effect of various mix proportions and aggregate types where investigated with the conclusion that;

i) the aggregate cement ratio has a significant effect on the initial slope of the $\sigma-\varepsilon$

curves. Leaner mixes indicate a smaller reduction of elasticity at higher temperatures.

ii) the aggregate type was the primary influencing factor. Concretes with hard aggregates (siliceous, basalt) and high coefficients of thermal expansion produce a steeper decrease in modulus of elasticity at elevated temperatures than soft aggregates (calcareous, expanded shale).

Transient tests have been conducted by Anderberg *et al* and presented by way of total deformation - temperature curves [8]. Figure 6.5 illustrates results for normal weight concrete tested under various load levels. Again, the critical temperature was lower for normal weight concrete than for lightweight, which is attributed to the type of aggregate.

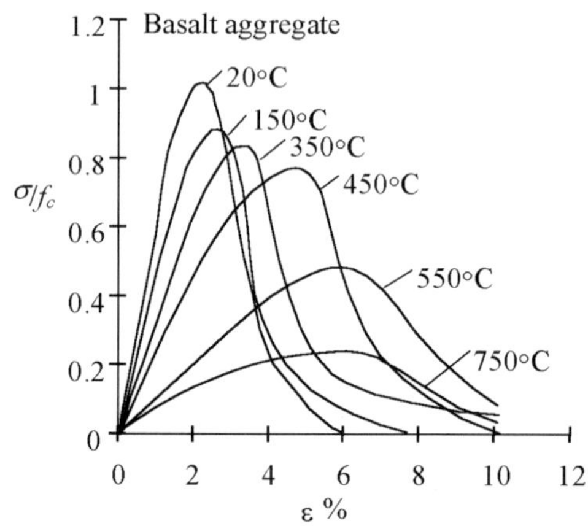

Figure 6.4: Schneider's stress-strain relationship.

In order for the $\sigma - \varepsilon$ relationship to be interpreted numerically it must be defined by several characteristics. The ultimate compressive stress is perhaps the most representative property of material strength, providing a simple criterion for material failure which is widely used in practice. The first high temperature compressive strength tests were carried out by Malhatro [9]. The specimens tested consisted of three mix proportions and four water cement ratios, with a siliceous gravel flint aggregate. Testing was conducted using three approaches;

i) stressed hot; where the specimen was subject to 20 percent of its original compressive strength, then heated to a constant temperature and loaded until failure.

ii) unstressed hot; where the specimen was heated to a constant temperature and loaded until failure.

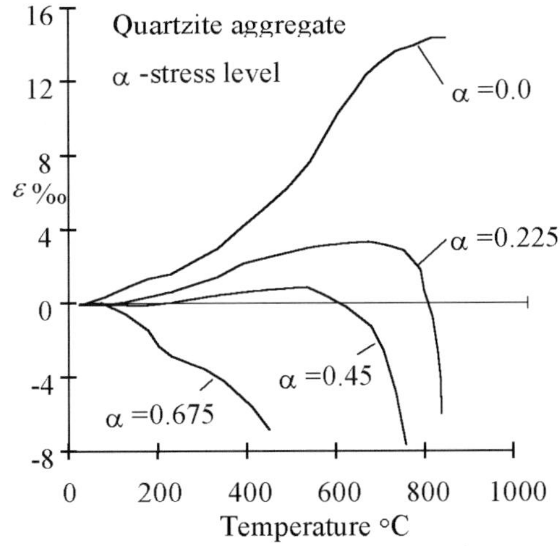

Figure 6.5: Transient temperature-strain results by Anderberg.

iii) unstressed residual; where the specimen was heated, allowed to cool, then loaded until failure.

The results of the investigation are shown in Figure 6.6. The most profound effect is that of test method. The stressed hot specimens experienced least reduction in compressive strength, which can be attributed to the initial stress preventing internal micro cracking between cement paste and aggregate. Mix proportions influenced results, with leaner mixes appearing to be proportionally stronger at the same temperature, thus being consistent with the findings of Schneider regarding the effect of aggregate-cement ratio on the $\sigma - \varepsilon$ relationship. The effect of aggregate type on the high temperature compressive strength of concrete has been reported by Schneider [7], concluding that calcareous and lightweight aggregate concretes experiences less strength reduction compared to siliceous aggregate concretes.

The ultimate strain ε_{ult}, which corresponds to the compressive strength, is another characteristic which defines the $\sigma - \varepsilon$ response. The temperature dependence of ε_{ult} for both loaded and nonloaded specimens has been investigated by Anderberg and Thelandersson [8]. In an unloaded OPC concrete, ε_{ult} typically ranges non-linearly from 0.002 at ambient temperatures to approximately 0.013 at 800°C. An increase in stress level reduces the rise of ε_{ult} with temperature, and at a stress level of approximately 20 percent, the ultimate strain may be assumed to remain constant at its initial value with increasing temperature.

Results from high temperature split cylinder tensile strength tests have shown reduction of tensile strength f_t, with increasing temperature (Figure 6.7). Elevated temperature tensile strength is also influenced by mix proportion, where a delayed reduction is portrayed by leaner mixes. The influence of aggregate type is seen to be

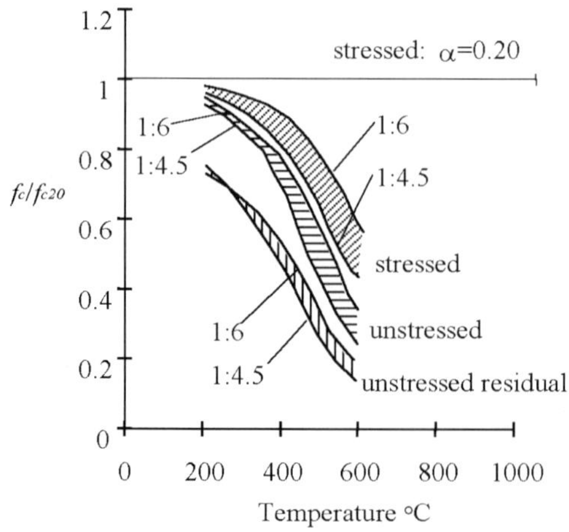

Figure 6.6: Compressive strength at high temperatures.

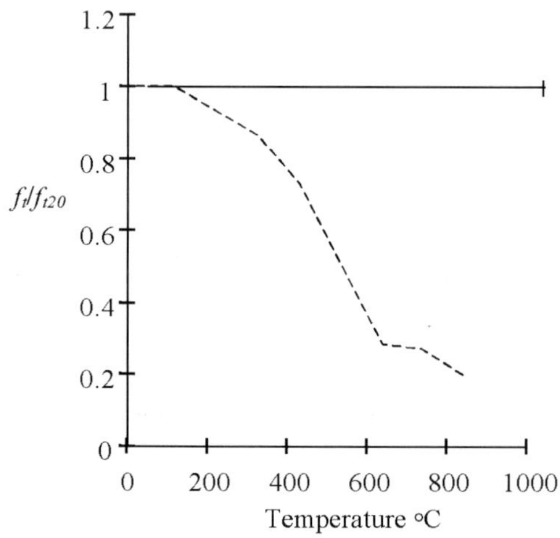

Figure 6.7: High temperature tensile strength.

the opposite of that on compressive strength, with calcareous aggregates exhibiting a decrease in f_t twice as high as siliceous aggregate concretes at 500°C.

The overall thermal expansion of concrete is determined by the individual thermal strain components in the cement paste and aggregate. While the cement paste exhibits shrinkage in temperatures above 300°C, the aggregate undergoes expansion at all elevated temperatures. The overall behaviour is one of thermal expansion, as shrinkage is overshadowed by aggregate expansion, simply because aggregate generally accounts for the largest proportion of the concrete volume. This claim has been substantiated by various authors, reporting that the thermal strain behaviour of concrete is largely determined by its aggregate type. Thermal strain - temperature relationships reported by Schneider [10] are shown in Figure 6.9, and the profound effect of aggregate type is most apparent. Siliceous aggregate concretes experience the highest thermal strain values due to the high content of quartz. Lightweight aggregates such as expanded shale/slag and pumice, produce the most thermally stable concrete, with thermal strains approximately half that of the siliceous concrete.

High temperature creep strain ε_{cr}, is determined by heating a specimen to a constant temperature, then applying a stress level and measuring the resultant strains over a period of time. At higher temperatures it is possible that the time dependent creep strains can be several times larger than instantaneous stress related strains, even after a relatively short period of two hours. Creep strain tests have been performed on quartzite aggregate concrete by Anderberg *et al* [8] over a period of three hours, under two stress levels of 45 and 22.5 percent, the findings of which are illustrated in Figure 6.8. High temperature creep tests have also been conducted over the longer period of seven days, and at $450°$C strains of 3.4% were recorded under a stress level of 0.35 at the end of testing. However, for practical fire situations a period of three hours provides sufficient information.

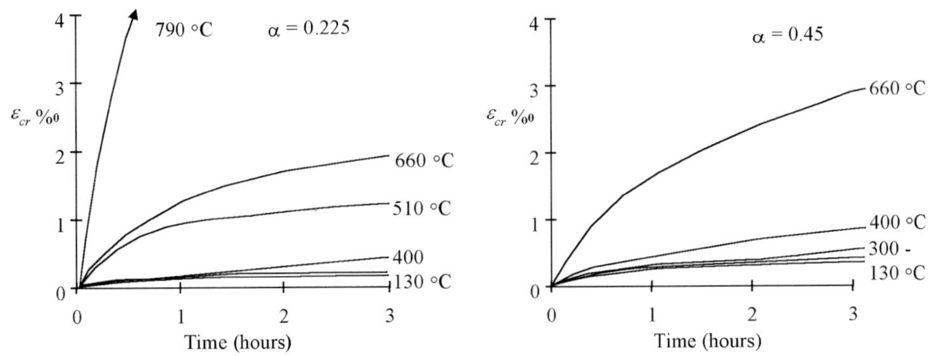

Figure 6.8: High temperature creep strains for load levels of 22.5 and 45 percent.

The influence of testing method to obtain elevated temperature material property results was noted by Schneider [7]. From investigation of total deformation data it became apparent that specimens tested under transient conditions experienced an additional component of strain than those tested under steady state conditions. Figure 6.10 shows the total strain - temperature curve produced by a specimen tested under transient conditions with a stress level of 35 percent. The thermal strain component of the same material was obtained from previous testing. From steady state test the stress related strains were estimated, along with creep strains from associated creep testing.

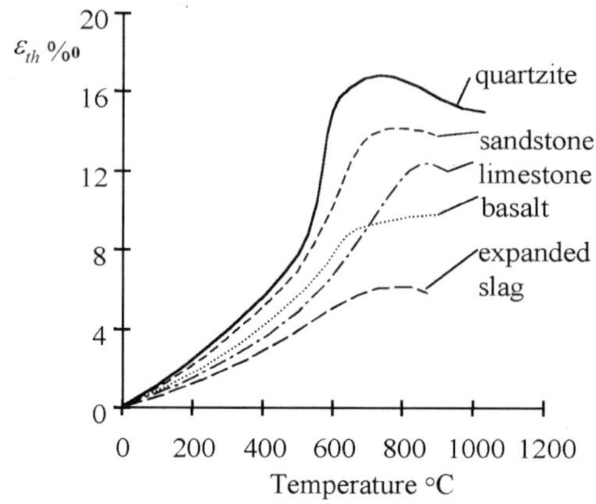

Figure 6.9: Thermal strains of various aggregate concretes.

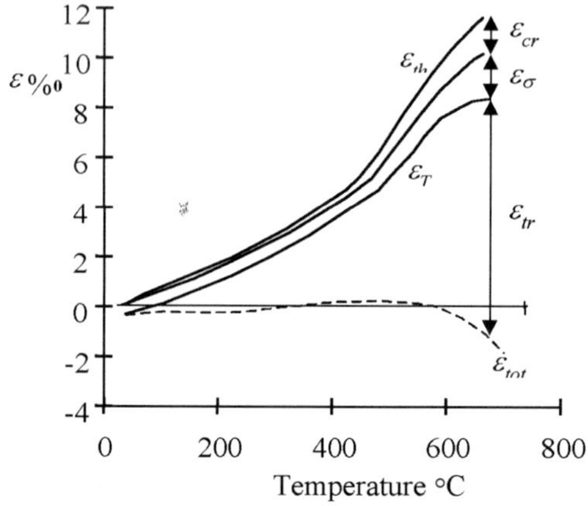

Figure 6.10: Total transient strains, showing individual strain components.

The sum of these three strain components were calculated by

$$\varepsilon_T = \varepsilon_{th} + \varepsilon_{\sigma} + \varepsilon_{cr} \tag{6.4}$$

and the values plotted in Figure 6.10. It is clear that an additional strain component must exist in transient conditions as the curves ε_T and ε_{tot} markedly differ. The transient strain is though to be a result of the initial stress level preventing internal micro cracking between cement paste and aggregate, and is reported to occur mainly in the cement paste. It must be noted that this behaviour is only present in concrete which is stressed and heated for the first time.

6.3.3 Elevated temperature clay and calcium silicate material behaviour

Clay masonry units may have variable properties dependent upon mineralogical composition. Tests, which have involved heating clay units up to 1000°C, have shown that clay bricks suffer no significant loss of strength with temperature. Therefore the same must be assumed for all other mechanical material properties since no evidence exists to suggest otherwise. The thermal expansion relationship of burnt-clay extruded bricks has been investigated by Sheehan [11], and the upper and lower boundaries of results obtained are given in Figure 6.11.

The thermal expansion of calcium silicate concrete with a quartzite aggregate was also investigated by Sheehan (as illustrated in Figure 6.12), and results are remarkably similar to those of an ordinary quartzite aggregate concrete. Tests on the residual compressive strength after heating to 1000°C showed a similar reduction in compressive strength as for concrete. No other information exists as to the behaviour of the elevated temperature mechanical material properties.

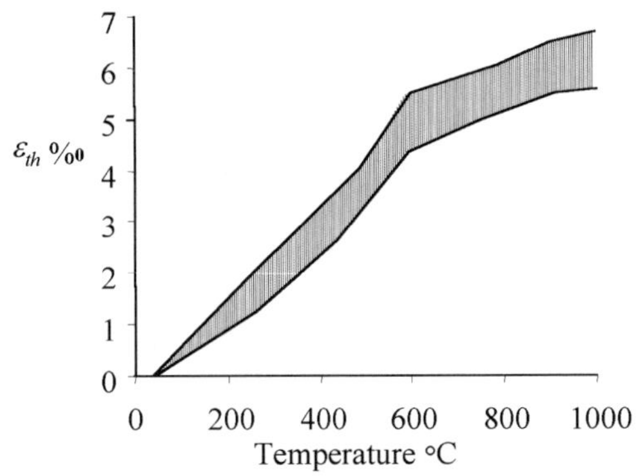

Figure 6.11: Clay thermal expansion.

6.4 Mathematical Modelling of an Elevated Temperature Masonry Material

Traditionally the numerical modelling of masonry materials has been approached with various levels of refinement. The modelling of brick unit and mortar joint separately provides an accurate macro level description between the two material constituents, thus enabling the failure modes and cracking patterns to be precisely predicted, however much at the cost of increasing computational expense. This approach is suitable for masonry elements of a small proportion, and its application to larger elements becomes uneconomical, moreover often unnecessary. Masonry material homogenisation

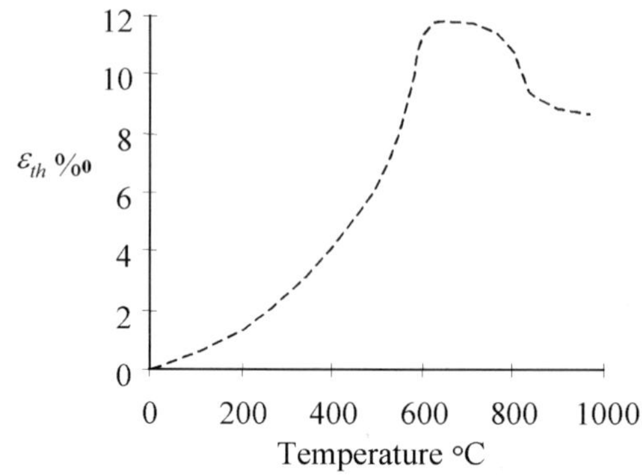

Figure 6.12: Calcium Silicate thermal expansion.

techniques have been developed to overcome this problem [12], where equivalent orthotropic material properties are determined for the masonry composite based on the individual properties of its constituents.

6.4.1 Elevated temperature concrete material model

Several high temperature concrete material models have been suggested in literature, however a complete description of each will not be given here and the reader is referred to [7]. Selective features from several models are presented, which are to be employed in the numerical model presented in Section 6.5. The formulation of models is based entirely on experimental data obtained from uniaxial test conditions.

The basic constitutive strain equation as agreed by several authors is given as:

$$\varepsilon_{tot} = \varepsilon_{\sigma} + \varepsilon_{th} + \varepsilon_{cr} + \varepsilon_{tr} \tag{6.5}$$

where; ε_{tot} = total strain
ε_{σ} = instantaneous stress related strain
ε_{th} = thermal strain
ε_{cr} = creeps strain
ε_{tr} = thermal transient strain

The total strain is derived from the field of displacements, resulting from both applied loading and thermal expansion.

The thermal strain relationship suggested by EC4 [13] is used; where for siliceous, quartzite and gravel aggregate concrete thermal strain is defined by:

$$\begin{aligned} \varepsilon_{th} &= -1.8 \times 10^{-4} + 9 \times 10^{-6}(T-20) \qquad (T \leq 700°\text{C}) \\ &\quad + 2.3 \times 10^{-11}(T-20)^3 \\ \varepsilon_{th} &= +1.4 \times 10^{-3} \qquad (700°\text{C} < T) \end{aligned} \tag{6.6}$$

and for lightweight concrete a linear thermal expansion relationship is suggested for all temperatures;

$$\varepsilon_{th} = 8.10 \times 10^{-6}(T - 20) \tag{6.7}$$

The transient strain component model is based on extensive transient test data performed by Anderberg *et al* [8]. The ε_{tr} values were arrived at by subtracting individual estimated strain components of the form:

$$\varepsilon_{tr} = \varepsilon_{tot} - \varepsilon_{th} - \varepsilon_{\sigma} - \varepsilon_{cr} \tag{6.8}$$

where the values of stress related strain are determined from associated high temperature $\sigma - \varepsilon$ curves, and creep strain determined from related creep tests. The total strain corresponds to the total measured deformation of the specimen in the transient test. A series of transient strains were obtained over a range of stress levels, and upon analysis of results the ε_{tr} component was seen to be uniquely a function of thermal strain and stress level, producing the following expression:

$$\varepsilon_{tr} = -2.35 \times \varepsilon_{th} \frac{\sigma}{f_{c(20)}} \tag{6.9}$$

where σ is the stress in the specimen, $f_{c(20)}$ is the material compressive strength at 20°C. The constant of 2.35 has proved to be sensitive to mix proportion. In leaner mixes the constant may be of the order 1.8, producing less ε_{tr}, since the cement paste content is reduced. Above temperatures of 500°C, transient strain departs from Equation (6.9), and the incremental change can accurately be represented by

$$\Delta\varepsilon_{tr} = 10^{-4} \times \Delta T \times \frac{\sigma}{f_c} \tag{6.10}$$

However the application of the above transient strain equations is somewhat difficult in structural analysis, since the level of stress at temperature increment j is dependent on the magnitude of transient strain. Therefore when determining transient strain at temperature increment j the stress level σ^j is generally unknown. To over come this difficulty the stress level of the previous increment σ^{j-1} is utilised in Equations (6.9) and (6.10) to obtain the current transient strain ε_{tr}^j. Inaccuracies may be avoided provided the temperature increment is small enough so the increase in stress level from increment $j - 1$ to j is significantly small.

From regression analysis of test data Anderberg *et al* have proposed a mathematical formulation for the creep strain ε_{cr}, at constant temperature and stress level.

$$\varepsilon_{cr} = \beta_0 \frac{\sigma}{f_{C(T)}} \left(\frac{t}{tr}\right) P e^{K_1(T-20)} \tag{6.11}$$

with the following constants suggested for granite aggregate concrete:

$$\begin{aligned} \beta_0 &= 0.53 \times 10^{-3} \\ tr &= 3.0\text{hrs} \\ P &= 0.5 \\ K_1 &= 3.04 \times 10^{-3}/°\text{C} \end{aligned}$$

Hence the creep strain is time, temperature and stress level dependent, and the above equation can be directly numerically applied to situations of constant temperature. However, in a transient temperature environment with variable stress and temperature, the situation is more complex and a similar approach is adopted to that of the strain hardening principle.

With all of the above strain components calculated, it only remains to determine the instantaneous stress related strain, which can be easily obtained by rearranging Equation (6.5):

$$\varepsilon_\sigma = \varepsilon_{tot} - \varepsilon_{th} - \varepsilon_{tr} - \varepsilon_{cr} \tag{6.12}$$

Internal stress can then be determined from the following temperature dependent $\sigma - \varepsilon$ relationship:

$$\sigma = f_{c(T)} \times \frac{\varepsilon_{sigma}}{\varepsilon_{ult}} \times \frac{n}{(n-1) + (\varepsilon_{sigma}/\varepsilon_{ult})^n} \tag{6.13}$$

where $f_{c(T)}$ is the temperature dependent compressive stress, n is 3 for normal weight concrete and 2.5 for lightweight concrete.

Compressive strength as a function of temperature and stress level is derived based on the findings of Malhatro [9], and given by the following formulation:

$$\begin{aligned}
f_{c(T)} &= f_{c(T)} \times 1.0 && (T < 100°\text{C}) \\
f_{c(T)} &= f_{c(20)} \times \begin{bmatrix} C1 \times 10^{-3}(T/100)^3 + \\ C2 \times 10^{-2}(T/100)^2 + \\ C3 \times 10^{-2}(T/100) + \\ C4 \end{bmatrix} && (100°\text{C} < T < 500°\text{C}) \\
f_{c(T)} &= f_{c(20)} \times \begin{bmatrix} B1 \times 10^{-4}(T/100)^3 + \\ B2 \times 10^{-2}(T/100)^2 + \\ B3 \times 10^{-2}(T/100) + \\ B4 \end{bmatrix} && (500°\text{C} < T < 1200°\text{C}) \\
f_{c(T)} &= f_{c(T)} \times 0.0 && (T > 1200°\text{C})
\end{aligned} \tag{6.14}$$

where for stress levels:

$\sigma^{j-1}/f_{c(20)} = 0.0$

$C1 = 2.95$ $B1 = -4.21$
$C2 = -4.88$ $B2 = 2.344$
$C3 = 8.87$ $B3 = 0.3806$
$C4 = 0.957$ $B4 = 1.919$

$\sigma^{j-1}/f_{c(20)} = 0.4$

$C1 = 1.76$ $B1 = -6.292$
$C2 = -4.9288$ $B2 = 2.52618$
$C3 = 20.1$ $B3 = 0.392$
$C4 = 0.87$ $B4 = 2.25$

The compressive strength degradation curve for the stress level of 0.4 is curtailed at 800°C, since the compressive strength has degraded to 0.4 of its original value at this point, and at temperatures above 800°C the compressive strength is taken as zero. A linear interpolation between the two curves may be made for intermediate stress levels, however, stress levels above 0.4 are assumed behave the same as 0.4.

The ultimate strain value is also dependent on stress history as well as temperature, and according to Schneider [14] may be determined by the expression:

$$\varepsilon_{ult(T)} = \varepsilon_{ult(20)} + \Delta\varepsilon_{ult(T)} \times f(\alpha) \tag{6.15}$$

where

$$\Delta\varepsilon_{ult(T)} = \left[4.2 \times 10^{-6} + 5.4 \times 10^{-9}(T - 20)\right] \times (T - 20) \tag{6.16}$$

$$\begin{aligned} f(\alpha) &= 1 \qquad \text{for } \alpha = 0 \qquad \alpha = \sigma^{j-1}/f_{c(20)} \\ &= 0.227 \qquad\quad\;\; = 0.1 \\ &= -0.095 \qquad\;\; = 0.3 \end{aligned}$$

For both compressive stress and ultimate strain the stress level is calculated based on stress history i.e. the stress in the previous increment $j-1$, essentially for the same reasons that this approach was adopted for determination of the transient strain.

To relate the uniaxial elevated temperature model to a continuum material, a biaxial stress interaction curve is employed based on the von Mises yield criterion. The biaxial compressive strength of concrete increases by approximately 15 to 20 percent from its uniaxial state [15], and the percentage increase is higher still at elevated temperatures with a reported increase of around 79 percent at 750°C [16]. It was decided to ignore the high temperature biaxial behaviour at this point since results would be further influenced by stress history which has yet to be investigated. Based on Kupfer's [15] results at ambient temperature and the von Mises yield criterion, the following expression has been suggested:

$$f(\sigma) = \sqrt{1.355\left[\left(\sigma_x^2 + \sigma_y^2 - \sigma_x\sigma_y\right) + 3\left(\tau_{xy}^2\right) + 0.355\sigma_0\left(\sigma_x\sigma_y\right)\right]} = \sigma_0 \tag{6.17}$$

6.4.2 Elevated temperature clay material model

As previously stated, test data on the elevated temperature material property behaviour of clay is somewhat meagre. Since no transient strain exists, and no high temperature creep information is available, the total constitutive strain equation is:

$$\varepsilon_{tot} = \varepsilon_\sigma + \varepsilon_{th} \tag{6.18}$$

A tri-linear thermal strain model is suggested based on Sheehan's [11] results:

$$\Delta\varepsilon_{th} = \alpha \times T \tag{6.19}$$

where

$$\begin{aligned} \alpha &= 7.0 \times 10^{-6} \text{ to } 11.0 \times 10^{-6} && (T < 575°\text{C}) \\ &= 4.0 \times 10^{-6} && (575°\text{C} < T < 1000°\text{C}) \\ &= 0.0 && (T > 1000°\text{C}) \end{aligned}$$

With total strain determined from the displacement field, the internal stress related strains are calculated as for concrete but excluding the transient and creep strain. The same parabolic stress-strain relationship of Equation (6.13) may be used to obtain internal stresses. All mechanical material properties are assumed to remain unaltered from their original ambient temperature values.

6.5 Description of the Numerical Model

6.5.1 Justification of Numerical Strategy

Several authors have attempted to numerically simulate the behaviour of masonry walls subject to fire on one side with various degrees of success. Gnanakrishnan *et al* [17] developed a simplified plane strain finite element model, using four-noded quadrilateral isoparametric elements. Dhanasekar *et al* [18] created a finite element model in the shell of LUSAS using layered thin shell elements. O'Meagher and Bennetts [19] produced a computer model based on the moment-curvature method of analysis.

The highly non-linear features of the problem prompt the use of the finite element method, and indeed it has been used to considerable effect in the past. Upon inspection of the problem it was perceived that for certain structural scenarios a fire wall may be accurately modelled in two dimensions as a cross-section through the thickness of the wall. In the standard fire testing of load bearing structural masonry walls, both the top and bottom boundary conditions are fixed, while the vertical edges are free. Therefore if a section through the thickness of the wall is considered, it may be assumed to be in a state of plane stress. Taking advantage of this situation it was decided to develop a two dimensional finite element model capable of predicting the thermo-structural behaviour of load bearing masonry walls.

The main disadvantage of the two dimensional model is its incapacity to simulate horizontal thermal bowing, however the effects of this are minimised in load bearing walls tested as described above. The axial load together with fixed top and bottom boundary conditions restrains horizontal thermal bow, hence limiting its contribution to central lateral displacements. The extent of horizontal thermal bow is further reduced with decreasing wall width. It must be noted that the presumption of plane stress in a section of the wall thickness is not entirely true, as the event of restrained horizontal thermal bowing over the width of the wall ensures a stress distribution perpendicular to that plane. In many fire wall tests carried out under the above conditions a number of vertical cracks appeared over the height of the wall, which are caused by the stresses produced from the restraining of horizontal thermal bow. The wall panels were effectively divided into a number of narrow column strips, hence making additional thermal bowing negligible.

The application of a two dimensional model to non-load bearing masonry walls of significant width is questionable as secondary horizontal thermal bowing may contribute equally as much to the structural response as vertical thermal bow, and is therefore most influential as testing has shown [5]. Consequently it is recommended that a two dimensional model be confined to the analysis of load bearing walls, or non-load bearing walls of a small width to height ratio.

6.5.2 The Finite Element Model Developed

Eight noded isoparametric quadrilateral elements were chosen to discretise the section through the thickness of the wall. The effects of both material and geometric non-linearity have been accounted for, and a smeared cracking model is incorporated to simulate the behaviour of a brittle masonry material. The fundamental structural related aspects when considering thermal effects are thermal expansion and material property deterioration, both of which have been included and their respective implementations are discussed in Sections 6.5.2.3 and 6.5.2.1 respectively. The program solution method adopts a standard or modified non-linear Newton-Raphson iteration procedure, in which the structure firstly receives an applied load, followed by successive temperature increments. The model is not coupled with a thermal analysis program for the calculation for temperature distributions, and depends on flux measurements made during testing. Temperatures are assigned to the nodal points by the user as input data.

6.5.2.1 Material model

The material model presented in Section 6.4 is fully implemented in the developed finite element model. Material non-linearity is approached using the tangent modulus method, to obtain a converged solution. The brittle nature of a masonry material is simulated by the use of a smeared cracking model with the total crack approach. If the maximum principal stress at any Gauss point exceeded the user specified tensile failure stress then a crack was assumed to have developed in a plane normal to the direction of the maximum principal stress over the associated Gaussian integration area. The crack was simulated by setting the Young's modulus in the direction normal to the crack as zero. The stress in the direction normal to the crack was not set to zero immediately but reduced according to a linear crack softening relationship. The effects of crack closure were also incorporated by restoring the compressive strength of the material in the relevant direction. A certain amount of shear capacity was retained, but reduced as the crack width increases. Figure 6.13 illustrates the stress-strain relationship of the cracking model. Crushing of a material was modelled by reducing all strength properties and stresses in an integration area to zero when the effective biaxial strain at that particular Gauss point exceeded the user specified crushing strain.

Upon the application of temperature the individual material properties are modified for each Gaussian integration area, according to the temperature at the Gauss sampling point. The temperature at the Gauss point, T_{gp} is determined from the nodal temperatures using the element shape functions:

$$T_{gp} = \sum_i T_i N_i \tag{6.20}$$

where T_i represents the nodal temperature at node i, while N_i is the associated shape function. Given that the material properties are altered for each integration area a sufficient number of elements, or more accurately, Gauss points must exist over the wall thickness to capture a representative material-temperature response.

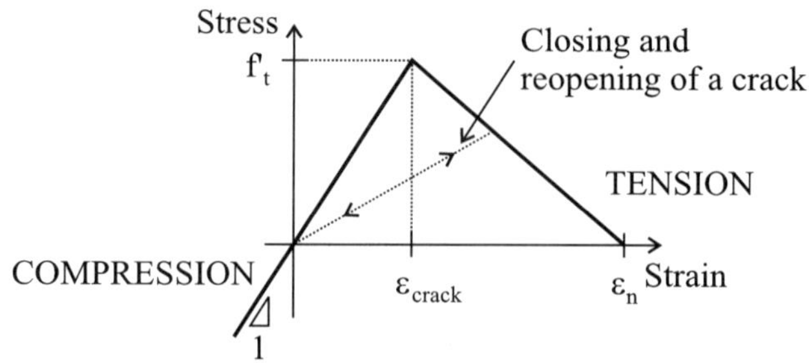

Figure 6.13: Tensile stress-strain relationship.

6.5.2.2 Geometric non-linearity

It was recognised that the problem at hand contained certain aspects of geometric non-linearity. Changes in geometry due to deformation significantly influence the predicted response of a masonry wall [19]. The total langrangian descriptor was used to redefine the structural geometry, and account for the additional stresses experienced in a laterally deforming axially loaded masonry wall. Green's strain tensor was used to include higher order strain terms. With the total Lagrangian approach the Piola Kirchoff stress field was used and referred to the original configuration. When calculating internal nodal forces the Piola Kirchoff stresses of the undeformed area were converted to the normal Cauchy stresses of the deformed configuration. The membrane stresses, which exist in an axially loaded slender structure, may actually have a stiffening effect and have been accounted for by the inclusion of the geometric stiffness matrix.

6.5.2.3 Implementation of thermal expansion

The effects of thermal expansion were included by the application of thermal strains ε_{th}. Thermal strains are indeed an initial strain component, and are therefore independent of stress. The nature of the finite element method written in displacement formulation prevents a direct explicit displacement-strain relationship. Therefore any deflection sought after (in this case thermal bowing) must be achieved from an equation of the form:

$$P^j = K^i a^i + f^j \tag{6.21}$$

where P is the vector of externally applied nodal loads, K is the stiffness matrix, a is the vector of nodal displacements and f is the vector of loads due to element body forces. Superscripts i and j denote iteration and increment respectively.

Temperature changes result in an initial strain vector of the form:

$$\varepsilon_{th}^j = \begin{Bmatrix} \varepsilon_{xth} \\ \varepsilon_{yth} \\ \gamma_{xyth} \end{Bmatrix} \tag{6.22}$$

where in an isotopic material $\gamma_{xyth} = 0$, as no shear strains are caused by thermal dilation. The thermal strains must be converted to equivalent nodal loads to act as body forces, f_{th}, which are capable of creating the state of thermal strains in the element, when applied to that element.

$$f_{th}^{j} = \int_{Ve} \left[B^{i}\right]^{\top} D_{T}^{i} \varepsilon_{th}^{j} \, \mathrm{d}V \tag{6.23}$$

The thermal stains are converted to imaginary thermal stresses, σ_{th}, using the current tangential elasticity matrix, D_T. The thermal stresses are then integrated over the element volume, therefore the thermal strains must be evaluated at the Gauss points using the ε_{th}-Temperature relationship defined by Equations (6.6),(6.7) or (6.19), to facilitate the Gauss quadrature integration.

Substituting Equation (6.21) into Equation (6.23) gives:

$$P^{j} + \int_{Ve} \left[B^{i}\right]^{\top} D_{T}^{i} \varepsilon_{th}^{j} \, \mathrm{d}V = K^{i} a^{i} \tag{6.24}$$

The equilibrium equations are solved for nodal displacements, a, and it is apparent that when the external applied nodal loads are zero, displacements are entirely due to thermal effects. Upon determining the nodal displacements, the total strains ε_{tot}, are calculated:

$$\varepsilon_{tot}^{i} = B^{i} a^{i} \tag{6.25}$$

Internal stresses can now be calculated as defined in Section 6.4, with the use of the constitutive strain Equation (6.12), and the stress-strain relationship of Equation (6.13) in conjunction with standard elasticity matrix for plane stress.

6.5.2.4 Interface elements

It was recognised that in both real structural situations and standard fire tests the end conditions of a masonry fire wall may not actually be bonded to adjacent surfaces, but simply in contact. A six noded interface element has been developed to simulate the contact-friction behaviour between non-bonded surfaces and its formulation is given elsewhere [20]. The interface behaviour is controlled by its relative shear and normal displacements, as indicated by Figure 6.14. and are directly related to the shear and normal stresses $\Delta\sigma_s$, $\Delta\sigma_n$ through a constitutive relationship of the form:

$$\begin{bmatrix} \Delta\sigma_s \\ \Delta\sigma_n \end{bmatrix} = \begin{bmatrix} K_s & 0 \\ 0 & K_n \end{bmatrix} \times \begin{bmatrix} \Delta s \\ \Delta n \end{bmatrix} \tag{6.26}$$

where K_s and K_n are the tangential and normal stiffness coefficients respectively. When the normal strain is in tension, separation is assumed to take place and the normal stiffness, K_n, is taken as zero. Since a separated interface cannot accommodate any shear the shear stiffness, K_s, is also set to zero. If the interface is in contact, i.e. $\Delta n \leq 0.0$, a very high value is assigned to K_n. The tangential stress-strain

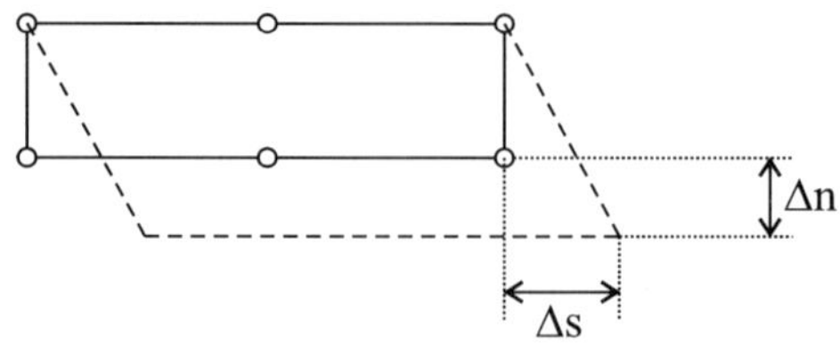

Figure 6.14: Interface element.

relationship is assumed to be elastic perfectly plastic, using a Mohr-Coulomb yield criterion with zero cohesion. When $\sigma_s \geq \mu\sigma_n$, where μ is the coefficient of friction between the two surfaces, firm contact is assumed and K_s is assigned a numerical value determined from testing. Slip between the two surfaces takes place when σ_s exceeds $\mu\sigma_n$, and in such a case σ_s is reduced to $\mu\sigma_n$ and K_s is set to zero.

6.5.3 Validation of the material model and its implementation

Before applying the model to structural fire wall analysis, the material model and its implementation were firstly evaluated against experimental results. Total restraint force and transient tests conducted by Anderberg [8] were used for this purpose. In the total restraint force test the specimen was heated while being fully axially restrained, and the restraint force generated over a temperature range recorded. Transient tests were performed at load levels of 0, 0.225, 0.35, and 0.675. The results yielded from both analysis and experiments are shown Figures 6.15 and 6.16, with reasonable agreement obtained. The initial restraint force generated in analysis is significantly higher than experimental, however peak values converge to approximately the same value. Concerns have been expressed regarding whether a true condition of full restraint was imposed during the initial stages of testing.

6.6 Numerical Examples and Comparison with Experimental Results

6.6.1 Experimental Investigation

6.6.1.1 Description of University of Ulster Tests

Two walls were selected from a range of tests conducted by Laverty [21]. The experiments employed $\frac{1}{2}$ scale model test methodology, since this allowed walls of greater slenderness ratios to be tested in the small furnace used. The model walls were initially subject to a factor of the normal service load defined by BS5628, and were then exposed to a model fire curve on one side. This model fire curve [22] was specifically developed to produce similar temperature profiles in a reduced scale wall that a

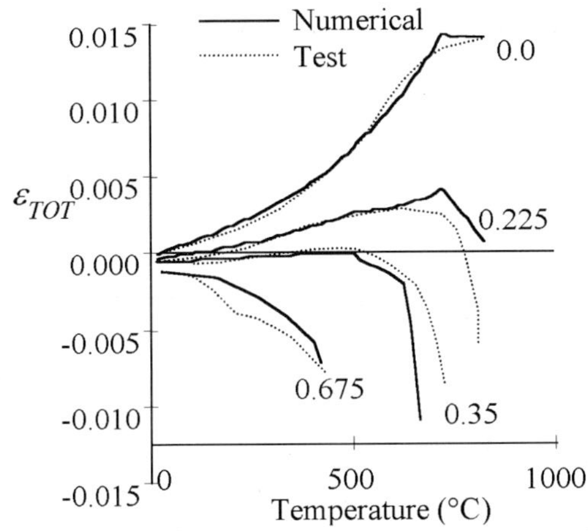

Figure 6.15: Total strain until failure - comparison between analysis and test.

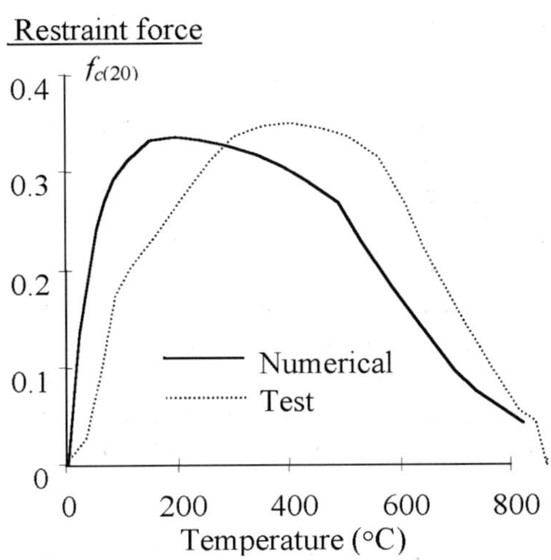

Figure 6.16: Total restraint force - numerical and experimental comparison.

standard BS476 fire curve would create in a full scale wall.

The first of the two walls (ref. FW1) was 430mm wide by 1330mm high, with a thickness of 50mm producing a slenderness ratio of 26.6. The second wall (ref. FW2) was 430mm wide by 700mm high, again with a thickness of 50mm producing a slenderness ratio of 14.0. Both walls were subject to a uniaxial applied load of 50kN. Masonry units consisted of concrete with a specified crushing strength of 20N/mm^2, bedded on a 1 : 3 mortar mix, using OPC. The base of the walls was built on a flat steel plate and the applied axial load was distributed over the width of the panel top through a steel loading plate which was restrained against rotational movements. The vertical edges of the walls were unrestrained with sufficient space for free thermal expansion.

6.6.1.2 Description of BRE tests

Testing on two single leaf non-loaded masonry cantilever walls was carried out by the fire research station at BRE [23]. Both the 215mm and 323mm thick walls were constructed using solid common clay bricks, and were unrestrained on both verticals and top boundaries. The dimensions of the walls were 950mm wide $\times$ 3000mm tall and exposed to the BS476 Pt. 20 temperature - time fire curve of three hours. However, for reasons of space, only the 215mm thick wall is chosen for numerical simulation herein.

6.6.2 Analysis of Experimental Results

6.6.2.1 University of Ulster Tests

The cross-section of the walls was modelled using eight noded isoparametric elements. Using three elements gave seven nodal points over the wall thickness enabling the curvi-linear temperature distribution to be simulated. The steel loading plate was also modelled and prevented numerical instabilities caused by local crushing of the top bricks. Interface elements were also introduced between the top of the wall and the loading plate, and also at the bottom of the wall.

The macro level modelling of the masonry composite enabled each constituent material to be given its individual material properties. The specified crushing strength of the masonry units was utilised, and an assumed value of 10N/mm^2 was taken for the mortar compressive strength. Tensile strength was taken as $0.1f_c$, and all other properties assumed typical values.

Temperatures were recorded during fire testing at dimensions of 0, 25.0 and 50.0 mm from the unexposed face, and estimated temperature profiles were interpolated by fitting a curvi-linear line through the three points recorded.

6.6.2.2 BRE Warrington Tests

The cross section of the masonry walls was discretised using a mesh of 5 elements wide and 80 elements high, with the mesh density increasing towards the fire exposed face. This resulted in 400 elements giving 1371 nodal points. The self weight of the wall was included by taking ρ_g as an applied body force on each element. The temperature distribution through the thickness of the wall was taken directly from experimental thermocouple measurements.

A homogenised masonry material was assumed with no material property degradation, as a clay material is seen to be thermally stable up to temperatures of 1000°C. Due to the uncertainty of fundamental material properties a sensitivity analysis was carried out to investigate the effects of various mechanical material properties on the analysis of testing. Two specific parameters were chosen: thermal strain-temperature relationship, and tensile strength. The thermal strain-temperature relationship was modelled on a trilinear form, as reported in section 6.4.2. Three different values of α_1 where chosen being, 11.0×10^{-6}, 9.0×10^{-6} and 7.0×10^{-6} all with $\alpha_2 = 4.0 \times 10^{-6}$, and the values were chosen to represent the upper and lower boundaries of the scatter of experimental results. A range of tensile strength values was considered primarily to examine the effects of thermal cracking. Impracticably high values of f_t were also chosen to minimise thermal cracking, hence making it possible to observe its influence.

It was appreciated that some transient strain may have been present in the mortar material, however it was considered negligible as the wall was unloaded and the material predominantly consisted of clay masonry units. Creep was also not included, again because the material was non-loaded, and no high temperature creep information was available for a clay material.

6.6.3 Discussion of Results

6.6.3.1 University of Ulster tests

Initial analysis of the problem was conducted with complete rotational restraint applied to the steel loading plate, which resulted in almost no thermal bowing. Little lateral displacements were experienced until temperature increased and the wall started to deflect away from the fire as a result of material property degradation. It was soon realised that complete rotational restraint was unlikely to be the actual situation in the test rig. Rotational restraint was removed therein, when it was then observed that rotations at the top of the wall in analysis, θ_{lr}, reached a magnitude of approximately 0.02 radians prior to structural collapse. Upon examination of the test rig it was considered quite feasible that rotations of this extent may have been allowed to occur freely.

The test duration for fire wall FW1 was 28 minutes, at which point complete structural collapse was experienced. During the test the wall clearly bowed towards the fire. The central deflection (Δ_{cx}) versus exposed face temperature (T_{ef}) relationship from both experiment and analysis is shown in Figure 6.17. The results are in reasonable agreement, where the onset of large lateral displacements occurred at approximately 812oC. As reported in Section 6.2.2, fire wall testing has indicated that failure due to buckling tended to occur at a central deflection of 90% the thickness of the wall. FW1 failed at a central deflection of 42.5mm (85% wall thickness), therefore the failure mode was almost definitely one of buckling instability. The large deflections created tensile stresses on the exposed face, which produced horizontal cracking in the mortar joints followed by almost immediate structural collapse. It was also observed during analysis that quite a pivotal event was the crushing of the mortar joint at the base of the wall on the fire side. This effectively relieved some of the rotational restraint at the bottom of the wall facilitating the onset of large displacements. This hypothesis could not be compared to experimental events, as the fire side of the wall could not be seen for obvious reasons.

As can be seen from Figure 6.18, the failure time in analysis was 19.5 minutes, which is considerably shorter than the test time of 28.0 minutes. This is attributed to a degree of rotational restraint during testing prolonging central displacements and delaying structural failure.

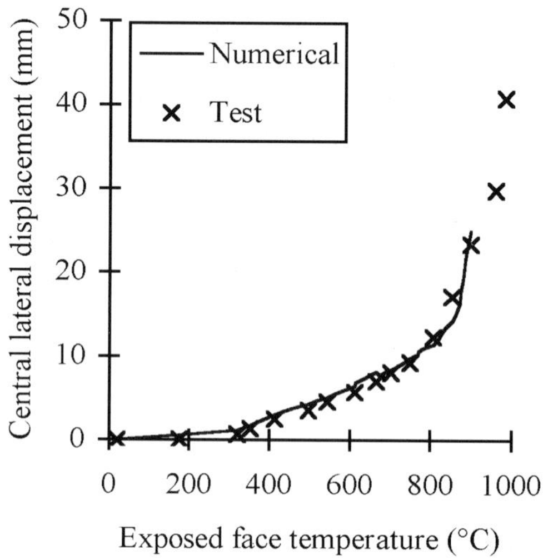

Figure 6.17: FW1- Central displacement versus exposed face temperature.

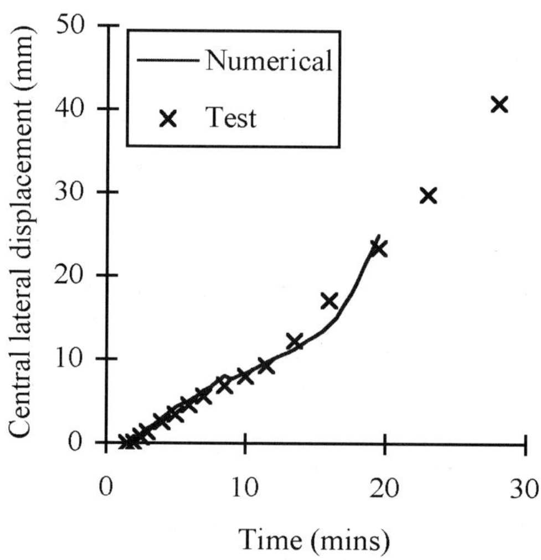

Figure 6.18: FW1 - Central displacement versus time.

During testing, fire wall FW2 was not brought to structural failure due to the stabilisation of the central lateral displacement. From Figure 6.19 reasonable agreement is achieved between test and numerical results up to approximately 11.0 minutes, after which numerical displacements are much more conservative than the experimental displacements. The exposed face temperature - central lateral displacement relationship is illustrated in Figure 6.20, and the presentation of results in this way appears to produce a better agreement between results. This indicates that the discrepancy between numerical data and experimental is not temperature related, but more likely to be a result of a structural mechanism which has occurred during testing. A possible mechanism was thought to be the crushing of the base mortar joint on the fire exposed face during testing, hence alleviating bottom rotational restraint, producing increased lateral displacements. Such behaviour was not experienced in analysis, however attempts to induce crushing of the mortar joint by lowering the crushing strain did produce increased lateral displacements, but a situation of numerical instability prevailed to produce early failure.

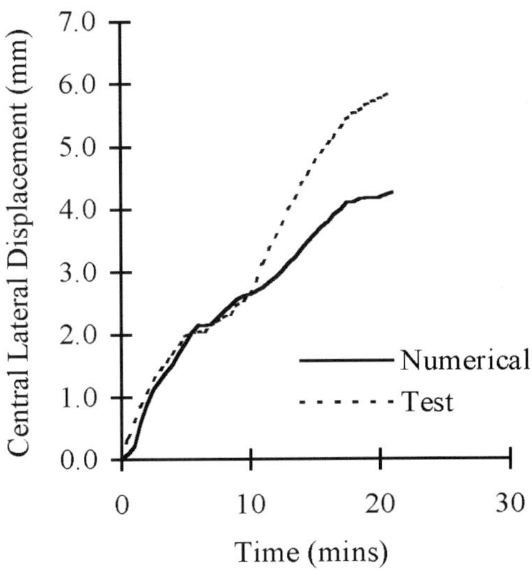

Figure 6.19: FW2- Central displacement versus exposed face temperature.

6.6.3.2 The BRE Cantilever Wall Results

The BRE results are presented by way of top deflection versus time, primarily because experimental measurements were made in this way.

The increase of α_1 was seen to be directly proportional to the increase in top displacements as Figure 6.21 shows. In the analysis of $C1$ and $C2$ no thermal cracking was experienced due to an impracticably high f_t value being deliberately employed.

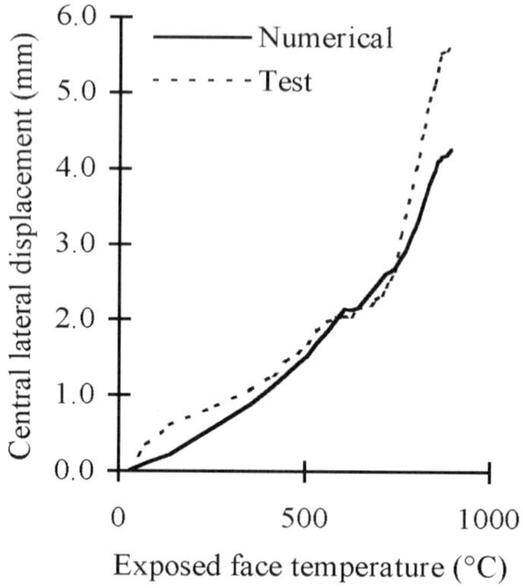

Figure 6.20: FW2 - Central displacement versus time.

However in $C3$ thermal cracking was experienced, even with the high tensile strength, due to a very profound curvilinear thermal strain distribution caused by the increased α_1 value of 11.0×10^{-6}.

The reduction of f_t increased the advent of thermal cracking, which was seen to dramatically increase deflections, however the final displacements at 180 minutes converge to the same point irrespective of f_t (see Figure 6.22). As thermal cracking was experienced this effectively reduced the shear strength of those particular elements, leading to considerable shear strains causing non-linear flexural strains, ε_{act} (Section 6.3.2). The flexural strains were therefore no longer constrained to linearity by shear and were free to occur in a profile similar to the applied thermal strains, thus leading to a reduction in the stress related thermal strains. This was verified by the investigation of the various strain components from first cracking (15 minutes) to considerable cracking (36 minutes). The convergence of the final displacements to a similar value came as a result of the thermal strain tending to a linear profile as it approached 180 minutes, therefore minimising shear strains and stress related thermal strains.

Upon examination of Figure 6.21 it was apparent that the final top displacement of test $C1$ was nearest to the experimental results. Considering the effects of a reduced tensile strength as shown in Figure 6.22, the correct tensile strength of 2.5N/mm^2 introduced with variables as in $C1$ yields results that are in excellent agreement with experimental figures, as illustrated in Figure 6.23.

Cooke [23] employed a computer program called SOSMEF to numerically simulate the structural response of the 215mm wall. SOSMEF is a finite difference based model and is capable of predicting the structural response of beams and columns subject to non-linear variations of temperature in all three directions. The material prop-

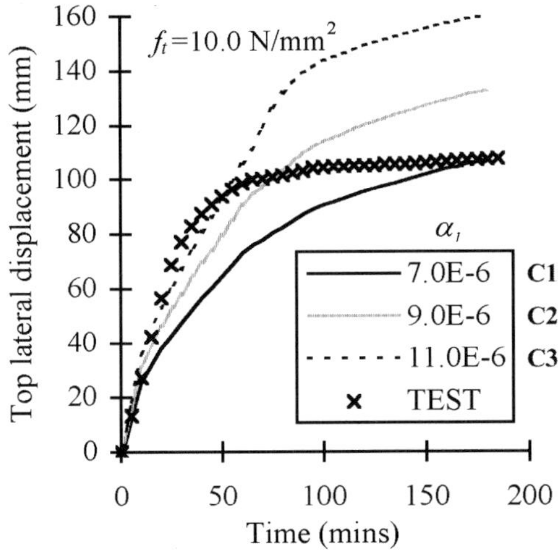

Figure 6.21: Effect of variable thermal expansion on the top lateral displacement.

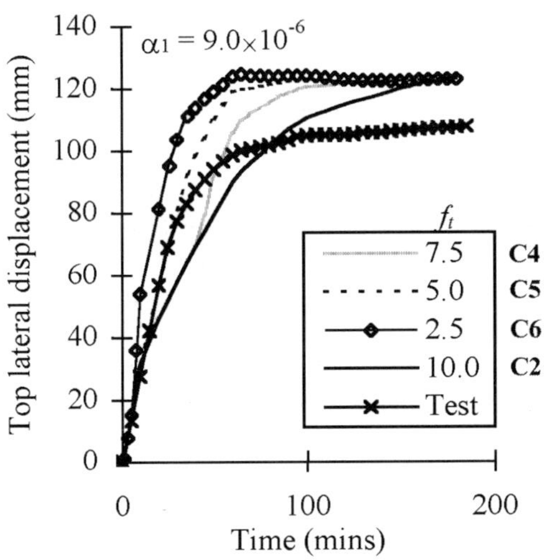

Figure 6.22: Effect of variable tensile strength on the top lateral displacement.

erties employed were:

$$
\begin{aligned}
f_c &= 14.0\text{N/mm}^2 \\
f_t &= 2.0\text{N/mm}^2 \\
\varepsilon_{act} &= 0.0028 \\
\alpha_1 &= 6.0 \times 10^{-6} \text{ (for all temperatures)}
\end{aligned}
$$

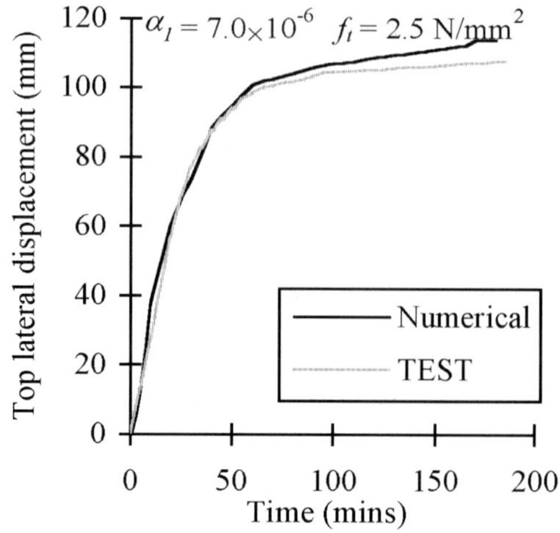

Figure 6.23: Assumed material properties giving best approximation to test results.

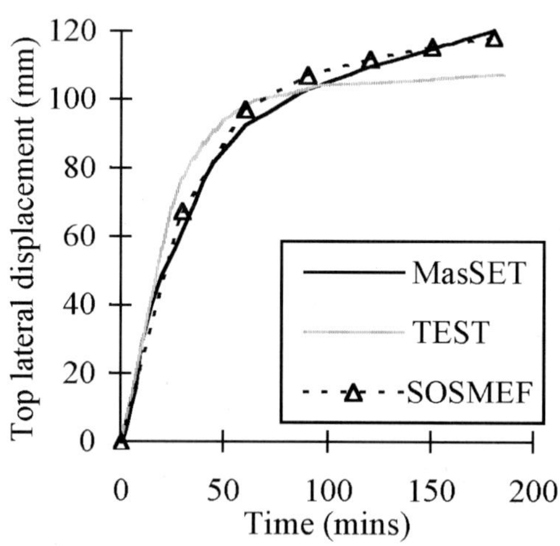

Figure 6.24: Comparison between results from SOSMEF, with MasSET and test.

These values were employed in MasSET for the purposes of evaluation and comparison with MasSET. Figure 6.24 illustrates the top displacement-time results from MasSET, SOSMEF and experimental data, and as can be seen MasSET shows reasonable comparison with SOSMEF, and indeed both numerical results tend to the experimental behaviour.

6.7 Concluding remarks

The chapter has provided an insight into the various factors affecting thermo-structural behaviour of masonry fire walls. The elevated temperature material properties of some masonry material have been discussed, which has facilitated greater understanding of the structural response in high temperature environments.

Both the representation of thermal strains and the elevated temperature material model used in the program have been validated by comparison with total strain transient temperature experimental results. The numerical model has then been used to simulate experimental data from standard fire testing on masonry walls, with considerable success. Anomalies between the University of Ulster tests and numerical analysis have largely been attributed to the uncertainty of the horizontal boundary conditions in testing. The BRE cantilever walls have highlighted the significant influence of thermal cracking, and also how the variation of the thermal expansion coefficient alters structural response.

However, although the accuracy of the model has been proved with respect to the current examples, its application must be limited to load-bearing walls of the type described in BS476, Pt.8, and to some non-load bearing walls, provided the vertical edges are free and the height to breadth ratio is large. It is recommended that the model be developed into three dimensions to accommodate all structural scenarios, however the initial 2-D model has provided an excellent basis for this future work.

References

[1] BS 5628, "Code of Practice for use of Masonry, Pt 1. Structural Use of Unreinforced Masonry", British Standards Institution, London, 1992.

[2] Eurocode No.6., "Design of Masonry Structures. Pt 10: Structural Fire Design", Commission of European Communities, 1990.

[3] Byrne S.M., "Fire Resistance of Load-bearing Masonry Walls", Fire Technology, 15 (3), p180-188, 1979.

[4] Lawrence S.J., Gnanakrishnan N., "The Fire Resistance of Masonry Walls", National Building Technology Centre, Technical record No. 531, Australia, 1988.

[5] Gnanakrishnan N., "The effect of end restraint on the stability of masonry walls exposed to fire", National Building Technology Centre, Australia, 1990.

[6] Khoury G.A., Grainger B.N. and Sullivan P.J.E., "Transient Thermal Strain of Concrete: Literature Review, Conditions Within Specimen and Behaviour of Individual Constituents", Magazine of Concrete Research, 37 (132), p131-144, 1985.

[7] Schneider U, "Behaviour of Concrete at High Temperatures", RILEM Committee Report. (44-PHT), 1983.

[8] Anderberg Y. and Thelandersson S., "Stress Deformation Characteristics of Concrete at High Temperatures", Division of Structural Mechanics and Concrete

Construction, Lund Institute of Technology, 15, 1976.

[9] Malhotra H.L., "The effect of temperature on the compressive strength of concrete", Magazine of concrete research, 8 (23), p85-94, 1956.

[10] Schneider U., "Behaviour of concrete under thermal steady state and non-steady state conditions", Fire and Materials, 1, p103-115, 1976.

[11] Sheehan P.K., "Measurement of linear thermal expansion", Technical record TR 44/153/432, Experimental Building Station, North Ryde, Australia, 1976.

[12] Pande G.N., Kralji B. and Middleton J., "Analysis of the compressive strength of masonry given by the equation fk = K(f'b)a (fm)b", The Structural Engineer, 71 (1), p7-12, 1994.

[13] Eurocode No.4., "Design of Composite Steel and Concrete Structures. Pt 1.2: Structural fire design (second draft)", Commission of the European Communities, 1992.

[14] Schneider U, Morita T, and Franssen J.M., "A concrete model considering the load history applied to centrally loaded columns under fire attack", in "Fire Safety Science, Proc. of the Fourth Int. Symp.", Canada, p1101-1121, 1994.

[15] Kupfer H., Hilsdorf K.H. and Rush H., "Behaviour of Concrete under Biaxial Stresses", Proc. ACI, 66 (8), p656-666, 1969.

[16] Kordina K., Ehm C. and Schneider U., "Effects of biaxial loading on the high temperature behaviour of concrete" in "Fire Safety Science, Proc. of the First Int. Sypm.", p281-290, 1984-1985.

[17] Gnanakrishnan N. and Lawther R., "Some Aspects of the Fire Performance of Single Leaf Masonry Construction2", in "International Symposium on Fire Engineering for Building Structures and Safety", Melbourne, The Institution of Engineers Australia. 89 (16), p93-99, 1989.

[18] Dhanasekar M., Chandrasekaran V. and Grubits S.J., "A numerical model for the thermal bowing of masonry walls", in "Proc. tenth international brick masonry conference", Canada, p1093-1102, 1994.

[19] O'Meagher A.J. and Bennetts I.D., "Modelling of Concrete Walls in Fire", Fire Safety Journal, 15, p315-335, 1991.

[20] Nadjai A., O'Connor D.J., O'Gara M.E., "Behaviour of Brickwork Structures Subject to Elevated Temperatures", in "Computer Techniques for Civil and Structural Engineering", Civil-Comp Press, Edinburgh, UK, 1999.

[21] Laverty D., Private Communication, University of Ulster at Jordanstown, 1998.

[22] O'Connor D.J. and Scotney B.W., "Determination of equivalent thermal response parameters for evaluating the structural response of beams subjected to transient thermal environments", International Journal of Mathematics, Education, Science and Technology, 26 (1), p111-130, 1995.

[23] Cooke G.M.E., "The thermal bowing of brick walls exposed to fire on one side" in "Proc. Interfam'96", Interscience, London, p915-919, 1996.

J.W. Bull, (Editor), Computational Modelling of Masonry, Brickwork and Blockwork Structures,
Saxe-Coburg Publications, Stirling, Scotland 177-196.

Chapter 7

Discontinuous Deformation Analysis of Masonry Bridges

N. Bicanic[†], D. Ponniah[‡] and J. Robinson[‡]

†*Department of Civil Engineering, University of Glasgow, United Kingdom*
‡*Department of Civil and Environmental Engineering, University of Edinburgh, United Kingdom*

Abstract

The application of the discontinuous deformation analysis (DDA) for an assessment of masonry arch bridges is considered in the context of various computational frameworks. Following a brief account of underlying DDA theory, the method is illustrated on a series of simple benchmark problems related to the stability of a masonry arch under its own weight. Benchmark problems build on the classical Couplet/Heyman interface conditions, by relaxing the non sliding interface condition. Two sets of sensitivity analyses looking at the influence of the interface friction angle are conducted for the Edinburgh University arch bridge and compared with experimental findings. Computational predictions using the DDA analysis was also compared with predictions from the nonlinear finite element analysis and the discrete element analysis. Satisfactory failure mode predictions, including formation of hinges were obtained.

7.1 Introduction

Preservation strategies and conservation of heritage masonry infrastructure and masonry bridges represent complex and sensitive processes, which require an understanding of an historical context and building traditions. Proper understanding of the original construction process may influence repair strategies and help to rationalise arguments related to the historic infrastructure asset management. Throughout history masonry arch bridges have continually been designed and constructed. Many of these are still standing, paying testament to the strength and durability of this type of structure (Figure 7.1). In parallel to the construction of masonry bridges, rational ways were sought to assess an existing arch bridge or design a new one. Many methods of arch analysis, some simple, others more complex, have been introduced over the years, most of them evolving, increasing or decreasing in use and popularity with time. Many of the currently available assessment methods are based on various embedded theories and assumptions of arch behaviour. For a very long time simple empirical rules [1] were widely used for the construction of masonry arch bridges. These methods did not consider material properties, but were based solely on observation of traditional bridges, which had proved to be safe over time. They represented actual observations, an undoubtedly important factor in the arch bridge design, but did not provide for any means of assessment of existing structures. A more scientific method was introduced with the development of early elastic methods of analysis, where a line of thrust is calculated inside the arch barrel from the dead load of the structure and the application of applied loads. Castigliano introduced a middle third rule, in order to ensure the no tension state within the masonry. The presently popular modified MEXE method [2] evolved from these early elastic methods. This method used

Figure 7.1: Typical masonry arch bridge.

various assumptions associated with arch shape and material properties in order to derive the maximum load which can be supported by an arch. Tables of this maximum load were constructed for different arch dimensions and failure loads for different arch shapes can be obtained with the use of modification factors. Heyman [3] refers to a very early alternative method introduced by Couplet, who considered the actual way in which arches failed by the formation of hinges, therefore forming a mechanism. Pippard [4] re-introduced this almost forgotten method of analysis as the new plastic analysis method. The method is based on the three basic assumptions - *masonry has zero tensile strength, masonry has infinite compressive strength and no sliding is possible between voussoirs*. The same assumptions were used in the development of the ARCHIE program [5], where the three main developments include assumptions for modelling of the fill (including load dispersal, dead load and horizontal pressure), the creation of zones of thrust in place of a line of thrust and the calculation of a required ring depth for a given magnitude and position of load.

7.2 Computational Frameworks for Masonry

A relatively recent development in the analysis of masonry arch bridges has been the use of finite element techniques. Towler and Sawko [6] showed the potential by computing load deflection curves and collapse loads for an arch modeled with 1-D beam elements to represent the arch barrel. The presence of fill was considered only as a dead load and no further soil-structure interaction effects were modelled. Crisfield [7] later introduced non-linear spring elements in order to try to model the resistance from the soil. The spring stiffness was termed the sub-grade modulus, and was initially pre-compressed to the equivalent at rest pressure. The maximum horizontal pressure was limited by the active or passive pressures depending upon the type of movement of the arch. The modelling framework CTAP is based on an elastic cracking and on Castigliano's original approach. The model assumes linear elastic behaviour in compression and zero strength in tension and good correlation with full scale tests to failure and finite element methods was observed [8]. Generalised matrix formulation for curved members with variable cross section is described in detail in the work by Molins [9]. The contribution of spandrel walls and fill is included by the use of tapered members joined to the arch ring and the method produced adequate estimates of failure loads and comparative mechanisms. Choo [10] modelled the arch ring differently, using tapered elements allowing failure in both tension and compression, resulting in zones having no further structural stiffness but still contributing to the arch compressive strength. The fill is modelled by springs and the dispersal angle is used to distribute the applied load onto the arch barrel. The MAFEA program, used by British Rail, is based on the above method, based on no tension in the arch ring and passive horizontal resistance from the soil fill. An alterative framework proposed by Ashour [11] utilises a numerical procedure by equating the internal energy dissipated in the arch by the masonry to the external work done by the loads in order to find the arch capacity. With this simplified model a computationally inexpensive method of analysis, which correlates well with previously conducted laboratory tests, has been produced.

Most recent computational approaches to the integrity assessment of structural masonry are conducted by a variety of methodologies, ranging from highly simplified methods to complex nonlinear finite element analyses using plasticity based material models, including joint and interface elements to model planes of weaknesses. Research into modelling of structural masonry mostly relies on the nonlinear continuum or homogenisation based techniques. Recently considerable attention has also been given to rational assessment methodologies [12], [13] which deal more directly with computational aspects and the discontinuous nature of structural masonry. Several international conferences have addressed more specifically the structural analysis of historic constructions [14].

It has been postulated that many aspects of a masonry arch bridge would make analysis with discrete element methods, DEM [15] particularly suitable. The joints between voussoirs represent the main weakness planes of arch bridges, allowing blocks to slip and the formation of hinges. Arch/fill interaction due to the load and arch movement is allowed for by these methods. The reliability of the simple numerical model, composed of rigid elements connected by frictional joints, has been verified by comparison with classic limit analysis. Early concepts of discrete element methods were proposed for the simulation of jointed rock masses in a combined discrete and finite continuum approach [16]. Most DEM based frameworks tend to use the same concept, solving equations of motion using an explicit time-stepping scheme. Contact force histories are typically obtained using a penalty method, where the contact force is proportional to the block inter-penetration. The resultant out of balance force is then used with new equations of motion for the next time step. Many improvements have been included, such as deformability of solid material, more complex contact interaction laws and failure or fracturing criteria controlling the solid material behaviour. Early discrete element methods treated a structure as a large number of interacting rigid part, but recent work allows for deformability of the blocks, leading to a combined discrete/finite element approach. The deformation of individual discrete elements was initially dealt with by subdivision of the element into triangular constant strain zones, allowing for a more accurate resolution of the strain field to be accounted for.

Owen *et al* [17] modelled an arch bridge with finite and discrete elements to represent the discrete and composite nature of actual bridges and the procedure is intended to assess the strength of damaged masonry and the efficiency of repair strategies. An idealised two-span bridge has been modelled, with masonry units represented by deformable discrete elements and the fill with spherical discrete elements. This structure was subjected to abnormal vehicular loading to determine the load distribution throughout the structure. Melbourne [18] used a finite element analysis with different brick/mortar elements as well as the interface elements, with a limiting interface strength, and a very good correlation between the numerical predictions and experimental results have been achieved.

The well known discrete element code UDEC [19] was often used to model arch structures, where the arch was divided into blocks connected by elastic spring contacts. The presence of backfill was typically represented by static loads acting on the arch extrados and the following qualitative conclusions were typically made

- *Block contact stiffness* was shown to have almost no influence on the failure load or mechanism, in agreement with the limit analysis theorems.

- *Interface friction angle* has a marked effect on the failure load and at lower values the failure mechanism involves some sliding, as well as the formation of hinges.
- *Number of blocks* (arch discretisation) in the model has a small effect on the failure load, believed to be occurring due mainly to a slightly different failure mechanism occuring.
- The progressive reduction of the thickness/span ratio produces a reduction of the failure load until the arch is no longer stable under self-weight alone.
- *Arch shape* was shown to have differing effects on the failure load for the various types of load applied.

The use of discrete elements to investigate arch bridges enables the investigation of many of the factors affecting the strength of arch bridges such as arch/fill interaction, failure mechanisms, spandrel walls and other 3-D effects.

Further discussion will concentrate almost exclusively on another computational framework, the discontinuous deformation analysis, DDA, developed by Gen-hua Shi [20] in the context of the analysis of jointed rock. Unlike most discrete element techniques the DDA method is an implicit method, where the contact constraints are imposed and rigorously satisfied. The main equilibrium equations are based on the minimisation of the potential energy, which stems from the deformability of arbitrarily shaped blocks, applied loads as well as from the contact between blocks (normal contact, shear sliding, stick and slip condition).

7.3 Discontinuous Deformation Analysis, DDA

A variety of discontinuous modelling frameworks (discrete element method, rigid block spring method, lattice modelling, discontinuous deformation analysis, combined finite elements) are increasingly utilised in modelling discontinuous, fractured and disjointed media. The applications range from modelling problems of an inherently discontinuous behaviour (granular and particulate materials, silo flow, sediment transport, jointed rocks, stone or brick masonry) to problems where the modelling of transition from continuum to discontinuum is also important (rock fracturing, rock blasting, concrete cracking). Typical applications include an analysis of the influence of backfill on the ultimate load capacity of historic masonry arch bridges , which require to be reassessed for a substantially increased axle load, where the whole range of different techniques was adopted.

Common to all discontinuous modelling frameworks is the treatment of a large number of distinct interacting domains, where the contact conditions are continuously updated and enforced as the solution progresses. Some modelling frameworks ignore the deformability of interacting domains (i.e. treat them as rigid) and concentrate predominantly on the contact nonlinearity (contact detection, separation, sliding). Although such a restriction simplifies the computational simulation and it may often be quite acceptable, the need to include the domain deformability is self evident, especially if some rational criteria for compressive failure and for further progressive fracturing of discrete domains (blocks, bricks) are required.

The inclusion of deformability is typically achieved by subdividing the discrete domain into finite elements, e.g. the "discrete finite elements" [16] or the "combined DEM/FEM" formulation [21]. Alternatively, the Discontinuous Deformation Analysis [20] is based on an assumed deformation field within distinct domains of arbitrary shapes and a rigorous imposition of contact constraints.

The DDA analysis appeared initially in the field of rock mechanics as an efficient framework of modelling jointed rock as a deformable blocky media and a way of introducing solid deformability into the discrete element framework, where block sliding and separation is considered along predetermined discontinuity planes. The displacement field is described in terms of centroid displacement and rotations, augmented by an assumed deformation field over the block, which together comprise the solution variables D_i for each block i in turn. Components of the system stiffness matrix and the load vector are obtained by the usual process of the minimisation of potential energy, and the distinction is made between the contribution to the potential energy of the whole system arising from the internal strain energy of the block itself and the potential energy associated with block contact constraints present.

Minimisation of potential energy leads to formation of a familiar set of equations with n mixed degrees of freedom for every block (displacement components of the block centroid and as many parameters as needed to describe the strain field).

The global stiffness matrix with $(NB * n) * (NB * n)$ terms, contains $n * n$ submatrices K_{ii} and K_{ij} for NB blocks

$$\begin{bmatrix} K_{11} & K_{12} & \dots & K_{1m} \\ K_{21} & K_{22} & \dots & K_{2m} \\ \dots & \dots & \ddots & \vdots \\ K_{m1} & K_{m2} & \dots & K_{mn} \end{bmatrix} \begin{bmatrix} D_1 \\ D_2 \\ \dots \\ D_m \end{bmatrix} = \begin{bmatrix} F_1 \\ F_2 \\ \dots \\ F_m \end{bmatrix} \tag{7.1}$$

$$K \quad D = F$$

where the nonzero submatrices K_{ij} are present only if and when the blocks i and j are in active contact, and D comprises deformation variables of all blocks considered in the system.

Depending on the inter-block contact stress state, the contact conditions either allow sliding with no penetration and no tension or impose a no sliding, no penetration and no tension condition. These conditions can be interpreted as constraints for the displacement variables of the two blocks in contact, which reduce algorithmically to an interaction problem between a vertex of one block with the side of another.

The simplest approach to impose the non-penetration condition is the *penalty format*, where the contact spring with a penalty stiffness p is placed between the vertex of one block and the side of the other. The presence of penalty springs, at all positions of active contact, contributes to the overall strain energy of the system, which in turn affects the system stiffness matrix, as well as the load vector. This formulation leads to a nonlinear iterative scheme, which proceeds until the global equilibrium is satisfied (norm of the out of balance forces is within some tolerance) while at the same time a *near zero* penetration condition is satisfied at all active contact positions. The advantage of the penalty format lies in its simplicity and the convergence of the solution algorithm depends greatly on the choice of the penalty term. The process may often lead to ill conditioned matrices, when a very large penalty term is employed to ensure the penetration remains close to zero. The penalty formulation also requires an auxiliary calculation of contact forces. In the *Lagrange multiplier method*, the Lagrange multipliers represent the actual contact forces at all active contacts which become additional unknowns of the problem. The system of equations is augmented by the presence of any active contact and the number of unknowns to be solved may change at every increment as the solution progresses, and contact are activated and deactivated. The major disadvantage of the method is that *every* additional active contacts leads to the introduction of an additional Lagrange multiplier λ_i resulting in a continuous change in the number of system equations. The contact forces are obtained directly as a part of the solution vector and no auxiliary computations are needed to evaluate the contact stress resultants. The most advanced treatment of contact constraints in the DDA context is the *Augmented Lagrangian Method*, which has been advocated by Lin [22] and Amadei *et al* [23], where an iterative combination of a Lagrangian multiplier and a contact penalty spring is utilised. The ensuing iterative procedure, to obtain the correct contact forces, proceeds until the penetration distance and the norm of the out of balance forces is smaller than some specified norm.

Consecutive iterates for λ are obtained from

$$\lambda_{i+1} = \lambda_i + pd(D_{i+1}) \tag{7.2}$$

where the penalty term acts as the solution accelerator parameter and $d(D_{i+1})$ is the penetration level at the previous iteration.

The contribution to the system potential energy from active contacts (both Lagrange multipliers and penalty springs) leads to changes both in the system stiffness matrix, as well as the right hand side load vector. The associated set of nonlinear equations can be written as

$$[K + K'(p, D_i)]D_{i+1} = F_{ext} + F'(p, D_i) \tag{7.3}$$

which requires an update for $\lambda_{i+1} = \lambda_i + pd(D_i)$ between iterations. As the value of the last iterate for λ_i is known and appears on the right hand side, the number of global unknowns does not change and the system size to solve for the new iterate D_{i+1} remains constant, comprising only the block deformation variables. The components of the current stiffness matrix are affected by the current solution vector D_i (which define the current penetrations $d(D_i)$ at all active contact positions), whereas the load vector comprises a contribution from the previous iterate of contact forces λ_i. The rate of convergence of the ensuing iterative procedure to obtain the contact forces is clearly controlled by the penalty stiffness term and the iteration process is ended when the norm of the out of balance forces in between the two iterates is less than a specified tolerance limit.

The interface constraints are subject to the Mohr Coulomb criterion, augmented with the tension cut off. Interface properties include therefore the friction angle, cohesion as well as the tensile strength and three types of contact conditions between blocks are continuously updated, i.e. purely normal contact under compressive forces, "stick" contact with normal and shear forces, as well as block sliding contact with normal and shear forces.

The DDA method has recently been reformulated as a subset of a more generalised framework, and the procedure is seen as an alternative approach along with a number of approximation procedures suitable for modelling discontinuous media. The most recent generalisation is the Manifold Method [24], which advocates very similar ideas to the ones used in the *meshless* methods [25] and in the method of *moving least squares* [26], where the development can be traced back to early irregular finite difference schemes. Similar to the meshless methods, the manifold method identifies the cover displacement function and the cover weighting function, where the geometry of the actual blocks is utilised for numerical integration purposes. The treatment of discontinuities is envisaged in the same way as with meshless methods, i.e. by introducing the concept of effective cover regions, where there is a need for n independent covers if a cover intersects n‘ disconnected domains. The commonality of the two approaches, where the meshless methods stem from the continuum side and the manifold method from the discontinuum end of the spectrum, indicates a possibility of a more rigorous treatment in modelling of progressive discontinuities in quasi brittle materials.

7.4 Couplet/Heyman Benchmark Problem

In the following, one of the well known classical solutions in the theory of masonry arches will be utilised as a benchmark problem for the DDA modelling framework. The benchmark problem relates to the minimum thickness t of a semicircular arch of a given mean radius a, required to support its own weight. The analytical solution (derived by Couplet in 1730, described in detail and corrected by Heyman) can be obtained under the assumptions of (a) *no sliding*, (b) *no crushing* and (c) *no tension* between voussoirs. The arch fails as a four hinge mechanism if the thickness of the arch is smaller than the critical value of $(t/a)_{crit} = 0.105869$ (Figure 7.2). For the discussion which follows, it is important to note that this solution is independent of the level of the specific self weight of the arch, as it is based on purely geometric arguments.

The DDA solution was conducted for the semicircular arch ($t = 0.103$m, $a = 1.0515$m, specific weight $\gamma_w = 15000$N/m^3, $E = 6 * 10^7$N/m^2, $\nu = 0.1$) (Figure 7.2). In order to make the DDA analysis correspond to the Couplet/Heyman assumptions, the DDA interface constitutive parameters are set to ensure that no sliding failure occurs (infinite cohesion, high friction angle, no tension). The solution confirms that the arch is unstable and that the four hinge mechanism is created, irrespective of the value of the selfweight. It is also clear that the material data referring to block deformability

(E, ν) are of no consequence for this comparison, they are included for completeness only.

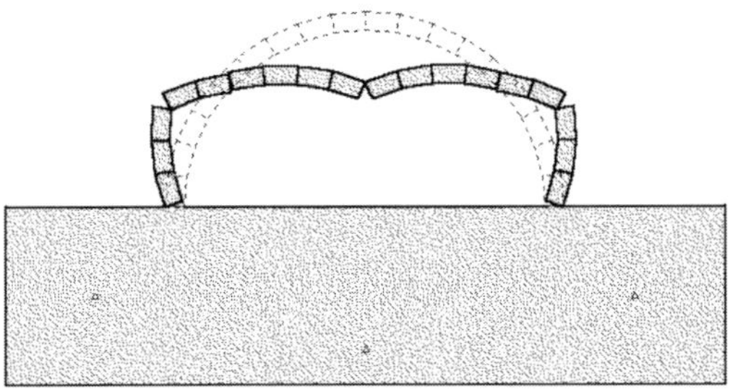

Figure 7.2: Couplet/Heyman benchmark problem A$t = 0.103$m, $a = 1.0515$m, $t/a = 0.098 < 0.10598$, unstable arch.

Slight modification of the arch geometry ($t = 0.12$m, $a = 1.06$m, specific weight $\gamma_w = 15000$N/m^3, $E = 6 * 10^7$ N/m^2, $\nu = 0.1$) leads to the $t/a = 0.113$, just above the critical ratio (Figure 7.3). Using the same interface constitutive parameters, the DDA solution confirms the analytical solution and the arch is found to be stable.

Clearly the DDA computational framework allows for the departure from the crucial assumptions for the interface constitutive law, as used by Couplet and Heyman - no sliding, infinite compressive strength (no crushing), no tension. A small scale sensitivity study was conducted by varying the interface constitutive parameters (cohesion, friction angle and tension) for the benchmark problem A (unstable arch) (Figure 7.2). The first sensitivity study was conducted to establish the level of the interface tensile strength required to stabilise the otherwise unstable arch, while maintaining the other

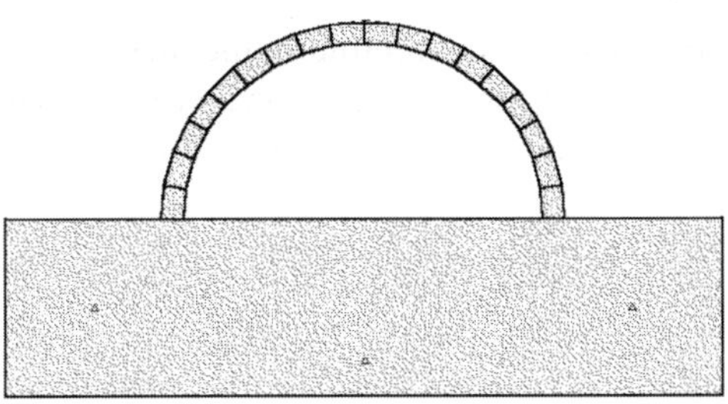

Figure 7.3: Couplet/Heyman benchmark problem B $t = 0.12$m, $a = 1.06$m, $t/a = 0.113 < 0.10598$, stable arch.

two original conditions of no sliding and no crushing. For the example above the analysis leads to the critical value of $f_t = 5250$ N/m^2. It is clear that this particular tensile strength level is problem dependent, associated with the level of selfweight (unlike the critical thickness/span ratio, which is purely geometrical). Indeed, another DDA analysis with the same arch geometry - but with a double selfweight - leads to doubling of the critical tensile strength value to prevent failure, i.e. $f_t = 10500$ N/m^2.

A further set of sensitivity analyses was conducted for the benchmark problem B (stable arch) by studying "cohesion only" (Figure 7.4) and "friction angle only" (Figure 7.5) interface properties. Two of the three Couplet/Heyman conditions were retained (no tension, no crushing) whereas the "no sliding" condition is replaced by searching for limiting values of cohesion and of friction angles respectively, which are still sufficient to secure arch stability.

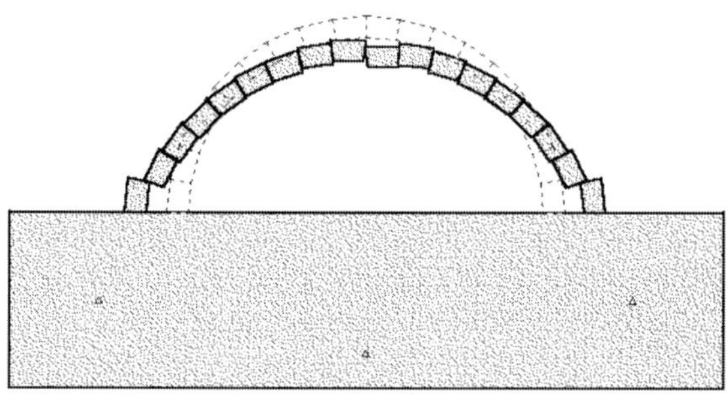

Figure 7.4: Couplet/Heyman benchmark problem B, $t = 0.12$m, $a = 1.06$m, $t/a = 0.113$ (Cohesion only interface ccrit $= 35000$ N/m^2, $f = 0$).

Fig 7.4 illustrates the failure mode and the limiting value of cohesion associated with the "cohesion only" interface i.e. for any cohesion above this critical value the arch remains safe. Note that the friction angle f is set to zero for this analysis.

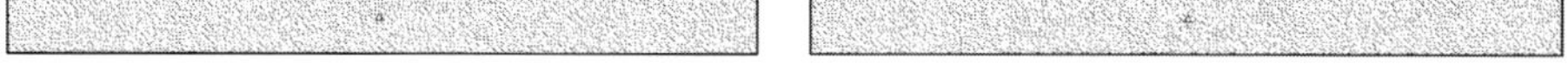

$\varphi = 35°$, $c = 0$ $\qquad$ $\varphi = 34°$, $c = 0$

Figure 7.5: Couplet/Heyman benchmark problem B, $t = 0.12$m, $a = 1.06$m, $t/a = 0.113$ Two "friction angle only" analyses, just above and just below φ_{crit}.

Again, these particular limiting values are for obvious reasons dependent on the specific weight of the arch material, unlike the original Couplet/Heyman criteria. Further confirmation comes from the DDA analysis of a thicker semicircular arch (hence stable in the Couplet/Heyman sense), where the critical cohesion (Figure 7.6) and the critical friction angle (Figure 7.7) can be lower than for the critical values obtained for a thinner arch and still ensure arch stability.

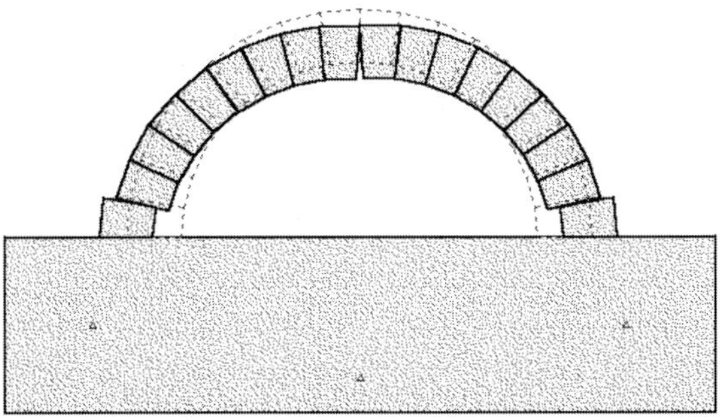

Figure 7.6: Couplet/Heyman benchmark problem B, $t = 0.303$m, $a = 1.1515$m, $t/a = 0.263$ (Cohesion only interface $c_{crit} = 21000$ N/m^2, $\varphi = 0$).

The DDA discontinuous modelling framework allows for the whole range of sensitivity analyses to be easily conducted, to establish the complete "interaction diagram" for (c, φ) interface parameters for the arch only problem.

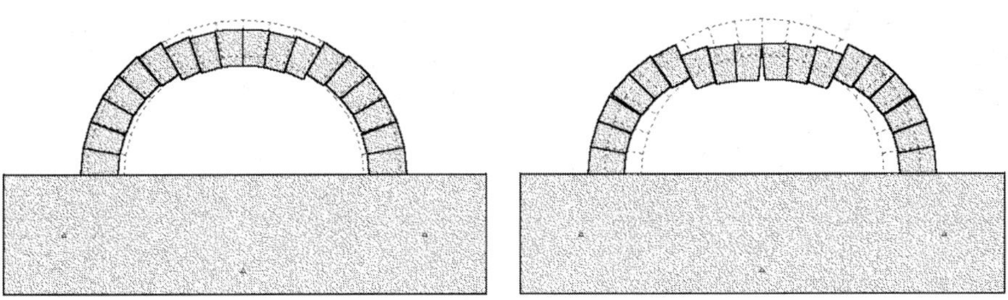

$\varphi = 15°\ , c = 0$ $\qquad$ $\varphi = 14°\ , c = 0$

Figure 7.7: Couplet/Heyman benchmark problem C, $t = 0.303$m, $a = 1.1515$m, $t/a = 0.263$ Two "friction angle only" analyses, just above and just below f_{crit}.

The more general application of the modelling framework is also indicated; inclusion of the backfill material, interface properties for backfill itself, as well as for the backfill/arch can be accomodated within a unified framework. Moreover, the inclusion of block deformability (inherent within DDA) allows for a further generalisation, where block material crushing or further fracturing may be accounted for.

7.5 Edinburgh Arch and Influence of Backfill

The arch modelled was constructed originally to conduct a series of instrumented tests on a idealised 2-D bridge structure [27]. The arch was constructed from a single ring of 41 class B Engineering bricks set on their sides to a total thickness of 102.5 mm with a nominal 10 mm mortar. The barrel was laid in English bond with half bricks cut every second course to face the arch ring; the bridge dimensions are shown in Figure 7.8. The fill material was medium uniform sand with a bulk density of 1517 kg/m^3 and an angle of shearing resistance of 34°. Stiffened timber walls were used to retain this fill material, separated from the arch barrel to ensure a minimal effect on the structural behaviour of the arch. Polythene strips were attached to the walls and arch also to prevent loss of fill and reduce friction effects. Deflection gauges were placed under the arch intrados to allow deflections to be compared with other measurements, although these were removed at higher loads to avoid equipment damage. The load was applied to the fill surface through a semi-flexible timber load platen of dimensions 0.18m by 1.79m. In the test to destruction the load was applied at 1/3 span and was applied at a steady rate. All instrumentation was read at the beginning of the test and after each load increment until collapse. Initial loading caused elastic compressive strain in the arch ring, allowing the arch to displace. Further movement and rotation formed visible cracks in the mortar joint or bond failure at the mortar brick interface, and the formation of hinges. Once four hinges had formed, alternating between the intrados and the extrados, the arch became unstable and collapsed. The failure load obtained from the experiment was 38.1kN (21.3 kN/m) and the failure occurred due to the formation of four hinges.

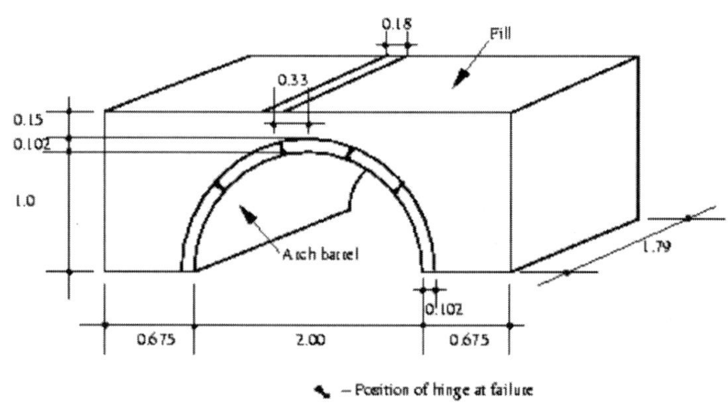

Figure 7.8: Edinburgh University Arch Bridge.

The illustrative DDA Analysis of the Edinburgh Arch problem follows as a direct extension of the Couplet/Heyman benchmark problem (Figures 7.9 and 7.10). In fact, the dimensions of the Edinburgh arch correspond to the earlier arch-only analysis (benchmark problem A) (Figure 7.2), where the arch was found to be unstable under its own weight. The investigation focuses on the influence of the backfill on the failure load of an otherwise unstable arch.

The DDA model includes two deformable blocks on the two sides to account for the stiffened timber walls. The backfill is modelled with deformable DDA blocks but the distribution of joints is arbitrary and for illustrative purposes. Joint interfaces conform to the Mohr-Coulomb constitutive law, with a tension cut-off. The sensitivity analysis conducted here focuses on the type of the interface behaviour, where, in addition to an infinite compressive strength, it is only the friction angle that controls the structure's response. The interface is unable of transmit any tensile forces and no cohesion is present. The arch is modelled with a smaller number of voussoirs than the actual number present in an actual arch, but it is believed that this does not take anything away from the illustrative value of this analysis.

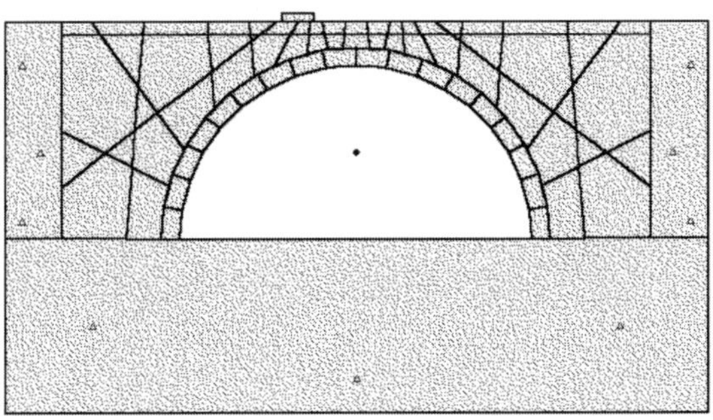

Figure 7.9: DDA Analysis of Edinburgh Arch Bridge Undeformed Block Geometry.

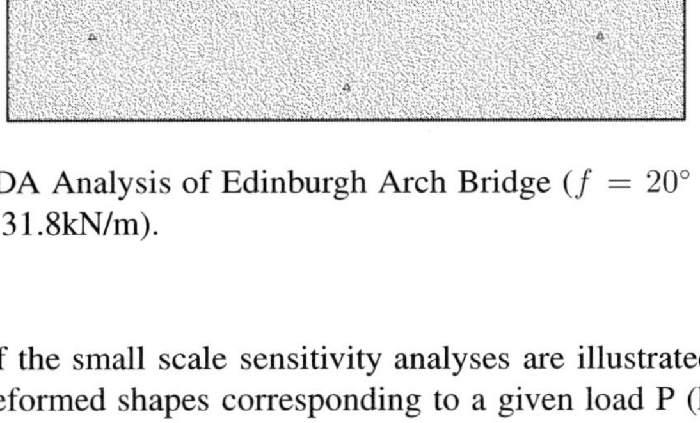

Figure 7.10: DDA Analysis of Edinburgh Arch Bridge ($f = 20°$ - Failure Mode at 31.8kN/m).

The results of the small scale sensitivity analyses are illustrated in Figures 7.11 and 7.12, with deformed shapes corresponding to a given load P (kN/m), presented against the interface friction angle φ. As expected the importance of the friction angle on the failure load is clearly visible. It is also interesting to observe structural behaviour under the loading which is equivalent to the experimental failure load (21 kN/m) - the load levels considered in this sensitivity study correspond to 80%, 100%, 120% and 150% of the experimentally observed failure load.

Despite the very crude discretisation, the formation of distinct hinges is clearly visible in very similar zones as observed during the experiment and the load level corresponding to formation of hinges also compares well. Although the discretisation of the backfill was arbitrary and another discretisation of the backfill would lead to somewhat different results, these are unlikely to be very different.

An important issue is also related to the actual definition of failure both in terms of experiments, as well as within an analysis. During the experiment, the observation of four hinges was deemed to constitute failure. On the other hand the DDA analyses presented were all conducted using the load control, and the absence of convergence can be seen as an indicator of failure.

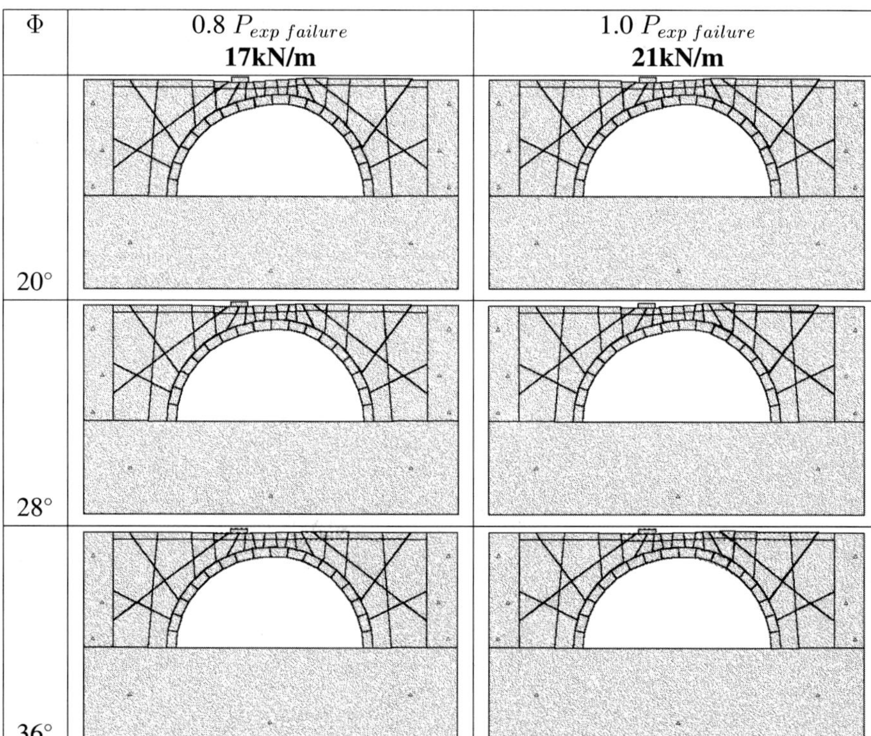

Figure 7.11: Interface Friction Angle φ Sensitivity Analysis for Edinburgh University Arch Bridge (Friction - No tension - No cohesion interfaces). Part 1.

Φ	1.20 $P_{exp\ failure}$ **25.5kN/m**	1.50 $P_{exp\ failure}$ **31.8kN/m**
20°	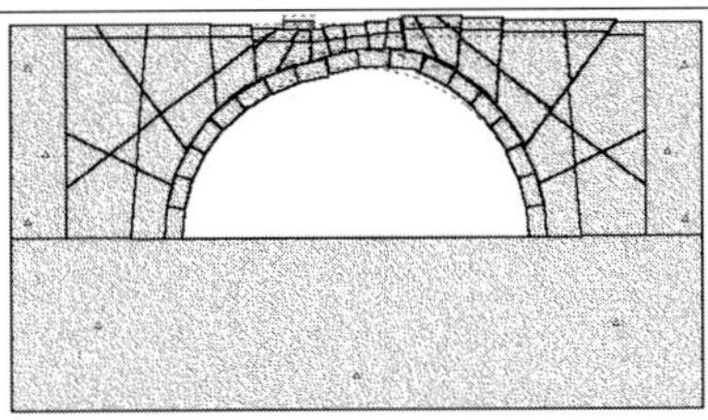	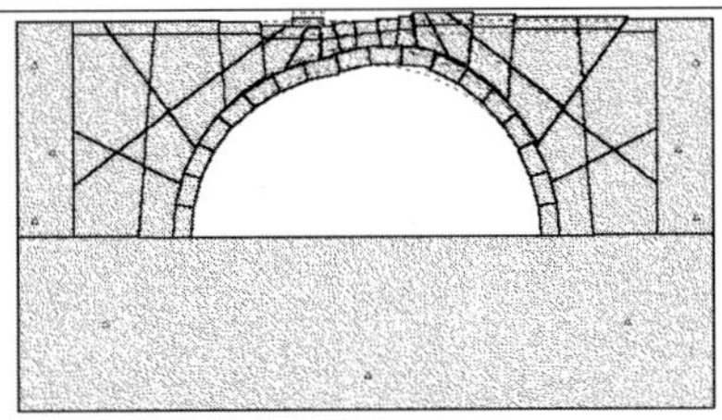
28°	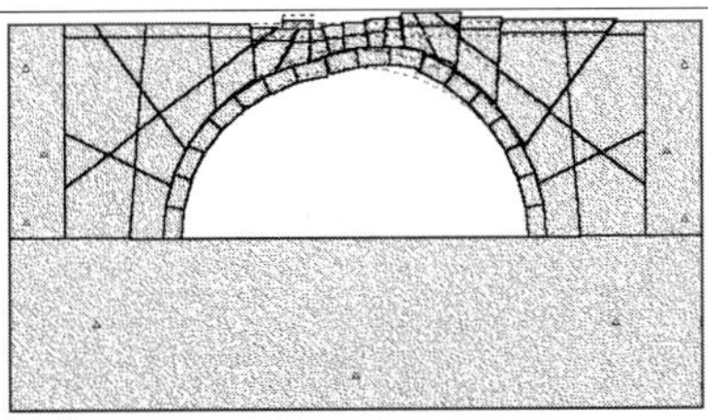	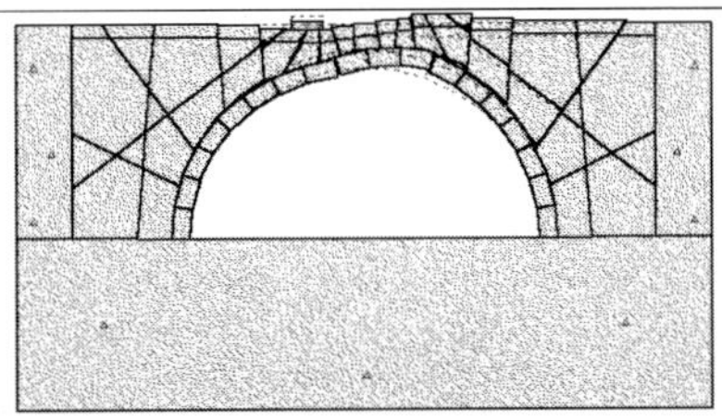
36°	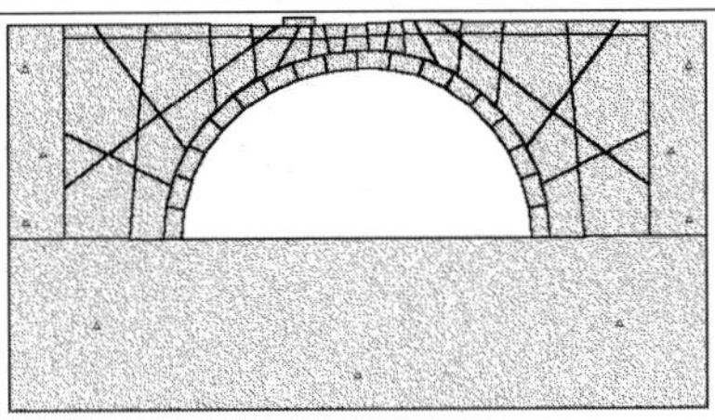	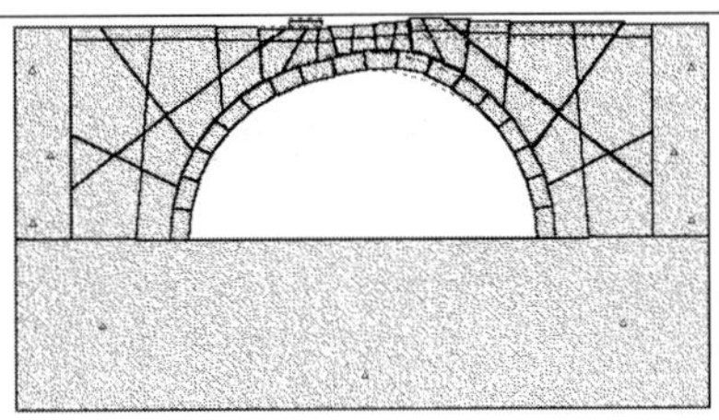

Figure 7.12: Interface Friction Angle φ Sensitivity Analysis for Edinburgh University Arch Bridge (Friction - No tension - No cohesion interfaces). Part 2.

Another sensitivity study was conducted for the Edinburgh Arch model, with a much finer backfill mesh, based on Voronoi tesselation, in order to allow for a more objective load dispersal and for more objective analysis of the influence of interface friction. Here, the voussoir interface friction was varied independently of the backfill interface friction angle. Both backfill and voussoir interfaces were assumed as no tension, no cohesion interfaces. Results of the DDA analysis were compared with comparative sensitivity analyses using the finite element method [28], using DIANA (Figure 7.13) (joint elements, Mohr-Coulomb law for the backfill) and the Rigid Disk-Spring (Figure 7.14) and Cluster Model [29], using PFC (Figure 7.15) (clusters of constrained particles forming voissoirs, backfill modelled as distinct rigid particles with elastic contacts). All analyses were performed using the displacement control, thereby allowing for an analysis to capture post peak response, leading to a better definition of the actual failure load level.

Sensitivity analysis was performed by varying the friction angle of the mortar joints and the fill material (Figure 7.16). Friction angles for the mortar joint were

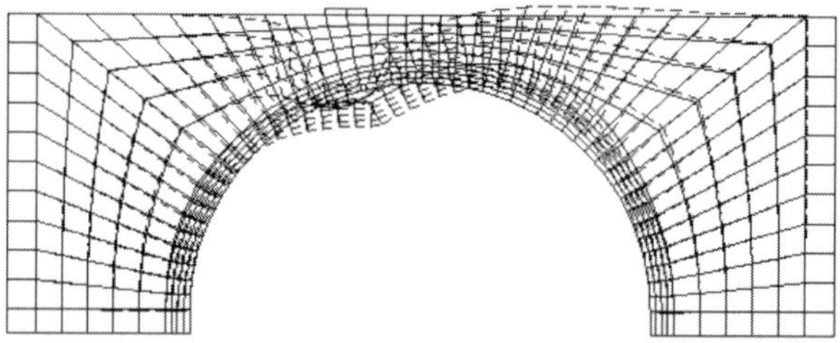

Figure 7.13: Edinburgh Arch Bridge Deformed Shape using DIANA Nonlinear FE Analysis, including Interface Elements.

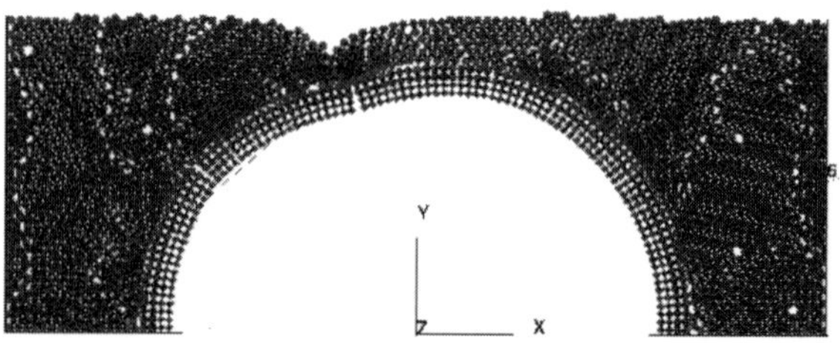

Figure 7.14: Edinburgh Arch Bridge Deformed Shape using Explicit DEM Analysis (PFC) Rigid Disk Particles -Spring and Cluster Model.

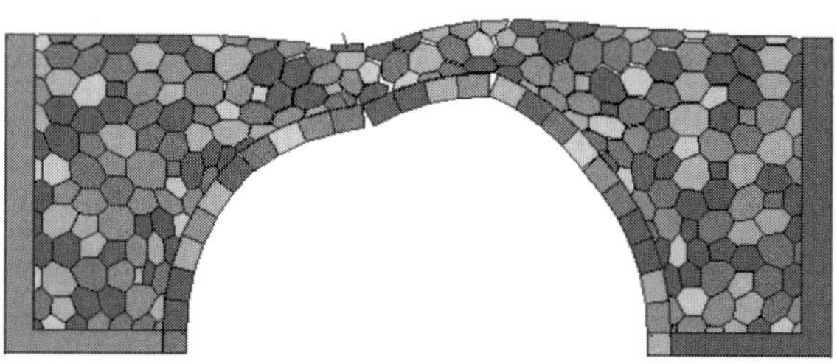

Figure 7.15: Edinburgh Arch Bridge Deformed Shape following DDA Analysis, with Simply Deformable Blocks.

varied between 30° and 40° and the fill angle was varied between 30° and 44°. The friction angle of the arch-fill interface was taken as 75% of the fill friction. Figure 7.16 illustrates the effect that changes in these values have on the failure load. Note that the values of the full failure load, rather than the failure load per unit width, are reported in these figures (the experimental failure load was 38 kN, corresponding to the failure load per unit with of 21.8 kN/m).

The results suggest that changing the friction angle for the mortar joints between voussoirs had little effect on the level of failure loads. On the other hand, the change of the fill friction angle has a significant effect on the failure load for all models, although the effect produced is not consistent.

The collapse loads predicted by each of the modelling frameworks shows a relatively good comparison between the different methods and the failure mechanism is closely replicated in all three cases. Comparing failure load predictions with "classical" assessment techniques suggest that the latter underestimates the strength of an arch with backfill included. The closest failure load predicted by MAFEA provided a failure load of 56% of the experimental data. The MEXE method gives the lowest failure load, only 28% of the experimental data, however this is not surprising since the size of the arch is outside the appropriate range for MEXE predictions.

7.6 Conclusions

Several computational and modelling issues related to the discontinuous modelling framework for masonry arch bridges were presented. In addition to the simple benchmark problems based on the analytical solution by Couplet and Heyman, two sensitivity studies were conducted and the analytical predictions compared with the experimental results for the Edinburgh University arch bridge. Analyses confirm the suitability of the discontinuous modelling frameworks to be used as assessment tools. The following conclusions may be drawn:

- Discretised assessment methods have been shown to produce realistic and accurate failure loads.
- Displacement controlled loading procedure has successfully been utilised to achieve the collapse load and failure mechanism.
- The modelling framework presented can give reliable and realistic failure predictions.
- Sensitivity analysis shows the significance of the backfill to the strength of the entire structure, not just acting as dead weight.
- Arch only analysis compares well with plastic failure methods.
- The effects of varying material parameters has been investigated and the results compare well with previous theory and model tests.

References

[1] Corradi M., "Empirical methods for the construction of masonry arch bridges in the 19th century", in Arch Bridges - History, analysis, assessment, maintenance and repair, Proc. Second Int. Arch Bridge Conference, p25-36, A.A. Balkema, October 1998.

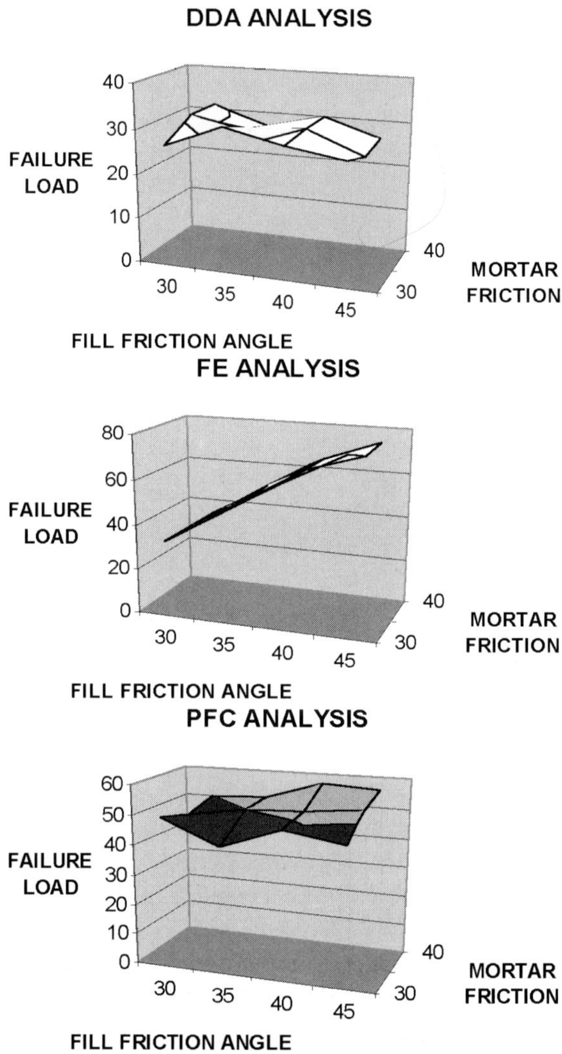

Figure 7.16: Edinburgh Arch Bridge - Failure Load Sensitivity Study with respect to Fill and Mortar Friction (DDA, FE and DEM Analysis).

[2] Military Engineering Experimental Establishment, "MEXE", in Classification of masonry arch bridges, Christchurch, 1952.

[3] Heyman J., "The masonry arch", Ellis Horwood Ltd., Chichester, England, 1982.

[4] Pippard A.J.S. *et al*, "The mechanics of voussoir arch", J. Inst Civ Eng, vol 4, p281, 1936.

[5] Harvey W.J., ARCHIE and MULTI operation manual, 1991.

[6] Towler K.D.S. and Sawko F., "Limit state behaviour of brickwork arches", in Proc. 6th International Brick Masonry Conference, Room, 1982.

[7] Crisfield M.A., "A finite element computer program for the analysis of masonry arches", Technical Report Laboratory Report 1115, Department of Transport, Crowthorne, 1984.

[8] Hughes T.G. and Blackler M.J., "A Review of the UK masonry arch assessment methods", Proc. Instn. Civ. Engrs Structs and Bldgs, 122, p305-315, August 1997.

[9] Molins C. and Roca P, "Capacity of masonry arches and spacial frames", Journal of Structural Engineering, ASCE, 124, 1998.

[10] Choo B.S., Coutie M.G. and Gong N.G., "Analysis of masonry arch bridges by a finite element method", in Proc. Of Forth Rail Bridge Centenary International Conference, p381-293, Edinburgh, 1990.

[11] Ashour A.F. and Garrity S.W., "An upper bound analysis for the strength assessment of masonry arch bridges", in Arch Bridges History, analysis, assessment, maintenance and repair, Proc. Second Int. Arch Bridge Conference, p139-146. A.A. Balkema, October 1998.

[12] Pande G.N. and Middleton J. (eds.), Proc Int Symp on Computer Methods in Structural Masonry, Swansea, 1993

[13] Rots J., "Structural Masonry – An Experimental/Numerical Basis for Practical Design Rules", A.A. Balkema, Rotterdam, 1997

[14] Roca P., Gonzales J.L., Mari A.R. and Onate E. (eds.), "Structural Analysis of Historical Constructions – Possibilities of Numerical and experimental techniques", CIMNE, Barcelona, p407, January 1997.

[15] Cundall P. and Hart R.D., "Numerical Modelling of Dicontinua", in Mustoe G. *et al* (edss), Proceedings 1st US Conference on Discrete Element Methods, CSM Press, Golden, Colorado, 1989.

[16] Ghaboussi J., "Fully deformable Discrete Element Analysis using a Finite Element Approach", Intl Journl of Comp and Geotech, 5, p175-195, 1988.

[17] Owen D.R.J. *et al*, "Finite/discrete element models for assessment and repair of masonry structures", in Arch Bridges - History, analysis, assessment, maintenance and repair, Proc. Second Int. Arch Bridge Conf., p173-180, A.A. Balkema, October 1998.

[18] Melbourne C., "Arch Bridges", Proceedings of the First International Conference on Arch Bridges", Thomas Telford Services Ltd., Bolton, England, September 1995.

[19] Cundall P., "UDEC – A generalised distinct element program for modelling jointed rock", US Army European Research Office, NTIS Order No AD-A087-610/2, 1987.

[20] Shi G.H., "Discontinous Deformation Analysis – a new numerical model for statics and dynamics of block systems", PhD thesis, University of California,

Berkeley, 1988.

[21] Munjiza A., Owen D.R.J. and Bicanic N., "A combined Finite/Discrete Element Method in Transient Dynamics of Fracturing Solids", Engng Computations, 12, p145-174, 1995.

[22] Lin C., "Extensions to the DDA for Jointed Rock Masses and other Blocky Systems", PhD Thesis, University of Colorado, Boulder, 1995.

[23] Amadei B., Lin C. and Dwyer J., "Recent Extension to the DDA Method", in DDA and Simulations of Discontinuous Media, Salami and Banks (eds.), TSI Press, Albuquerque, p1-30, 1996.

[24] Shi G.H., "Numerical Manifold Method", in Ohnishi Y. (ed.), Proceedings ICADD-2 Conference on Analysis of Discontinuous Deformation, Kyoto University, p1-35, July 1997.

[25] Belytschko T., Lu Y. and Gu L., "Element Free Galerkin Methods", Intl Journal for Numerical Methods in Engineering, 37, p329-356, 1994.

[26] Nayroles B., Touzot G. and Villon P., "Generating the finite element method by diffuse approximation and diffuse elements", Journal of Computational Mechanics, 10, p307-318, 1992.

[27] Fairfield C.A., "Soil-structure interaction in arch bridges", PhD thesis, University of Edinburgh, 1994.

[28] DIANA - Finite element analysis, User's Manual. TNO Building and Construction Research, Delft, Netherlands, 6.1 edition, 1999.

[29] Itasca Consulting Group, "PFC-2D, Particle flow code in 2 dimensions", Volume 1: User's Manual, Minneapolis, USA, 1995.

©2001, Saxe-Coburg Publications, Stirling, Scotland
J.W. Bull, (Editor), Computational Modelling of Masonry, Brickwork and Blockwork Structures,
Saxe-Coburg Publications, Stirling, Scotland 197-220.

Chapter 8

Modelling Masonry Arch Bridges

C. Melbourne[†] **and M. Gilbert**[‡]

†*Department of Civil and Environmental Engineering, University of Salford, United Kingdom*
‡*Department of Civil and Structural Engineering, University of Sheffield, United Kingdom*

Abstract

This chapter reviews some of the basic problems associated with modelling masonry arches. The masonry arch has been used since antiquity. Its aesthetic appeal remains unchallenged but the analysis of such structural forms remains fertile ground for the analyst. Prior to considering global arch bridge modelling approaches, some of the basic characteristics of masonry materials are discussed. The development of a conceptually simple 'rigid block' mechanism method of analysis is then described. This method permits estimates of the ultimate load carrying capacity of arches (including multi-span and multi-ring brickwork arches) to be rapidly obtained. However, in some circumstances it is clearly not realistic to assume the constituent blocks in a masonry arch are rigid, and hence there will always be a place for (non-linear) elastic methods of analysis; issues relevant to this are discussed.

8.1 Introduction

The history of the masonry arch can be traced back to antiquity. Its development as a structural form has embraced the changes in technology and taste with many fine examples standing as evidence of its durability not only structurally but also aesthetically. One use of the masonry arch which has attracted much recent interest is the masonry arch bridge. There are over 40000 such structures in the UK alone ranging from small remote bridges over mountain streams to major viaducts carrying arterial mainline railways. All were built when the nature and volume of modern traffic could not have been envisaged by their designers. The assessment and maintenance of this bridge stock demands a fundamental understanding of their behaviour. This is true for all types of structure but can become fogged by the longevity of masonry arch bridges leading to the assumption that little needs to be done in response to the changing loading regime and the passage of time. This would be a risky path to tread.

It is important before embarking on any structural modelling that the analyst has an appreciation and understanding of the behaviour of the structure that is to be modelled and of the experimental data which are going to be used to validate the computer models. Table 8.1 presents some of the masonry material parameters that may influence structural behaviour - whether taken into account in a given mathematical model or not. It is also important to realise that the construction of masonry arch bridges can be diverse and it is this diversity which can present the analyst at the macroscopic level with many problems. At the microscopic level different problems arise. These will be discussed first.

Compressive strength	Elastic modulus
Flexural strength	Shear modulus
Tensile strength	Poisson's ratio
Shear strength	Fracture energy
Bond strength (at mortar/unit interface)	Area of mortar/unit adhesion
Friction coefficient (at mortar/unit interface)	

Table 8.1: Material parameters (unit, mortar and masonry) to be considered in a masonry arch bridge analysis.

8.2 The influence of masonry materials

Masonry (whether comprising bricks or stone blocks) may be considered to be a composite where units are held together (or kept apart depending on one's perception) by mortar joints. In any event, the properties of the unit, mortar, and the unit/mortar interface must all be robustly represented in the model. This may be in a form which allows the computer program to decide the nature and extent of the response or it may be greatly influenced by predetermined constraints. It is this range of interventionist modelling (this also applies to experimental strategies) which can influence the model response and hence the predicted behaviour and carrying capacity.

Figure 8.1 shows a range of possible modelling strategies that may be used for masonry structures. Ideally, the analyst would wish to represent each physical unit with elements possessing the mechanical properties of the unit material [Figure 8.1 (a)]. Similarly, the mortar would be appropriately represented. The two materials are then joined at the interface [Figure 8.1 (b)]. This level of 'accuracy' may be misplaced given the variability in the constituent materials' mechanical properties, which are significantly influenced by the workmanship.

An alternative strategy is to replace the mortar and units by an equivalent material and the joint by an equivalent interface (this is often termed a 'discrete crack' modelling strategy). Using this strategy individual cracks are modelled and predicted crack patterns can be correlated with actual structural behaviour [Figure 8.1 (c)].

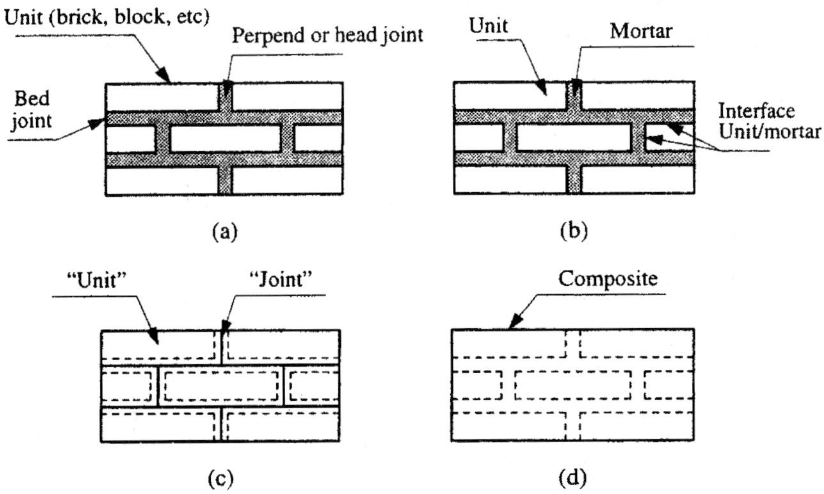

Figure 8.1: Modelling strategies for masonry structures (from Lourenço, 1996 [3]).

Finally a general composite may be provided which has the overall properties of the brickwork or stonework which is 'smeared' over the structural component. Crack criteria can be set but these may result in a crack pattern which is oblivious of the masonry bonding [Figure 8.1 (d)].

At a microscopic level both the units (whether brick, concrete or rock) and the mortar display softening that is indicative of a deterioration of mechanical performance under monotonic or cyclic increase in deformation. Such behaviour is usually attributed to the nature of and defects within ceramic materials and the complexity of their interaction. By the very nature of the construction process, masonry arches contain many flaws. As suggested above, bricks contain flaws that are consequential upon the manufacturing process and which in engineering quality bricks are stable at low stress levels. Similarly, the stone blocks may contain flaws initially formed during the geological process that created the rock from which the blocks were quarried. The latter process can reveal and/or aggravate defects.

Initially, the micro-cracks are stable and propagation is only caused by increased stress levels. A durable unit will be one where cyclic loading within a known stress regime does not cause crack propagation. Although this sounds quite straightforward to quantify, the situation is significantly complicated by the interaction between the mortar and the units - deterioration of any of the mechanical properties e.g. the interface bond, may change the 'internal' response to a given 'external' stress regime and thus allow crack propagation. This can be an important consideration when assessing arches containing macroscopic defects (like ring separation in multi-ring brickwork arch bridges). The effect of the presence of cracks is to cause stress redistribution with the crack surface attracting zero normal stress and thus locally releasing the strain energy (this phenomenon is utilised in the construction of mass concrete arches containing inclusions [1] which 'protect' the destressed material). Typical stress displacement graphs for ceramic materials are shown in Figure 8.2. In each case there is a well-defined softening which under soft loading systems would result in 'brittle' failure. Recent work [2], [3] has involved determining experimentally values for the tensile fracture energy, G_f^I, the compressive fracture energy G_c and the mode II fracture energy G_f^{II}. Clearly, the behaviour of masonry subjected to shear loading is significantly influenced by the amount of pre-compression. This leads to the definition of mode II fracture energy G_f^{II} as the integral of the shear stress vs. shear deformation graph in the absence of a stress normal to the bedding plane.

Notwithstanding the above there is a strong correlation between the strength of the masonry and the strength of its constituent parts. Although the unit-mortar interface may be considered a 'weakening' feature of masonry, it must be remembered that it is this very 'weakness' which provides masonry with the ability to articulate and accommodate movement without loss of structural integrity.

Van der Pluijm [4] carried out a series of controlled laboratory tests on small solid clay and calcium silicate units. Mode I fracture energies G_f^I were obtained which ranged from 0.005 to 0.02 Nmm/mm^2 for tensile bond strengths ranging from 0.3 to 0.9 N/mm^2. Interestingly, it was observed that the bonded area was concentrated in the inner part of the specimen; this was probably a consequence of shrinkage resulting from surface moisture loss. The effect of this was to reduce the bonded area for the small square interfaces to about one third of the total area with the implication that the bonded area for a long wall would be about 60%. This of course makes no allowance for the effects of workmanship and presupposes that the mortar-unit is engaged under optimum conditions if the unit is neither totally dry nor saturated (Hendry [5] suggests about 75% full saturation) and there is 100% contact at the time of laying.

Mode II failure is notoriously difficult to simulate under experimental conditions [6]. Van der Pluijm [7] studied the behaviour of clay and calcium silicate units in a specially designed test apparatus which allowed shear stress measurements under confining compressive stresses of 0.1, 0.5 and 1.0 N/mm^2. The work raises several interesting observations which indicate dependency on the nature of the units and mortar.

Figure 8.2 shows a plot of shear stress against shear displacement. The area under the curve is defined as the mode II fracture energy G_f^{II} with values ranging from 0.01 to 0.25 Nmm/mm^2 for initial cohesion, c, values ranging from 0.1 to 1.8 Nmm/mm^2 respectively. The test also allowed the determination of the initial friction angle ϕ_0 associated with a Coulomb friction model which ranged from 0.7 to 1.2 for the unit/mortar combinations tested. Significantly the residual internal friction angle remained approximately constant and equal to 0.75. The dilatancy angle, ϕ, was recorded as one unit sheared over another. What is important is its dependency upon the confining pressure, the unit surface roughness and the amount of slip. At low confining pressures $\tan\phi$ ranged from 0.2 to 0.7 depending on the unit surface roughness. As the confining pressure increased $\tan\phi$ decreased, an obvious consequence of local crushing of the sand particles as they attempted to pass over each other. Similarly, at large amounts of slip the shearing surfaces 'smoothed' with $\tan\phi$ decreasing to zero.

The writers are unaware of any reliable test data on Mode III behaviour of masonry. In the absence of these data it has to be assumed that extrapolation of the Mode II data is the best that can be offered. The dependency on confining pressure, unit roughness and amount of slippage can be assumed, although the contact area and shear distribution is open to some speculation.

Having considered in some detail the nature of the mortar/unit interface it is now necessary to consider the overall behaviour of the composite. There is the traditional dilemma between standardised test data and 'real' structure performance - for example in a typical series of tests the strength of two-unit wide specimen may be about 10% larger than that of one-unit wide specimens. Under uniform compressive loading failure usually occurs by the development of tensile cracks parallel to the axis of loading. The degree to which the development of such cracks is prevented due to restraint will positively influence the carrying capacity. Similarly, it is the lack of restraint of the unit by the mortar which results in the masonry having a lower strength than the unit. The converse is true for the mortar which is constrained by the units and consequently achieves a much higher strength although its relatively high Poisson's ratio leads to a propensity to tensile cracking of the units as stated above. It follows that the thickness of the mortar beds (and the volume of mortar per cubic metre of masonry in the case of random rubble masonry) has a significant effect upon strength.

Figure 8.4 shows the modes of failure of solid clay units of masonry under biaxial loading. Although the behaviour of a masonry panel subjected to uniaxial loading perpendicular or parallel to the bedding plane may be well understood and well documented, the response is anisotropic and highly dependent on the bonding and unit/mortar strengths which makes the extrapolation of a continuous failure surface somewhat circumspect. Page [8,9] demonstrated the strength enhancement for biaxial loading compared with uniaxial loading. It is believed that the failure mechanism is the same *viz.* tensile cracking resulting from the tensile stresses induced at a flaw as the compressive stress flows round it. It is important to notice that in the case of biaxial compression the expected mode of failure is splitting in the plane of the applied loads.

8.3 Development of modelling strategies for masonry arch bridges

Having considered the behaviour of masonry as a material subjected to idealised loading conditions it is now necessary to marry our knowledge of the behaviour of masonry

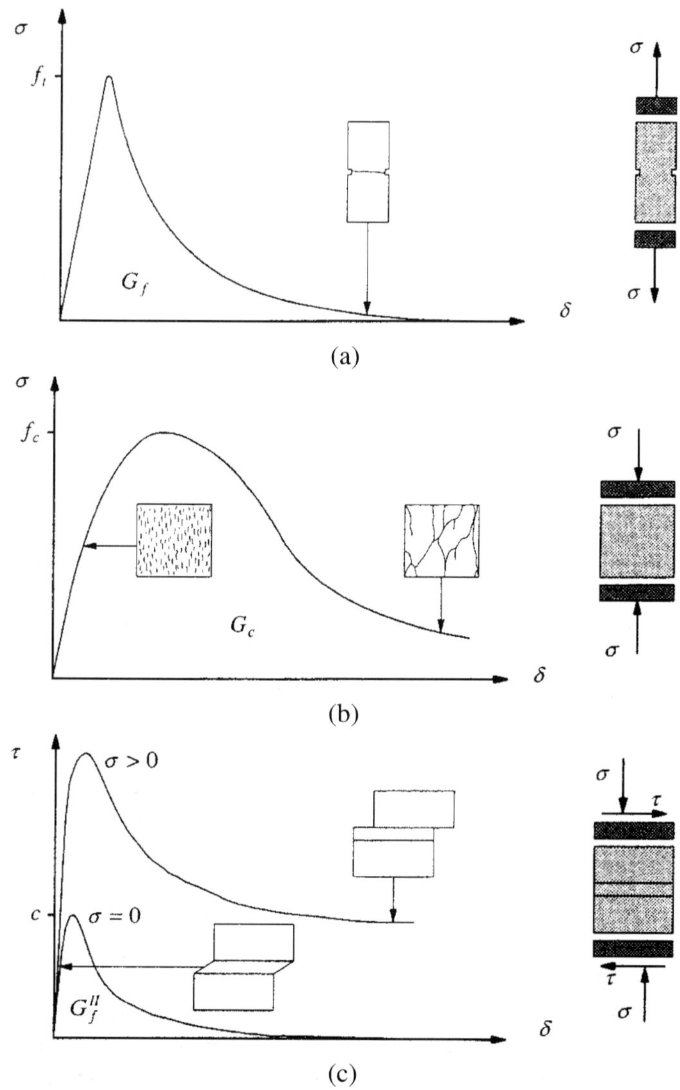

Figure 8.2: Typical behaviour of quasi-brittle materials under uniaxial tensile, compressive and shear loading (from Lourenço, 1996 [3]).

arch bridges to the available mathematical models. It must be remembered that most masonry arch bridges were conceived as gravity structures for which mass and geometry were the design criteria. Certainly, the proportions passed down from antiquity had no thought of stress criteria and were probably based upon bitter experience. It was Galileo in 1638 in his 'Two new sciences' who first considered the analysis of

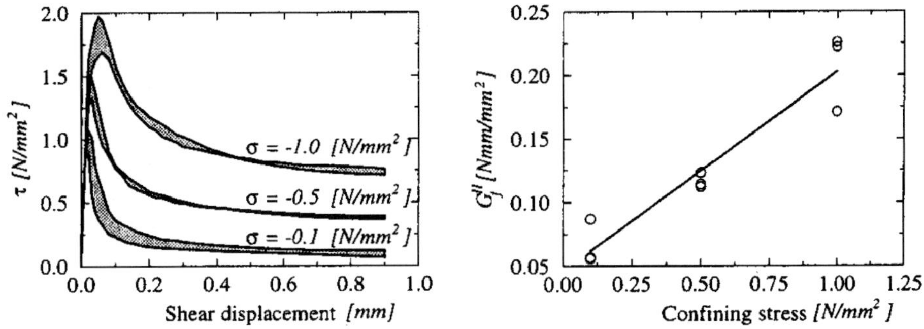

Figure 8.3: Typical shear bond behaviour of the joints for clay bricks (after Van der Pluijm, 1993 [7]).

the strengths of a structure - a cantilever beam. He showed that the strength was not a simple proportion of size - doubling the beam depth more than doubled the strength. At about the same time the strength of various materials were being quantified. It was a logical step in the interests of safety and economy to try to relate the calculated stress for the material strength and apply an acceptable factor of safety. Thus was born modern structural engineering.

It was left to Robert Hooke in 1675 to publish a Latin anagram, which translated states "As hangs the flexible line, so but inverted will stand the rigid arch". The inference is that a hanging simple cable once frozen will carry its own weight if inverted (this ignores stability problems). If the inverted cable is symmetrically increased in thickness then it continues to carry its own self-weight without inducing any bending. Clearly the application of an external load will induce bending which can be idealised internally as an eccentric thrust. Provided this thrust line everywhere lies within the arch, then the arch will remain stable. Barlow [10] demonstrated that there was no unique thrust line associated with a stable arch but that there were many possibilities. Navier's work [11] showed that for linear elastic materials where plane sections remained plane, tension could be avoided by ensuring that the thrust line lay within the middle third of the section. Combining these two pieces of work led to the well-known middle third rule of design which aimed to eliminate tension in the masonry and thus avoid any cracking. It also afforded analysts the luxury of modelling the arch as an elastic continuum.

Castigliano [12] applied the theorems of minimum strain energy to the arch. The position of the thrust line was determined and then checked to see that it lay within the middle third; if this was not the case then the tensile zone was 'removed' and the calculation iterated until no tension was present at any point in the arch.

The main advantage of an elastic analysis is that stress levels and deflections can be calculated - how meaningful these are is open to much debate but it has to be conceded that they provide a 'feel' for their probable values. However, it is universally accepted that masonry arch bridges are cracked - even before the centering is removed! This is a very important observation which should not be obscured by the sophistication of some of the currently available modelling software. Thus at the opposite extreme of considering masonry as an elastic continuum is the analytical model of regarding masonry as a particulate assemblage of stones or bricks, the irregularities between them being filled with a mortar that will permit the transfer of compressive force whilst preventing sliding and not allowing tensile stresses to be transmitted. This model had been shown experimentally to replicate the behaviour of full-scale structures tested by Danyzy in 1732; these tests demonstrated that the blocks within the arch accommodated settlement and collapsed by rotating in kinematically admission loci about hinge points. However, it was only recently that 'plastic' theory was applied to such assemblages [13, 14] with a robustness that permitted analysis.

Thus in its simplest form the voussoir arch may be considered to be a two-dimensional assemblage of wedge shaped voussoir stones between unyielding supports subjected to vertical loading. An arch with fixed supports has 3 redundancies; consequently, for the structure to collapse there must be 4 releases. Based upon Pippard's work [15] which showed that one mechanism involved the formation of 4 'hinges' Heyman (and others) applied plastic theorems to develop the 'mechanism' methods of analysis which form the basis of several commercial analysis programs. At the very least the 'mechanism' method provides a useful means of rapidly estimating bridge strength prior to carrying out a more detailed and, perhaps arguably, accurate analysis.

The basic 'mechanism' method assumptions are that:

1. the masonry in the arch has no tensile strength;

2. the masonry in the arch is incompressible;

3. sliding between the masonry units cannot occur.

Figure 8.5 shows a single span 2-dimensional arch that complies with the above assumptions. Failure can occur only by the formation of sufficient hinges to develop a mechanism within the arch.

Livesley [16] has presented a general method of analysis for masonry arches based on an 'equilibrium' formulation; a 'mechanism' formulation is presented here. However, Charnes and Greenberg [17] showed that when 'mechanism' and 'equilibrium' formulations of limit analysis are linearised, they give rise to dual linear programming (LP) problems.

8.4 The 'mechanism' method of analysis: a linear programming formulation

8.4.1 Basic method

A brief description of a modern (rigid block) formulation of the upper bound or 'mechanism' method of analysis follows. The method described is of general applicability

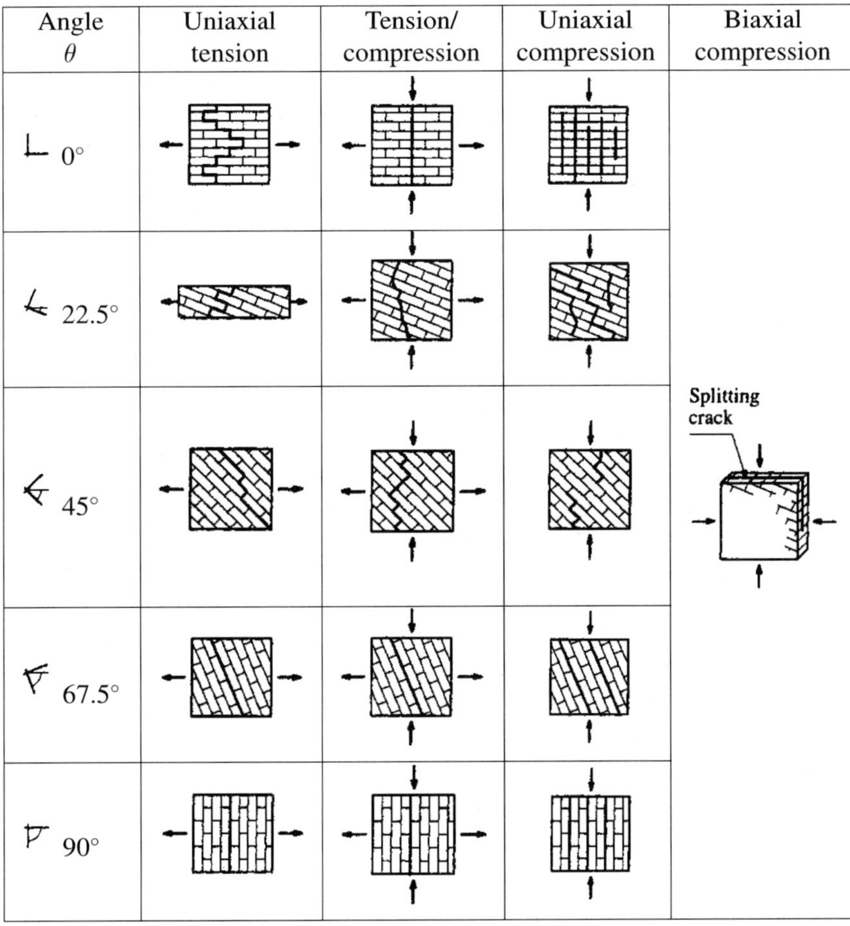

Figure 8.4: Modes of failure of solid clay brickwork under biaxial loading (after Dhanaseker, 1985).

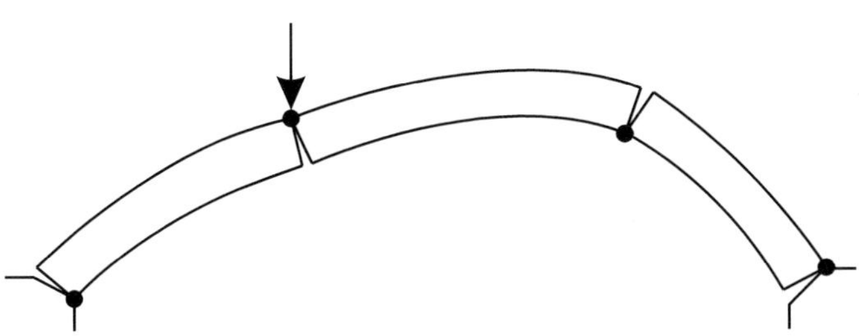

Figure 8.5: Four-hinge failure mechanism.

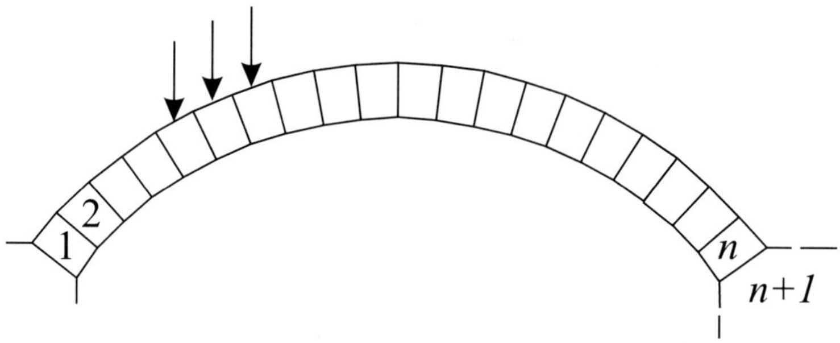

Figure 8.6: Loaded arch rib containing n voussoirs.

and is particularly suitable for computer use.

Consider a loaded arch rib comprising n voussoirs [Figure 8.6], which are rigid, infinitely strong and have surfaces which are rough enough to prevent sliding failure occurring. Assume that constituent blocks within the arch may be subject to both dead loading p and live loading q. At the ultimate limit state the problem to be solved is: what load factor λ applied to the live loading will lead to global collapse of the arch?

Using an approach based on virtual work, the problem becomes:

Minimise

$$\lambda|q| = p^T d \tag{8.1}$$

Where the *whole structure* live load, dead load and displacement vectors are denoted respectively $q^T = \{q_1, q_2 \cdots q_{n+1}\}$, $p^T = \{p_1, p_2 \cdots p_{n+1}\}$ and $d^T = \{d_1, d_2 \cdots d_{n+1}\}$. And where the *block* live load, dead load and displacement vectors are denoted respectively $q_j^T = \{q_x, q_y, m_q\}_j$, $p_j^T = \{p_x, p_y, m_p\}_j$ and $d_j^T = \{u, v, \phi\}_j$; individual block displacement and force components are shown on Figure 8.7.

Subject to constraints which:

(a) stipulate that the right hand abutment (notional block $n + 1$) is fixed in space. i.e:

$$d_{n+1} = \begin{bmatrix} 0 \\ 0 \\ 0 \end{bmatrix} \tag{8.2}$$

(b) displace the structure according to the magnitude of the live load. Thus in general the required constraint is:

$$\frac{1}{|\,q\,|}\sum_{j=1}^{n} q_j^T d_j = 1 \tag{8.3}$$

Alternatively displacements of block j, d_j, can be expressed as a function of the relative rotations between adjacent units. i.e:

$$d_j = A_j r \tag{8.4}$$

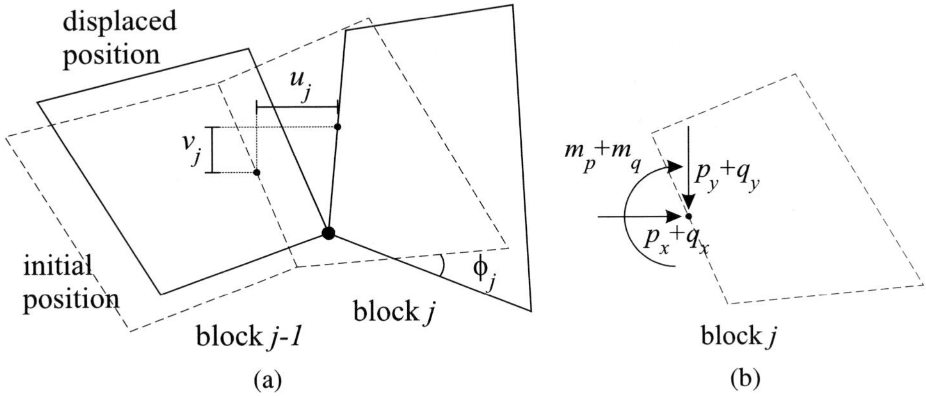

Figure 8.7: Block displacement components (a), and force components (b).

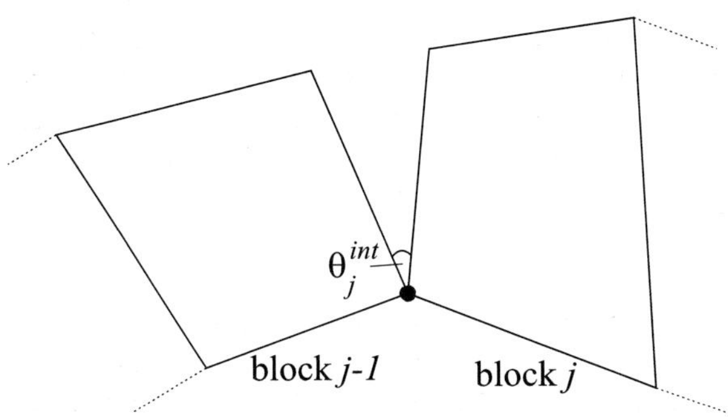

Figure 8.8: Rotation of adjacent blocks (about intrados).

Where A_j is a transformation matrix derived from the geometry of the structure [see Appendix for derivation], $r^T = \{\theta_1^{int}, \theta_1^{ext}, \theta_2^{int}, \theta_2^{ext}, \cdots \theta_{n+1}^{int}, \theta_{n+1}^{ext}\}$, and where θ_j^{int}, θ_j^{ext} represent rotations about the intrados and extrados respectively (e.g as shown on Figure 8.8).

In fact it is convenient to express the displacement terms in equations (8.1) to (8.3) in terms of the relative rotations between blocks, θ_j^{int},θ_j^{ext}, and to select these as the problem variables. This is because the necessary stipulation that these variables must take on only non-negative values is a standard feature of linear programming solution algorithms (note that θ_j^{int},θ_j^{ext}, are not restricted from simultaneously taking on positive values; this represents separation). Any standard linear programming technique, such as the Simplex method [18], can then be used to obtain a solution for λ. There are, however, two important reasons why it may not be possible to obtain a value for λ:

(a) There might be *no feasible solution to the problem*, i.e. there may be no combination of values for the problem variables (θ's) that satisfy all constraining equations. If this is the case then it follows that the arch is *geometrically locked* - the arch will not fail in a hinged mechanism. For example, this outcome might arise in the case of a very thick or flat arch.

(b) The solution for λ may be *unbounded.* This will occur if the arch is already *unstable under its own dead weight.* For example, this outcome might arise in the case of a very thin semicircular arch.

As an upper-bound problem is the linear-programming dual of the corresponding lower-bound problem, the values of the variables involved in the dual (lower-bound) problem are obtainable from the final LP tableau. Thus in the case of an arch bridge the magnitude of the abutment thrusts at failure become available without further calculation.

8.4.2 Removing 'no-sliding' assumption

As stated earlier, a 2-dimensional arch has 3 redundancies and hence requires 4 'releases' to produce a mechanism; so far these have been constrained to rotational releases. Removal of the 'no-sliding' restriction increases the generality of the method by permitting adjacent blocks greater freedom of movement.

Inclusion of Coulomb frictional sliding is difficult from a mathematical point of view. However, these difficulties disappear if the principle of normality is assumed to hold true (i.e. if an associated flow rule is assumed). In this case sliding movement between blocks is treated as 'plastic shearing', also termed dilatant or interlocking friction. Drucker [19] has shown that results from an analytical model in which frictional interfaces are modelled as plastic shearing will be upper bounds on the 'exact' values.

In terms of the formulation of the rigid-block method, additional variables can be introduced to allow sliding movements to be controlled. Thus variables y_j^+, y_j^- can be defined, respectively, as the positive and negative shearing displacements of block j relative to block $j-1$ parallel to the interface between the blocks ($y_j^+, y_j^- \geq 0$). Movement perpendicular to the interface (separation) can be stipulated to be $s_b(y_j^+ + y_j^-)$ where s_b is the coefficient of dilatant friction between blocks. Clearly the parameter s_b must normally always take on some (small) non-zero value, otherwise the arch will be found to be unstable under its own dead weight.

To include the additional variables in the analysis (8.4) is rewritten as:

$$d_j = A_j r + B_j s \tag{8.5}$$

Where B_j is a suitable transformation matrix [see Appendix for derivation] and where $s^T = \{y_1^+, y_1^-, y_2^+, y_2^1, \cdots y_{n+1}^+, y_{n+1}^-\}$.

Figure 8.9 shows the full range of feasible movements of adjacent blocks when the sliding capability is added. Figure 8.10 shows the influence of the value of the coefficient of dilatant friction on the predicted failure load and associated failure mechanism for the 23-block arch tested originally by Pippard and Ashby [13].

Figure 8.9: Feasible relative movements of adjacent blocks.

8.4.3 Including crushing of the masonry in the analysis

The assumption that the masonry is incompressible implies that infinite stresses can be sustained by the masonry at hinges. Although this is clearly unrealistic, the assumption will often lead to only small overestimates in collapse load, depending of course on the compressive strength of the masonry and on the bridge span and shape. Removing the assumption from the analysis transforms the problem from a linear to a non-linear one. However, using a simple iterative solution strategy, such as the one given below, it is typically found that convergence can be obtained in three or four iterations [20]:

1. Compute the collapse load of the bridge, and, from the final LP tableau, deduce the magnitudes of the thrust transmitted across each joint.

2. Determine the depth of masonry required to transmit the thrust across each joint and then locally reduce the depth of masonry in the arch by half this depth[1] (from the intrados or extrados as appropriate).

3. Update the transformation matrices A_j and B_j.

4. Repeat from step 1 until specified convergence criteria are met.

[1]This is the method put forward by Livesley [21]. Alternatively the masonry depth can be reduced by the full depth required to transmit the force. In this case the work equation (8.1) must then be modified to take account of the work done in subsequently opening the hinge, which now has an effective plastic moment of resistance. This latter method was suggested by Crisfield and Packham [22].

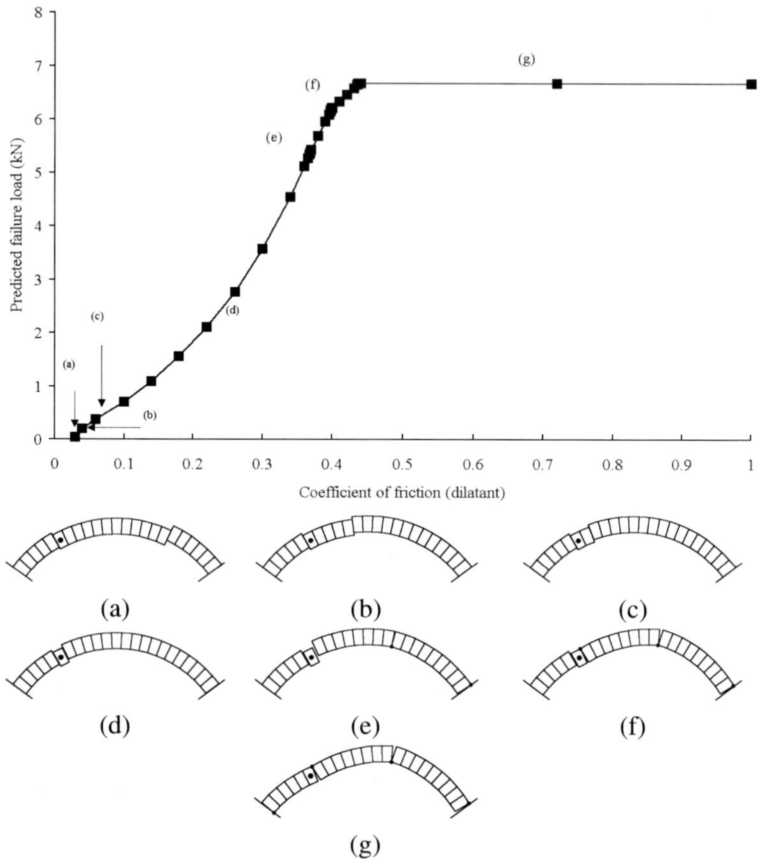

Figure 8.10: Predicted influence of coefficent of friction on collapse mode and load.

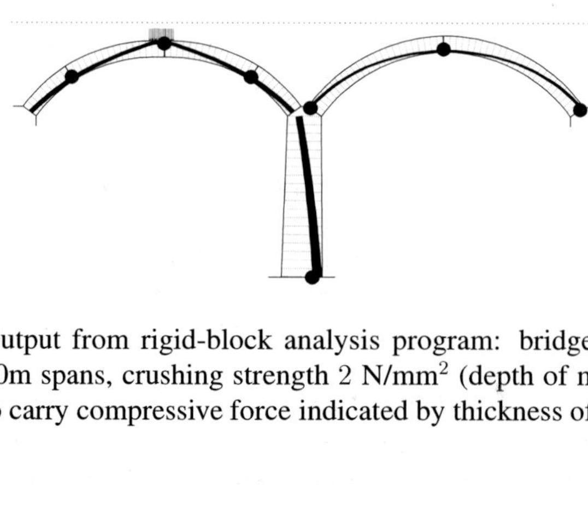

Figure 8.11: Output from rigid-block analysis program: bridge containing two 10m spans, crushing strength 2 N/mm^2 (depth of masonry required to carry compressive force indicated by thickness of thrustline).

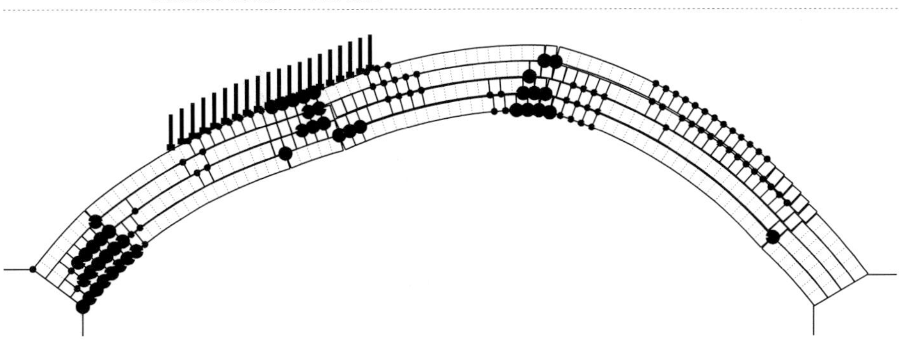

Figure 8.12: Output from rigid-block analysis program: Bolton model single-span model bridge 5-2.

8.4.4 Multi-span arches

Due to the general nature of the rigid block analysis method modelling multi-span masonry arches present no special difficulties. In a multi-span problem each span will have its own prescribed boundary constraints (8.2), leading to a total of $3q + 1$ constraining equations and a single objective function, where q is the number of spans in the structure under consideration. Figure 8.11 shows typical output from the rigid block analysis for a twin-span arch bridge (with finite masonry crushing strength).

8.4.5 Multi-ring arches

In cases where an arch barrel is built up from a number of individual rings of bricks, connected only by a weak layer of mortar, it will frequently be found that these rings become debonded; consequently bridge carrying capacity will be reduced [23].

Bridges with debonded arch rings can be modelled using the rigid block analysis method. Firstly, each ring is modelled as a separate arch with its own prescribed boundary constraints (8.2). Secondly only the top ring will normally be subject to live loading q. Thirdly, as adjacent rings of blocks will in reality interact with one another, additional constraints must be added to those already described in order to prevent blocks overlapping and, if required, to properly account for friction at the interface between rings. Further details of the method are provided in [24]. Figure 8.12 shows typical output from a multi-ring arch analysis, in this case the predicted collapse mechanism for Bolton bridge 5-2 (the bridge actually failed at a load of 500 kN).

Because of the number of additional constraints required to avoid inter-penetration of adjacent blocks it is found that multi-ring arch problems are considerably more computationally expensive than single ring problems (although many problems still take only a matter of a few seconds to solve on a modern desktop PC).

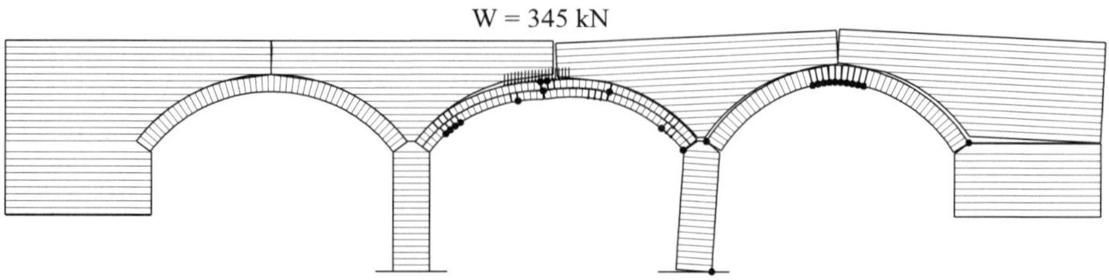

Figure 8.13: Output from rigid-block analysis program: Bolton model multi-span bridge no. 3 (experimentally observed partial ring separation in centre span back-substituted into model).

8.4.6 Masonry in the spandrel zone

The rigid block analysis procedure already described can be used to model a wide variety of configurations of masonry blocks, including internal spandrels, haunching and spandrel walls. The computational effort required to perform an analysis may be reduced if several physical units are represented by an equivalent block; some engineering judgement is required and reference to published large-scale test data is advised (e.g. Page [25], Melbourne [26]). At the same time it must be remembered that using fewer rigid blocks than physical masonry units will be unsafe, the solutions obtained being upper bounds on the actual collapse load.

In the case of spandrel walls (or masonry backing), additional constraints must be introduced to ensure that the interaction between blocks in the wall and blocks in the arch is geometrically admissible. These latter constraints can be formulated in the same way as those formulated for the arches with ring separation, if it is assumed that the shear-bond strength possessed by the joint between the barrel and the wall is negligible. This approach assumes the structure to be 2-dimensional - engineering judgement is needed to apportion the effective weight of the superincumbent spandrels. On narrow bridges (span to width ratio greater than one) it is often the case that at the springings the spandrel wall might occupy 25% of the width of the bridge.

Figure 8.13 shows the output from a 2-dimensional analysis of one of the Bolton model multi-span bridge containing attached spandrel walls (the bridge actually failed at a load of 325 kN). For the purpose of this analysis the walls have been modelled using blocks which extend up to the crown of each of the arch spans (where the wall is of minimum cross-sectional area and cracking in full scale tests was observed [27]). It is interesting to note that the hinges shown in Figure 8.13 are somewhat diffused in the vicinity of the wall/barrel contact positions, indicating the influence of the walls on the barrel failure mechanism.

8.4.7 Soil-structure interaction

In equation (8.1) dead load forces p that are either vertical or horizontal can be accommodated. Thus horizontal resistance forces from the backfill material may easily be included in the analysis (field and laboratory tests carried out over the last decade in the UK have shown that ultimate bridge strength is considerably enhanced by the presence of horizontal backfill pressures). However, some care must be taken in representing backfill pressures. Firstly it should be noted that the horizontal pressures applied to the extrados cannot exceed the concomitant frictional resistance. Secondly it should be appreciated that significant 'passive' pressures are normally only mobilized by relatively large structural displacements. Thus the assumption made in most mechanism analysis programs - that peak backfill pressures can be mobilized by infinitesimal structural displacements will clearly lead to non-conservative results (this problem also applies to geometrically-linear elastic analyses).

To address the problem a gross-displacement mechanism analysis can be performed. The suggested procedure [28] involves carrying out 'standard' mechanism analyses sequentially. i.e:

1. Perform 'standard' mechanism analysis of structure.

2. Magnify 'infinitesimal' displacements to determine new positions of blocks.

3. Modify soil-structure pressures as necessary.

4. Repeat from step 1.

In principle the same type of approach could be used to model a complete bridge, including abutments, so that the influence of abutment movements during loading could be accounted for. However, short of linking the rigid block analysis to a complex (non-linear elastic) soil model, this approach is likely to be problematic as prior to an analysis it would be necessary to specify the magnitude and distribution of the pressures to be applied to the back of the arch and abutments for all likely patterns of block displacements.

8.5 The use of elastic methods of analysis for masonry arch bridges

There are a number of types of problem for which the simple rigid block 'mechanism' analysis approach described in the previous section is clearly not well suited. Such problems include:

1. Determining in-service deflections and/or stresses.

2. Modelling elastic instability (e.g. snap-through).

3. Accurately modelling the soil for realistic soil-structure interaction studies.

4. Modelling fracture (e.g. between rings in multi-ring brickwork arches).

Numerous elastic analysis techniques (e.g. the finite element method [29], the discrete element method [30], etc.), have been developed over the past few decades and most can in principle be applied to solve one or more of the above problems. Because of the plethora of methods available, basic principles rather than any particular method will be considered here.

Two of the key issues associated with elastic analysis are:

(a) In order to model hinge formation and subsequent collapse the analysis must be (materially) non-linear. [Note that this is in contrast to the rigid block analysis method, where a collapse analysis is a linear problem].

(b) To properly model phenomenon such as elastic snap-through the analysis must also be geometrically non-linear (i.e. geometry from iteration n must be used for the calculations in iteration $n + 1$).

Many of the non-linear solution algorithms used in general purpose finite element packages are not robust; hence convergence may be erratic and difficult to achieve. A simple non-linear solution procedure that is particularly suitable for masonry arches, and does not require the use of complex material models, was proposed by Castigliano [12]. This procedure involves performing several linear elastic analyses and, after each, progressively removing areas of masonry containing tensile stresses until hinges and, eventually, a mechanism forms. This procedure is sometimes called a 'thinning elastic analysis' and has been widely used [31, 32]. Care needs to be taken that the algorithm implemented properly allows hinges to move freely prior to failure.

Assuming a robust solution scheme is being used, practical modelling difficulties are likely to stem from the fact that: (a) the initial stress-state is unknown, and (b) the material properties are not known. Whilst parametric studies can be performed to investigate the sensitivity of the analysis to changes in the boundary conditions and in material properties, this process can be very time-consuming, particularly when performing a fully 3-dimensional analysis. This can make an elastic analysis of a masonry arch bridge analysis tedious to perform in practice.

8.5.1 Modelling cracking

Analysts should be aware that when a non-zero value for the masonry tensile or shear strength is specified serious modelling difficulties can arise. More than two decades ago Hillerbourg [33], whilst attempting to model fracture in concrete, demonstrated that objective (mesh-size independent) results could only be obtained if suitable fracture parameters were included in the analysis (i.e. the fracture energy, shown in Figure 8.2, must be included in analysis). This finding appears to have gone largely unnoticed in the masonry community, at least until recently. The ambitious programme of experimental work carried out in the Netherlands in the 1990's (referred to in 8.2) for the first time provided fracture parameters that could be used for modelling masonry.

However, in the context of masonry arch bridges, whilst ever basic material properties such as compressive and tensile strength are unknown (and must be estimated using engineering judgement) it may appear to be incongruous that additional - more poorly understood and difficult to measure – parameters should also be required for analysis. Hence it is perhaps understandable that even now constitutive models which do not include these important additional parameters are used for masonry arch bridge analysis (e.g. [34]). However, because the results from these models will always be highly mesh size dependent, any conclusions drawn from such studies must be treated with due scepticism.

As mentioned in 8.2 there are essentially two main approaches to the modelling of cracking in masonry: 'smeared crack' models and 'discrete crack' models. In general 'discrete crack' models will provide much more reliable and realistic results, but these can be prohibitively expensive computationally when applied to large structures.

8.5.2 Example

Figure 8.14 shows output from a fully 3-dimensional 'discrete crack' model of a masonry bridge parapet subjected to an out-of-plane impact loading (it is often found when modelling masonry arch bridges that it is not the arch barrel that gives the greatest cause for concern). The general-purpose finite element code DYNA3D [35] was used, but a number of modifications [36] were required in order to make the code suitable for modelling masonry. The modifications included:

1. Inclusion of a Mohr-Coulomb type pressure dependent joint failure criterion.

2. Inclusion of an exponential softening relationship for joint failure, in tension and shear.

3. Inclusion of dilatancy during shearing at the interface (this means that during dynamic loading there is effectively an inertial clamping force normal to each fracturing joint; this allowed the experimentally observed dynamic enhancement of the bond strength to be simulated by the model).

The explicit (time-stepping) solution strategy used by DYNA3D means that the convergence problems associated with implicit non-linear finite models are not experienced. A similar (explicit) modelling approach has been successfully used recently for studies of the arch barrel [37].

8.6 Conclusions

It is important that the analyst takes a holistic approach to the modelling of masonry arches. The relative importance of each element of the structure should be carefully considered when creating a mathematical model. The contribution of spandrel walls and backfill, the rigidity of foundations and the influence of the stiffness of the road construction may all need to be included. Additionally values for the basic material

(a)

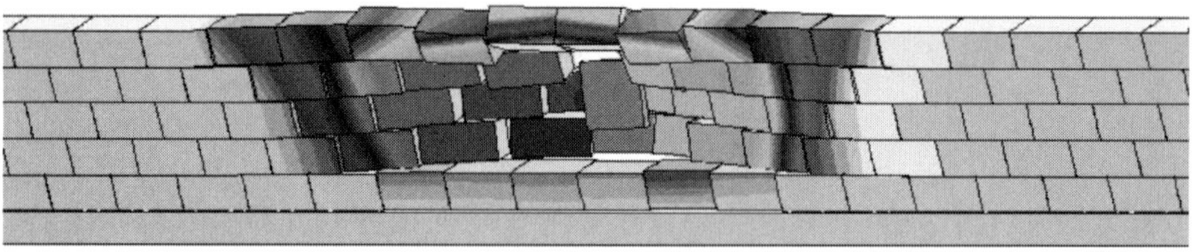

(b)

Figure 8.14: Discrete cracking models of masonry parapet wall subjected to impact loading: (a) 'large panel' failure mode of well mortared wall; (b) 'loose block' failure mode of weakly mortared wall.

properties must be estimated and an appropriate analytical idealisation made. Most 'elastic' methods suffer from the fact that the internal properties and initial stress state are uncertain whilst 'mechanism' analyses give upper bound solutions and provide no indications about in-service response. In spite of the advances in analytical tools and computer power the analysis of the masonry arch requires much engineering judgement.

APPENDIX

8.A Derivation of the transformation matrices for rigid block analysis

Constituent block j of an undisplaced arch is shown on Figure 8.15. Suppose some point is defined as the local origin of the block, for example the midpoint of the interface closest to the left-hand abutment. Using polar co-ordinates the position of O_j relative to O_{j+1} can be defined by $[b_j, \beta_j]$.

When an arch is subjected to a small (virtual) displacement, a number of hinges will form, at the intrados and extrados of the arch. The angle opened by a hinge on the extrados or intrados between block nos. $j-1$ and j of a given span are denoted as θ_j^{ext} and θ_j^{int} respectively (θ_j^{ext},$\theta_j^{int} \geq 0$). Using the assumptions of small deflection theory, and noting that the left-hand abutment is fixed in space, it can be shown that

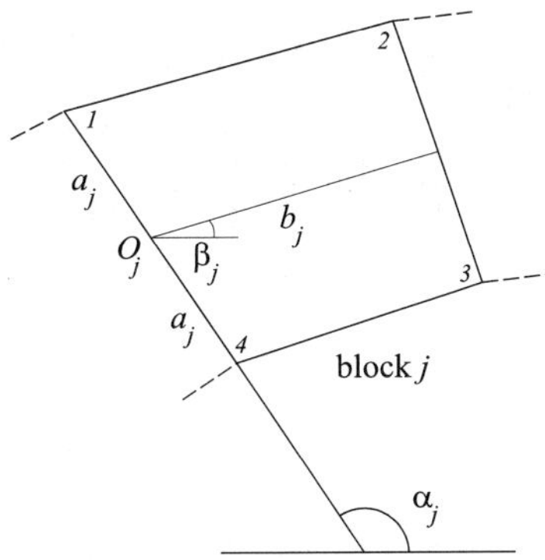

Figure 8.15: Geometry of block in undisplaced arch.

the horizontal displacement of the local origin of block j can be written as:

$$u_j = \sum_{k=1}^{k=j} a_k \left(\theta_k^{ext} + \theta_k^{int}\right) \sin \alpha_k - b_{k-1}\phi_{k-1} \sin \beta_{k-1} \tag{8.6}$$

Where:

$$\phi_k = \sum_{i=1}^{i=k} \left(\theta_l^{ext} - \theta_l^{int}\right) \tag{8.7}$$

Similarly the vertical displacement of point O_j can be written as:

$$v_j = \sum_{k=1}^{k=j} a_k \left(\theta_k^{ext} + \theta_k^{int}\right) \cos \alpha_k + b_{k-1}\phi_{k-1} \cos \beta_{k-1} \tag{8.8}$$

Thus it follows that the transformation matrix A_j will be of the form:

$$A_j = \begin{bmatrix} e_1 + f_1 + \cdots + f_{j-1} & c_1 + d_1 \cdots + d_{j-1} & 1 \\ e_1 - f_1 - \cdots - f_{j-1} & c_1 - d_1 - \cdots - d_{j-1} & -1 \\ \cdots & \cdots & \cdots \\ e_{j-1} + f_{j-1} & c_{j-1} - d_{j-1} & 1 \\ e_{j-1} - f_{j-1} & c_{j-1} - d_{j-1} & -1 \\ e_j & c_j & 1 \\ e_j & c_j & -1 \\ 0 & 0 & 0 \\ \cdots & \cdots & \cdots \\ 0 & 0 & 0 \end{bmatrix}^\top \tag{8.9}$$

Where $c_k = a_k \cos \alpha_k$, $d_k = b_k \cos \beta_k$, $e_k = \alpha_k \sin \alpha_k$, $f_k = b_k \sin \beta_k$

Figure 8.16: Relative sliding between blocks (assuming associated flow rule).

Suppose that sliding can also occur. Variables y_j^+, y_j^- control respectively the positive and negative shearing displacements of block j relative to block $j-1$ parallel to the interface between the blocks ($y_j^+, y_j^- \geq 0$); refer to Figure 8.16. Movement perpendicular to the interface (separation) is given by $s_b(y_j^+ + y_j^-)$, where s_b is termed the coefficient of dilatant friction (or separation) between blocks. The additional horizontal displacement of block j due to sliding can be shown to be:

$$u_j^{/} = \sum_{k=1}^{j} \left(\left[y_k^+ - y_k^- \right] \cos\alpha_k + s_b \left[y_k^+ + y_k^- \right] \sin\alpha_k \right) \tag{8.10}$$

Similarly the additional vertical displacement of block j due to sliding can be shown to be:

$$v_j^{/} = \sum_{k=1}^{j} \left(\left[y_k^+ - y_k^- \right] \sin\alpha_k - s_b \left[y_k^+ + y_k^- \right] \cos\alpha_k \right) \tag{8.11}$$

Thus it follows that the transformation matrix B_j will be of the form:

$$B_j = \begin{bmatrix} \cos\alpha + s_b \sin\alpha_1 & \sin\alpha_1 - s_b\cos\alpha_1 & 0 \\ -\cos\alpha_1 + s_b\sin\alpha_1 & -\sin\alpha_1 - s_b\cos\alpha_1 & 0 \\ \cdots & \cdots & \cdots \\ \cos\alpha_j + s_b\sin\alpha_j & \sin\alpha_j - s_b\cos\alpha_j & 0 \\ -\cos\alpha_j + s_b\sin\alpha_j & -\sin\alpha_j - s_b\cos\alpha & 0 \\ 0 & 0 & 0 \\ \cdots & \cdots & \cdots \\ 0 & 0 & 0 \end{bmatrix}^{\top} \tag{8.12}$$

References

[1] Melbourne C., Njumbe S.K., "Mass concrete arches", 2nd International Arch Bridges Conference, Venice, 331-340, 1998.

[2] Rots J.G., "Structural Masonry" (CUR Report 171), Balkema, Rotterdam, 1997.

[3] Lourenco P.B., "Computational strategies for masonry structures", Delft University Press, Delft, 1996.

[4] Pluijm R.Van Der, "Material properties of masonry and its components under tension and shear", Proc. 6th Canadian Masonry Symposium, Saskatoon, Saskatchewan, Canada, 675-686, 1992.

[5] Hendry A.W., "Structural Masonry", Macmillan, Basingstoke, 1990.

[6] Riddington J., Jukes P., "A review of masonry joint shear strength test methods", J. Brit. Mas. Soc., Masonry International, 11(2), 37-43, 1997.

[7] Pluijm R.Van Der, "Shear behaviour of bed joints", Proc. 6th North American Masonry Conf., Drexel University, Philadelphia, 125-136, 1993.

[8] Page A.W., "The biaxial compressive strength of brick masonry", Proc. Instn. Civ. Engrs., Part 2, 71, 893-906, 1981.

[9] Page A.W., "The strength of brick masonry under biaxial compression-tension", Int. J. Masonry Constr., 3(1), 26-31, 1983.

[10] Barlow W.H., "On the existence (practically) of the line of equal horizontal thrust, and the mode of determining it by geometrical construction", Minutes of the Institution of Civil Engineers, 5, 162-182, 1846.

[11] Navier L.M.H., "Resume des lecons donnees a l'Ecole des Ponts et Chaussees sur l'application de la mechanique a l'etablissement des construction et des machines", Part 1, 1826.

[12] Castigliano A., "Theorie de l'equilibre des systemes elastiques et ses applications", Torino, 1879.

[13] Kooharian A., "Limit analysis of voussoir (segmental) and concrete arches", Journal American Concrete Institute, 24, 317-328, 1952.

[14] Heyman J., "The stone sketeton", Int. J. Solids and Structures, 2, 1966.

[15] Pippard A.J.S., Baker, J.F., "The analysis of engineering structures", Arnold, London, 1943.

[16] Livesley R.K., "Limit analysis of structures formed from rigid blocks", International Journal of Numerical Methods in Engineering, 12, 1853-1871, 1978.

[17] Charnes A., Greenberg H.J., "Plastic collapse and linear programming", Bulletin American Mathematical Society, 57, 480, 1951.

[18] Dantzig G.B., "Computational algorithm of the revised simplex method", Rand Corporation reprint RM-1266, 1953.

[19] Drucker D.C.: "Coulomb friction, plasticity and limit loads", Transactions American Society Mechanical Engineers, 76, 71-74, 1953.

[20] Gilbert M., "On the analysis of multi-ring brickwork arch bridges", 2nd International Arch Bridges Conference, Venice, 109-118, 1998.

[21] Livesley R.K., "The collapse analysis of masonry arch bridges", Proc. conf. applied solid mechanics 4, Elsevier, 1992.

[22] Crisfield M.A., Packham A., "A mechanism program for computing the strength of masonry arch bridges", TRRL research report 124, DoT, UK, 1987.

[23] Melbourne C., Gilbert M., "The behaviour of multiring brickwork arch bridges", The Structural Engineer, 73, 39-47, 1995.

[24] Gilbert M., Melbourne C., "Rigid-block analysis of masonry structures", The Structural Engineer, 72(21), 356-361, 1994.

[25] Page J., (Editor), "Masonry arch bridges", HMSO, 1993

[26] Melbourne C., (Editor), "Arch bridges", Thomas Telford, London, 1995.

[27] Melbourne C., Gilbert, M., Wagstaff M., "The collapse behaviour of multi- span brickwork arch bridges", The Structural Engineer, 75(17), 297-305, 1997.

[28] Gilbert M., "Gross displacement mechanism analysis of masonry bridges and tunnels", Proc.11th International Brick/Block masonry conference, Shanghai, 473-482, 1997.

[29] Bathe K.J., "Finite Element Procedures", Prentice Hall, New Jersey, 1996.

[30] Cundall P.A., Strack, O.D.L., "A discrete numerical model for granular assemblies", Geotechnique, Vol 29, 47-65, 1979.

[31] Choo B.S., Coutie M.G., Gong N.G., "Finite element analysis of masonry arch bridges using tapered beam elements", Proc. Instn. Civ. Engrs., Part 2, 91, 755-770, 1991.

[32] Bridle R.J., Hughes T.G., "An energy method for arch bridge analysis", Proc. Instn. Civ. Engrs., Part 2, 89, 375-385, 1990.

[33] Hillerbourg A., "Analysis of crack formation and crack growth in concrete by means of fracture mechanics and finite elements", Cement and concrete research, 6, Pergamon, 773-782, 1976.

[34] Ng K.H., Fairfield C.A., Sibbald A., "Finite element analysis of masonry arch bridges", Proc. Instn. Civ. Engrs Structs & Bldgs, 134, 119-127, 1999.

[35] Hallquist J.L., Whirley R., "DYNA3D User Manual (v.4)", LLNL, California, 1994.

[36] Gilbert M., Molyneaux T.C.K., Hobbs B., "A dynamic finite element modelling approach for masonry structures", Proc. British Masonry Society, 8, 182-187, 1998.

[37] Owen D.R.J., Peric D., Petrinic N., Brookes C.L., James P.J., "Finite/discrete element models for assessment and repair of masonry structures", 2nd International Arch Bridges Conference, Venice, 173-180, 1998.

©2001, Saxe-Coburg Publications, Stirling, Scotland
J.W. Bull, (Editor), Computational Modelling of Masonry,
Brickwork and Blockwork Structures,
Saxe-Coburg Publications, Stirling, Scotland 221-272.

Chapter 9

Numerical Analysis of Old Masonry Buildings

F. Genna and P. Ronca

Department of Civil Engineering, University of Brescia, Italy

Abstract

We illustrate some of the alternatives existing, and difficulties to be expected when studying the behaviour of old masonry buildings numerically, especially in the non-linear range. Even if, nowadays, several quite sophisticated, continuum, discrete or interface constitutive models for masonry/mortar/dry blocks exist, the difficulty of properly applying them to real situations suggests the use of simpler approaches, such as elasticity or elastoplasticity as a starting point. Even in these simplified contexts, however, the analysis of old masonry poses several problems, some of which are illustrated by means of application examples, first with reference to the analysis of masonry walls, then with reference to the analysis of arches and vaults. The importance of properly taking into account the effects of design and construction details on the overall behaviour is stressed, as well as that of complementing the numerical results with analytical, simplified estimates of the structure behaviour, such as those given by the methods of the thrust line or limit analysis, when applicable.

9.1 Introduction

The structural members of old masonry buildings are essentially either plane walls, loaded in their own plane, or arches and vaults; both types of structures are usually designed to support vertical loads only. Today's engineers are often confronted with the task of studying, to the best of their ability, the behaviour of structural members of old masonry buildings, and, in doing so, they must face at least two very different situations. The first is the need to check the safety of the structural parts against their original design loads; the second occurs when an old, possibly historical building happens to be in a seisimc zone. Then its safety against horizontal loads must be checked; engineers are often called upon to undertake the task of assessing the causes of dramatic collapses.

The features of masonry, seen as a material, make it necessary to resort to numerical analysis to try and understand its actual behaviour. In this chapter we will try and summarise what can be done by applying computational techniques to study old masonry buildings; in doing so, we will make reference to the Finite Element Method [1] as the standard numerical tool, even if, with the aim of modelling the complex texture of masonry, other techniques might be employed, such as, for instance, the Boundary Element Method [2, 3]. The Finite Element Method (hereafter shortened to FEM), however, has become quite popular and is much more general than most other techniques. Here we will assume that the reader is familiar with the essentials of such a method, which can be found discussed in great detail in [1].

We must underline that, in any case, when dealing with the task of interpreting the result of a FEM analysis of an old structure, a mosaic of different assumed structural models and analysis methods have to be compared with each other. Our experience in this field strongly suggests that there are great structural resources behind the apparent weakness of the material. As a consequence, masonry structures can reach different, sometimes unpredictable limit states, with different safety coefficients. Such complexity of behaviour requires a broad spectrum of numerical analysis tools, including extensions of simplified methods proposed in the past both for the elastic and for the limit analysis [4].

If one wants to categorise the different structural problems, with reference to the relevant analysis techniques, one might obtain the following picture.

1. Masonry walls loaded in their plane.

   Very little can be done, analytically, to understand their elastic behaviour, whereas limit analysis tools can usually be employed to obtain lower bound estimates of the collapse load [5]. Numerical methods may provide information both about the elastic behaviour and about the inelastic and limit behaviour, at the cost of meeting several difficulties, some of which will be illustrated in this chapter.

2. Arches and vaults.

   Here it is possible to perform several types of analytical investigations, both in the elastic and in the inelastic ranges. It is in fact possible to study the elastic and post-elastic behaviour of arches by means of the "thrust line" method, which gives information about the structural behaviour up to collapse. The techniques of limit analysis can be employed to obtain, with relatively little effort, upper bounds to the collapse load. However, it is often difficult to take into account, in the analytical models, details such as the presence of spandrel walls. The numerical analysis is a useful alternative, which should complement whatever results can be obtained from hand computations.

3. There are other types of structural problems which may be encountered when dealing with old buildings, but which are usually not very significant, owing to the design and construction modalities. One of them is the analysis of box-like structures, in which the geometry and loading conditions force the engineer to consider an assembly of walls as a fully three-dimensional structure. In most cases the walls are studied as isolated from each other, under the assumption that the main loading acts in their own plane. However, this might not be true; although we will not consider such cases here, the reader can find some information about the modelling of masonry walls loaded as plates in [6].

4. Other topics of interest in the analysis of masonry structures are the stability of compressed columns (discussed, with reference to new masonry, in [7–9]), and the structural strength under seismic actions. This last topic has a special relevance in the context of this chapter, as hinted before, since masonry buildings, be they old masterpieces such as gothic cathedrals or of minor architectural relevance but of great historical importance anyway, were not built to withstand horizontal actions. Finally, a very specific but important field of work is the assessment of the load bearing capacity of old masonry when the restoration design implies a new use for the buildings (such as in the case of museums, etc.), and therefore heavier loading than for the original structure.

One can never stress enough the difficulty of performing numerical analyses in the nonlinear range, especially in the presence of brittle behaviour, with the aim of obtaining reasonably accurate information and possibly useful from the engineering viewpoint. As already said, the results of numerical analyses should always be complemented by results given from both different structural models and analytical computations, even if the latter can provide only a crude estimate of the desired results.

This warning is particularly important in the investigation about an old building, whose behaviour depends in a very complex way on all the usual basic ingredients – most notably the material constitutive properties – but also on the construction details, very often unknown to us. Both the difficulty of defining accurate constitutive laws for masonry, and the difficulty of modelling geometrical details which have great importance for the overall structural behaviour, are reasons which have so far substantially limited the progress of research and understanding in this field. When dealing with old masonry buildings, both the researcher and the practising engineer find very little reference in the literature, very little experimental data, and substantial difficulties in modelling the problem. Here we will try and examine the possible alternatives among which one has to choose when one has to analyse an old masonry building by means of a FEM code. We will then illustrate, by means of examples concerning both walls and arches, some results obtained by means of the numerical analysis methods of old masonry buildings.

9.2 Issues in the numerical modelling of old masonry

9.2.1 Geometry of the numerical model

When dealing with any real solid, and in particular with masonry walls, the first problem is whether to use a two-dimensional or a three-dimensional model. When studying arches or vaults there is a double alternative: between a continuum (with the choice between solid two- or three-dimensional elements) or a structural (choice between beam and shell elements) approach.

If one considers the extremely large amount of simplifying assumptions underlying any such analysis, one is led to use the simplest possible model, but this might not always be a feasible alternative. In the writers' experience, continuum elements are to be preferred over structural ones, even if they provide information sometimes difficult to interpret.

9.2.2 Discrete numerical model

Masonry is a heterogeneous material, made by two components exhibiting very different material properties. Often old structures are built using either blocks or something closely resembling dry masonry, in which shear forces are transmitted between blocks by friction only. The numerical analysis of similar structures can be approached from several different viewpoints.

The first is to consider masonry as a homogeneous material, at the scale of the full structure, governed by a suitable number of elastic constants (see, for instance, [10] and references therein quoted) and a suitable, more or less "standard" elastic-plastic model for describing its "homogenised" behaviour in the nonlinear range. We will come back later to this last topic, crucial to this chapter (see references [11–16]).

An approach for nonlinear analysis, similar to the first, but conceptually more refined, takes into account the physical sources of nonlinearity and rupture in masonry, even if, from the numerical viewpoint, the discrete model still considers masonry as a homogeneous material. Examples of models based on this approach can be found in [17].

More refined techniques imply the discretisation of the full geometry (so called "micromodels"), taking into account the single blocks and mortar beds. Such an approach seems not to be feasible for practical purposes, owing to the burden of the discretisation and to the size of the resulting problem. However, several researchers have studied the behaviour of joints between masonry blocks, and have tested their models against experimental results by means of fully discretised models, in which each single homogeneous part is discretised independently. Some references to work of this type can be found in [18–24]. The main application of these models is clearly limited, at the present stage of hardware development, to the calibration of the models actually used in the analysis of large structures.

Finally, some researchers have developed specific models for dry masonry, in which the blocks are treated as rigid (so called "discontinuous deformation analysis"). This way of working might have a strong relevance in the field of old structures; examples of these models can be found in [25–27], with the main reference to limit analysis.

9.2.3 Loading and boundary conditions

Both the analysis of masonry walls and the analysis of arches and vaults often imply difficult decisions concerning the modelling of boundary conditions and loading. We don't even consider the usual difficulties that arise, for instance, when reducing a three-dimensional structure to an assembly of plane parts. Specific problems arise, for instance, from the modelling of the usually unknown foundation structures – if there are any – which may have a really important effect on the supported structure. It is very difficult to know exactly how the structure transmits loads to the ground, if foundation settlements have taken place, and, if they seem to have occurred, to estimate their significance. Similar problems arise when trying to understand what types of loading a structure has withstood during the centuries; in particular, it is almost impossible to take into account the effect of atmospheric agents, such as wind, snow, temperature and humidity changes, etc.

9.2.4 Choice of formulation and finite elements

Once the engineer has decided his model, he must also decide which FEM code he wants to use. The choice of a specific FEM code implies both the choice of the problem formulation and the available options about finite element and material model choices. We will discuss the topic of the material model later; here we want to point out that there are several different finite element formulations and implementations, both with reference to the basic variational approach and with reference to solution techniques, especially in the nonlinear range.

The "standard" approach – the one most commercial codes adopt – is based on a displacement formulation, and the use of Newton-type solution schemes for tackling the nonlinear cases. Here the advantage lies in the generality of the approach, which can handle various different cases (although if one wants to perform limit analysis or to follow unstable equilibrium paths one needs to resort to some modifications of the basic technique). Such formulation, however, encounters problems in dealing with vector-valued yield surfaces (or with singular points in the yield surface, such as the vertex of the Drucker-Prager cone in the principal stress space); other problems arise when trying to do limit analysis, unless specific solution algorithms, such as the Riks method, are available. Also, a displacement formulation is defined, in the nonlinear range, in such a way as to prescribe only the satisfaction of the equations written in finite form at discrete intervals of the loading history; it never deals with the rate problem, which is the only one capable of giving accurate information about both the actual equilibrium path and all the possible critical points lying on it.

Other approaches have been tried, such as, in the linear elastic range, a stress formulation ("equilibrated" finite elements), or mixed formulations; these do not seem to have had much success. In the nonlinear range, one formulation particularly suited to deal with the rate plasticity equations is that proposed by Maier [28], based on plastic strain increments. Such a formulation allows one to obtain really accurate results, both for the standard elastic-plastic case, whatever be the yield function, and for the softening case [29,30]. Among its advantages is the possibility of performing a "real" limit analysis (as opposed to what is usually done with a standard formulation, in which collapse is defined as lack of convergence of the global equilibrium equations), as well as "real" tests on the stability of the equilibrium path [31]; as hinted, such an approach enables one to deal with vector-valued yield functions with no difficulty.

These aspects suggest that such a formulation should be preferred to a standard formulation, when dealing with structures made of brittle materials, usually in the range of small displacements and often with the aim of estimating the safety factor against a particular set of loads. Codes based on this formulation are commercially available, even if in a very small number [32].

Once the geometrical model and the formulation to be used are defined, one has to decide what type of finite element is best suited for the analysis. Here we are concerned mainly with nonlinear behaviour, and care must be taken, when choosing finite elements, in several respects. If masonry is modelled as a homogeneous continuum, there is a large choice of elements; however, if one is interested in performing an analysis up to collapse, one must prefer a selected group of elements, i.e., those with the ability to correctly catch the collapse mechanism. There are several examples of very poor performances of standard finite elements in situations of diffused yielding, even in the relatively simple case of metal plasticity. In [33] an example of poor element performance is given for the case of the so-called "no tension" constitutive model. In general, for continuum elements, the basic four-noded elements, either fully integrated or with reduced integration, should not be used in analyses pushed close to structural collapse. Even constant stress triangles may prove troublesome. In the authors' experience, a good element, in this respect, is the underintegrated eight-noded quadrilateral, able to correctly represent most collapse modes, even those implying strain rate concentration along thin bands.

Analogous care must be exerted both in the modelling of bricks and mortar as separate constituents and in the choice of interface elements. Also the use of structural elements implies several sources of possible troubles which, however, are too specific to be treated in detail here.

9.2.5 Choice of the constitutive model

The choice of the constitutive model, together with the choice of the material parameters, discussed later on, is the most difficult problem in the case of the analysis of old masonry buildings. Here one has to deal with several different, and conflicting, alternatives, and must decide whether one prefers simplicity and low cost, or sophistication – whatever this can mean in this field – coupled with high cost and uncertainty about the results.

Already in the linear elastic range, if masonry is treated as a homogeneous material, one has to decide whether or not to consider orthotropy. It is not difficult to include this feature into the elastic analysis; the real problem is the evaluation of the required material parameters, whose uncertainty might nullify whatever improvement one seeks by considering orthotropy.

In the nonlinear range there are several categories of constitutive laws from which to choose, namely (i) standard plasticity, (ii) damage, possibly coupled with plasticity, (iii) viscoelasticity, (iv) softening and (v) nonlinear fracture mechanics models. The choice of a very sophisticated constitutive model has meaning when the main objective of the analysis is to follow with the best possible accuracy the post-failure behaviour at all points of a structure. Usually softening or brittle fracture models are developed to simulate results of uniaxial tests on a very simple geometry; in some cases the main interest is focused on the joint behaviour alone. In dealing with a full scale old structure, however, it is impractical to try and catch all the equilibrium path details, putting too much emphasis on post-cracking behaviour. One should consider oneself happy if one can correctly predict the onset of global failure and, to this purpose, a standard plasticity model may be accurate enough. As we will comment later, the use of sophisticated models usually implies the definition of several material parameters which are anyway difficult to obtain from experiments, and almost impossible to determine for old masonry. Therefore, even if we give a brief summary of the recent contributions to the field of constitutive models for masonry, we feel that, for the analysis of old buildings, one should start with the simplest, that is elastic-plastic laws.

Several standard elastic-plastic constitutive laws have been developed to model "brittle behaviour", starting with the simple "no tension" law, described and used, for instance, in [11] and [34]. Other plasticity models commonly used in analysing masonry are the Drucker-Prager and Mohr-Coulomb laws, possibly with modifications, such as those described in [17]; the simple Galileo-Rankine model, used in [35]. Even the von Mises constitutive model has sometimes been used, such as, for instance, in [21]. Quite popular is the use of a generalised "no tension" model, with limited compressive strength, described in terms of bending moment and axial force in [4] with reference to the limit analysis of arches. More refined plasticity models have been proposed in the framework of the homogenisation theories, such as those described in [13, 25].

Damage constitutive models have only recently been used for studying masonry. Examples can be found in [12, 14] and [20]; in the last work the damage equations have been introduced to describe the behaviour of joints between blocks.

Creep has been recognised to be a possible cause of collapse by some researchers. We are not aware of any constitutive model explicitly developed for the numerical simulation of creep in masonry, but we can suggest reading reference [36], where the sudden collapse of the Civic Tower of Pavia, Italy, has been ascribed primarily to strength degradation due to viscous effects, triggered by the high permanent stresses acting upon the structure. Other work on the effect of aging on the mechanical behaviour of masonry can be found in [37–39].

Softening models, as a tool to describe fracture in a plasticity framework, have received attention in recent times. Their use is in any case quite difficult, owing to the inherent sources of instability; moreover, their careless use might produce pathologically wrong results. The theory underlying these difficulties is at the present stage well understood, and equations explicitly derived for masonry, and implemented into FEM codes, are being published in increasing numbers. Examples of these models can be found in [15, 16, 19, 40].

The importance of fracture mechanics concepts, when trying to reproduce the phenomenon of brittle rupture, which fully characterises the ultimate behaviour of masonry, has been recognised since the early seventies. It was the emergence of this discipline that led to the understanding of the puzzling results sometimes arising from the more standard softening analyses. Nowadays several FEM codes incorporate suitably designed elements and constitutive models able to follow crack initiation and propagation in a number of significant situations. A recent review of these models can be found in [41]. Some work has already been done to investigate the significance of fracture mechanics concepts in the field of the analysis of masonry structures; reference to such work can be found, for instance, in [42–44].

There is another category of constitutive models proposed for the study of masonry, based only on so-called "phenomenological" considerations, and which simply try to somehow fit experimental data with suitable analytical expressions. We feel that such approaches lack both a strong mechanical foundation and imply the use of several material parameters quite difficult to identify; therefore we will not treat them in any further detail.

As one can easily appreciate, there is indeed a huge variety of proposed models from which to choose. Each has advantages and disadvantages; it is safe to say that the most sophisticated constitutive laws are, at the present stage of development of the available hardware, essentially suited for academic work only. The task of following the initiation and propagation of several micro and macrocracks in a large structure, where every detail can be a source of cracking, is obviously out of reach of today's computers, even if such a goal is of primary importance. This consideration, coupled with the extreme difficulty one encounters in trying to define the material parameters (see next subsection), leads to the use of less refined, but more feasible choices. Standard elastic-plastic analyses clearly cannot model brittle fracture, but can nevertheless give useful indications about the state of the structure; we will see that even in the case of the simplest material models the choice of material parameters is very difficult.

In the following sections we will therefore illustrate some analyses performed on masonry structures by means of standard plasticity models only, with the aim of showing what type of results one can obtain, and what is the order of magnitude of the various material parameters involved by the models used. To conclude this subsection, however, we will point out two further sources of choice, with reference to constitutive modelling. The first concerns the flow-laws to be used in rate plasticity equations: the choice here is between associated or non-associated flow-laws, the latter being more realistic but more difficult to implement and use, owing to the loss of symmetry of the governing equations and to the introduction of further sources of instabilities into the equilibrium path.

The second is the alternative between the use of a rate formulation and the use of a nonlinear elastic (holonomic) one, assuming that the available code gives the possibility of choosing between the two. The use of a reversible model, written in terms of finite quantities, is definitely convenient when performing limit analysis calculations, since the collapse load factor in limit analysis is independent of the actual loading path, and a holonomic analysis is much cheaper than the integration of the rate equations. Also, if one is interested in estimating cracking by means of elastic-plastic equations, one might argue that fracture can be modelled – although only very crudely – by means of *reversible* plastic strains, therefore concluding that a holonomic model would be fully adequate to the purpose.

9.2.6 Choice of the material parameters

Once the analyst has decided that a specific constitutive law offers the best compromise for his purposes, he must face the problem of feeding into the model the material parameters data. One has to bear in mind that, in any case, either in the analysis of masonry vertical walls or in that of curved elements, the variety of construction modalities, such as the ratio between the thickness of bricks and mortar layers, or the inhomogeneity of the masonry texture in the thickness of the structural element, has a great influence on the overall "macroscopic" values of material parameters (see, for instance, [45,46]).

In the field of old masonry there is very little available information, for the obvious reason that, in order to test the material, one must inflict permanent damage to old, possibly historical buildings. There are some non-destructive tests which might be employed to obtain data about the stiffness of old masonry; the most used is the flatjack technique, originally applied to measure the *in-situ* stress level, but then extended to determine the Young modulus of the material. The results given by this technique are unfortunately extremely difficult to interpret; recent work has suggested a way to depurate the effect of plastic strains induced by the test modalities [47,48], but the conclusions still need standardisation.

A review of both constitutive models and of material parameters data for masonry, although in most cases with reference to new masonry, can be found in [49], and in the references herein quoted. We are aware of several works concerning the non-destructive analysis of chemical properties of old masonry, with reference to both brick and mortar deterioration (see, for instance, [50–52]) but of almost no work at all about the experimental analysis of mechanical properties of old masonry. A few exceptions are to be found when, as said, structural collapse allows researchers to find material to test, such as in [36], or when extensive restoration both requires and is a justification for extracting samples of material from old buildings, such as illustrated, for instance, in [53].

Even in the case where material samples are available for testing, it would be very difficult to obtain most of the parameters required by sophisticated constitutive models, such as, for instance, those illustrated in [14] or in [40]. To this purpose one might resort to identification techniques, such as discussed in [54], with specific reference to the model described in [14]. Such techniques, however, are still in too early a stage of development and testing, at least with reference to masonry and masonry-like materials, to be considered as a feasible tool for use by practising engineers. This alone is a sufficient reason to justify resorting to standard elastic-plastic models, whose strength parameters – usually uniaxial strengths, or cohesion and internal friction angle – are difficult enough to obtain from experimental tests.

The few examples illustrated in the following sections will show that, in order to obtain predictions comparable to observed fracture patterns, both the stiffness and the strength parameters to be adopted can have very low values, especially so if compared to values obtained from laboratory tests on masonry samples. An example of this can be found in [55], where a FEM analysis of the dynamic behaviour of the Parma Cathedral Bell-Tower is reported. There a Young modulus $E = 300$ MPa is used, together with a Drucker-Prager plastic constitutive law, with strength in compression $\sigma_c = 10$ MPa and strength in tension $\sigma_t = 0.5$ MPa. In [55], however, no indication is given about the source of such data.

9.2.7 Other issues

The reader may have noticed that a good portion of the available literature concerning old masonry buildings comes from Italian researchers. This happens for the good reason that Italy has plenty of such structures always undergoing the risk of annihilation due to earthquakes, if nothing else. The Italian School has studied the behaviour of masonry-like materials for a long time, both from the mechanical and the chemical viewpoint. While the latter is outside the scope of this chapter, about the first we might observe that the Italian School seems to have insisted on the issue of studying old, cracked masonry by means of the no-tension constitutive law (see, for instance, [3, 11, 34, 56–58]. In later work it has been shown, however, that the use of no-tension models can lead to significant errors in the prediction of the cracked zone, both if compared to results given by different constitutive models and if compared to surveyed cracking patterns [33, 59, 60]. We will show some of the results reported in [60] in section 9.3.1.

It can also be pointed out that often FEM analyses results are difficult to interpret from a practising engineering viewpoint. The situation is relatively simple for arches and vaults, whose structural nature allows one to define generalised variables such as bending moments and axial forces which can be easily represented even as a result of a continuum FEM model. For two-dimensional structures, such as masonry walls, one usually sees a lot of colour contour plots of stress components, or simply deformed shapes. These pictures give a qualitative idea of where stresses concentrate, but are hardly easy to use as a design tool. A better idea can be obtained from illustrations of the so called isostatic lines, showing the direction of principal components of tensors at some points of the structure. Not all the computer codes can draw such lines.

9.3 Analysis of masonry walls

In this section we summarise results obtained from the analysis of two different masonry walls, one taken from a XVI century cloister in Brescia, Italy, the other taken from a XIV century church in Verolanuova, Italy. Both walls had structural problems, which led to extensive cracking and, eventually, to the need for restoration. Both analyses have been done considering the situation both before and after the consolidation work; the first analysis has also been done with the specific purpose of comparing results given by several different elastic-plastic constitutive models.

Here there is not enough space to write down and explain all the equations describing each model; for this reason the reader is encouraged to go to the original published work which, at least for the first case, is in English [60]. Here we will try and give a brief summary of the main relevant results, and the conclusions which can be drawn from them.

9.3.1 A wall of the San Faustino cloister in Brescia, Italy

The wall is part of the monumental complex of the San Faustino Maggiore Monastery in Brescia, Italy.

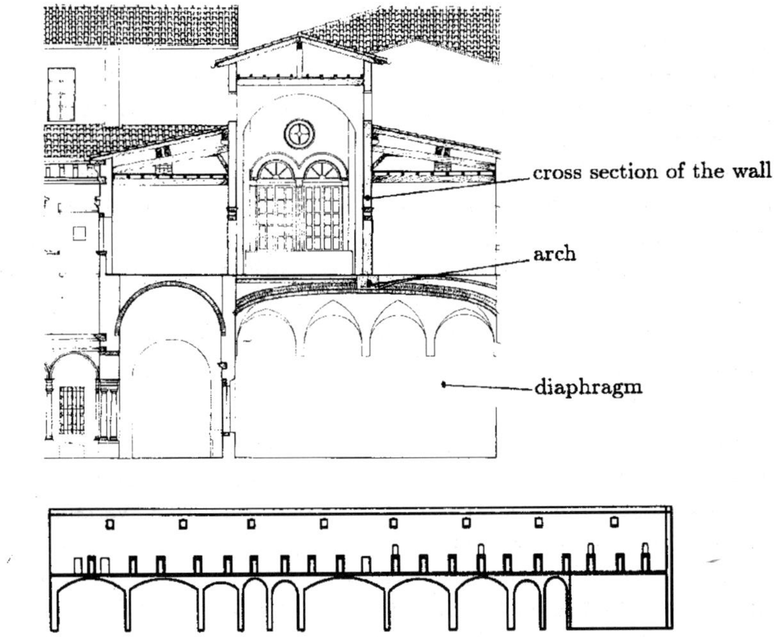

Figure 9.1: Cross section and lateral view of the San Faustino cloister wall.

Figure 9.1 shows a view of the structure, a masonry wall about 87 m long, 8.25 m high and 30 cm thick. The wall exhibits a static anomaly, in that it does not reach the ground, but it rests on both transverse diaphragms and on low rise arches, 60 cm thick, which also support the ceilings of the friars' cells at the ground floor. All these components are made of clay masonry; the average brick size is $28 \times 12 \times 6$ cm and the thickness of the mortar layers ranges from 1 to 2 cm. The main construction for this part of the complex was completed by the end of the XVI century [53].

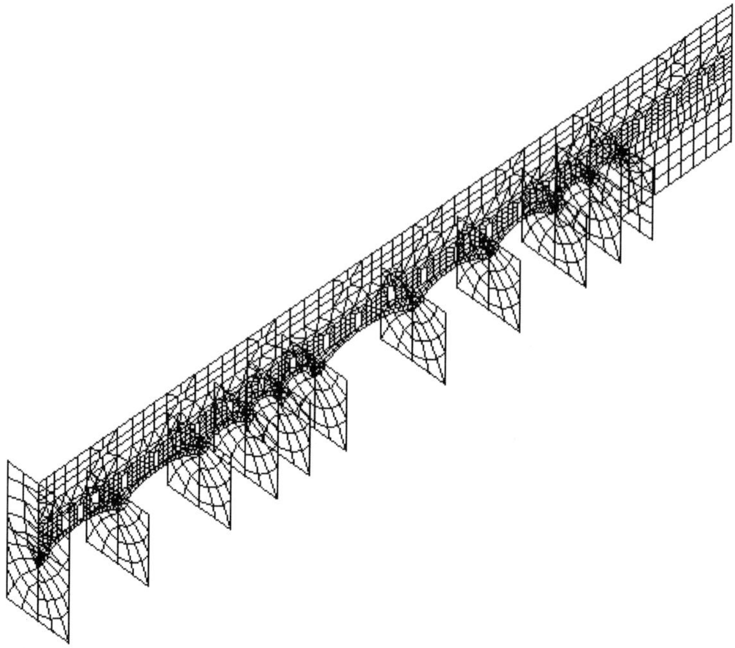

Figure 9.2: Three dimensional mesh of the San Faustino cloister wall.

Figures 9.2, 9.3 and 9.4 show different models analysed. The first two models describe the full wall; the third model has been chosen as representing the portion of the structure exhibiting the worst cracking pattern, and at the same time being of a size which allowed the performing of several nonlinear analyses. In the model of Figure 9.4 the three main structural elements – wall, arch and supporting diaphragms – have been physically separated, and suitable finite elements have been introduced at their interfaces, in order to allow the simulation of the interface behaviour. The real interface between the elements is made by either a simple support with interposed mortar (wall on arches and wall on diaphragms) or by clamping, brick by brick, in the lateral joints between wall and diaphragms. Such modelling is essential to avoid the

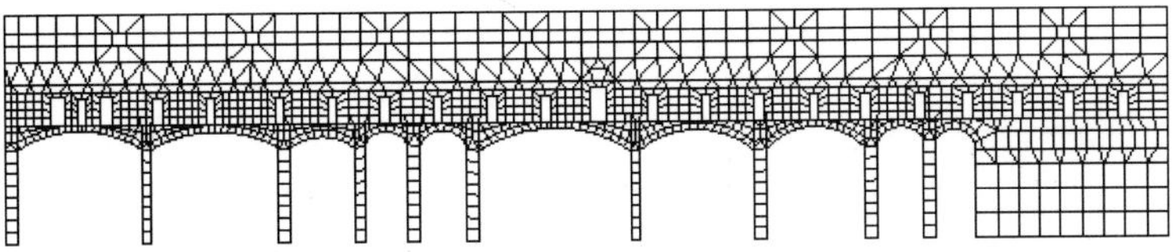

Figure 9.3: Two dimensional mesh of the San Faustino cloister wall.

prediction of an excessive collaboration between the structural components.

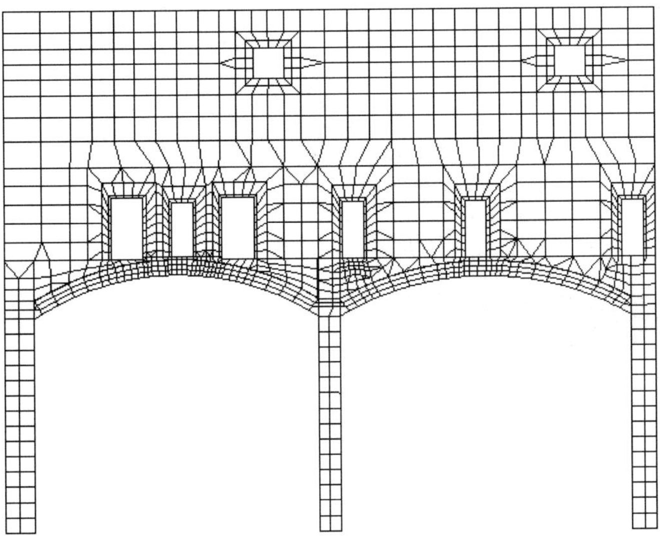

Figure 9.4: Two dimensional mesh of the studied part of San Faustino cloister wall.

The material has been assumed to behave as elastic-perfectly plastic, with the exception of the supporting diaphragms, whose main stress components lie in their plane, i.e., in the plane orthogonal to that containing the plane model of Figure 9.4. With such a model it is worthless to try and catch any nonlinear phenomenon in the diaphragms, which have therefore been considered as linear elastic. The flow-laws have always been considered as associated; this should not create many problems for most analyses, whereas it might lead to inaccurate results in the case of limit analyses.

The loading condition has been dead load (self weight and weight of the various roof slabs resting on the studied wall), as well as live load given by snow. All the details concerning these data can be found in [60].

The material models used in the analyses are as follows:

1. Mohr-Coulomb friction law with cohesion, for the interface elements;
2. Galileo-Rankine elastic-plastic model;
3. no-tension elastic-plastic model;
4. Drucker-Prager elastic-plastic model;
5. a tension cracking constitutive model, as described in [30];
6. modified De Felice elastic-plastic constitutive model, as described in [25] and modified as explained in [60].

The choice of material parameters in the elastic range has been somewhat guided by experimental results, performed on small portions of the building obtained during the consolidation work. A typical force-displacement laboratory curve is shown in Figure 9.5a, with the corresponding stress-strain curve shown in Figure 9.5b. From this data one can obtain a linear regression fit for the Young modulus, giving the value $E \approx 4550$ MPa, and an estimate of the compressive strength as about $\sigma_c \approx 2.5$ MPa.

In the analyses we have used the following sets of data.

- Interface elements. In a first group of analyses, they have been treated as linear elastic, in order to check only the effect of modelling of the continuum elements. After these analyses we have reduced to two the set of plastic models to be used for the continuum, and have modelled also the interfaces as nonlinear. After several preliminary tests, we have decided to use a zero cohesion value, and a friction angle $\Phi = 50°$.
- Wall masonry. Young modulus $E = 5000$ MPa, Poisson coefficient $\nu = 0.2$; the material has been treated as isotropic.
- Galileo-Rankine model. Strength in tension $\sigma_t = 0.3$ MPa; strength in compression $\sigma_c = 1.0$ MPa.
- Drucker-Prager model. Cohesion $k = 0.23$ MPa, friction angle $\Phi = 56°$; these data correspond to $\sigma_t = 0.2$ MPa and $\sigma_c = 1.5$ MPa;
- Tension cracking model. Uniaxial strength in tension $\sigma_t = 0.3$ MPa; fracture energy $\mathcal{G}_f = 0.02$ N/mm. This last value has been guessed, in the absence of other evidence, as about 1/5 of that suggested for plain concrete. The choice of a low value for the fracture energy, together with the large size of the structure, should magnify possible local instability phenomena due to sudden elastic energy release when cracking occurs.

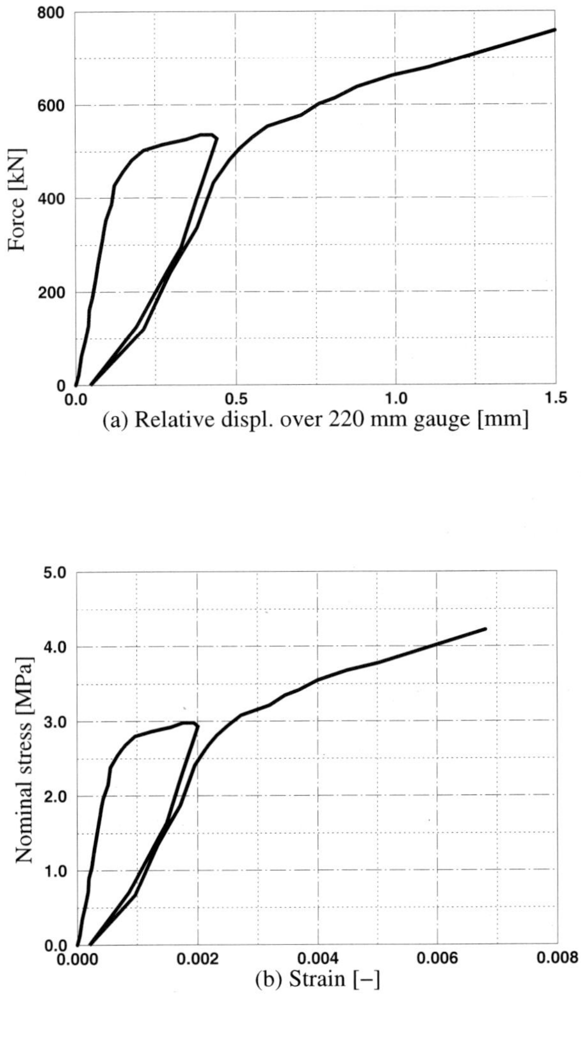

Figure 9.5: Experimental results of a uniaxial compression test on a portion of the San Faustion cloister wall. (a) shows force vs. displacement; (b) shows stress vs. strain.

- De Felice model. In the wall: cohesion $k = 0.2$ MPa, friction angle $\Phi = 45°$,

$\sigma_t = 0.2$ MPa, $\sigma_c = 1.5$ MPa, ratio m between height and width of a block $m = 0.2$. In the arches: cohesion $k = 0.18$ MPa, friction angle $\Phi = 45°$, $\sigma_t = 0.2$ MPa, $\sigma_c = 1.5$ MPa, ratio m between height and width of a block $m = 5.0$.

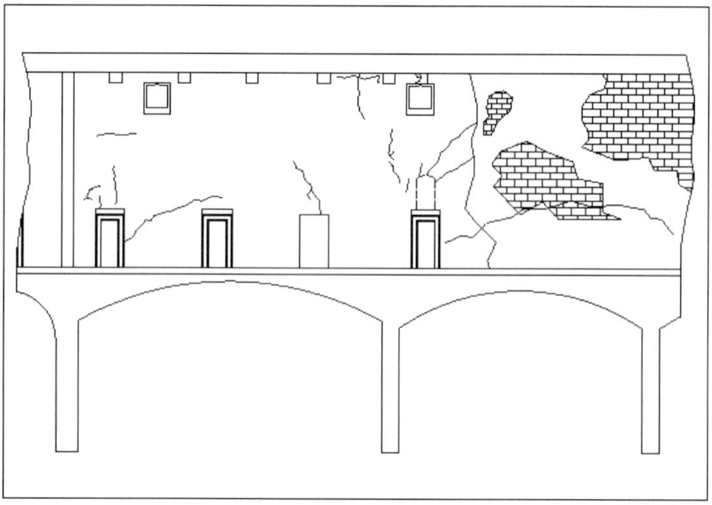

Figure 9.6: Surveyed crack pattern in a portion of San Faustino cloister wall.

All these values have been individuated after several numerical tests, and after comparison of the predicted cracked zone with surveyed results, such as those shown in Figure 9.6.

The numerical results are summarised in Figures 9.7 to 9.10. Figure 9.7 compares results of the analyses performed considering linear elastic interfaces, for the 4 constitutive model used for the continuum elements. Figure 9.7a shows plastic strains for the Galileo-Rankine model, Figure 9.7b for the Drucker-Prager model, Figure 9.7c for the no tension model and Figure 9.7d for the cracking model. The most striking feature here is the great difference between the no-tension results and the three others, which exhibit more or less similar features. The no-tension model predicts displacements about three times larger than the other models, and compressive stresses about two times larger than the other models. Also the geometry of the predicted "cracked" zone is quite different, in Figure 9.7c, from all the others. These results confirm what has already been found in [33, 59] about the possible lack of meaning of results given by a no-tension model when applied to the analysis of large brittle structures. Similar conclusions, but based on completely different reasons, have been reached in [43].

Also apparent from the results shown in Figures 9.7 is the excess of collaboration predicted at the interface between wall and arches, which gives rise to unrealistic "yielding". Figures 9.8 and 9.9 refer to an analysis in which wall and arches are mod-

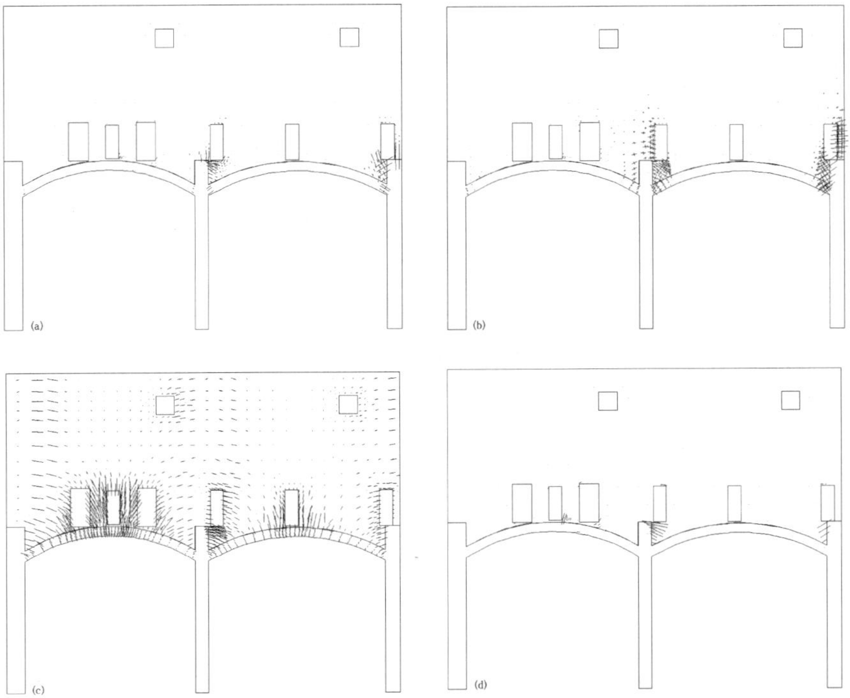

Figure 9.7: Predicted plastic strains for 4 constitutive models: (a) Galileo-Rankine; (b) Drucker-Prager; (c) no tension (d) cracking model.

elled by means of the Drucker-Prager law, and the interfaces are treated as nonlinear, as described above. Figure 9.8 displays the displaced configuration (9.8a), which shows the detachment between some portions of the wall and arches, the predicted plastic zone (9.8b), and the isostatic lines of tension (9.8c) and compression (9.8d). These last confirm, in terms of stress distribution, what is visible in Figure 9.8a. The load applied just above the arch in the left portion of the wall cannot be supported by the arch itself, and the wall above behaves like a simply supported slab subjected to distributed load applied to its lower edge. On the contrary, the right portion of the wall still rests for the most part on the arch, and its stress pattern resembles that of a slab with distributed load applied on the top. Figure 9.9 shows contour plots (originally in colour) of the principal stresses. Figure 9.10 shows, in the same way as before, the results obtained using the De Felice model for the continuum elements. The predicted plastic zone and stress distribution are quite similar to those given by the Drucker-

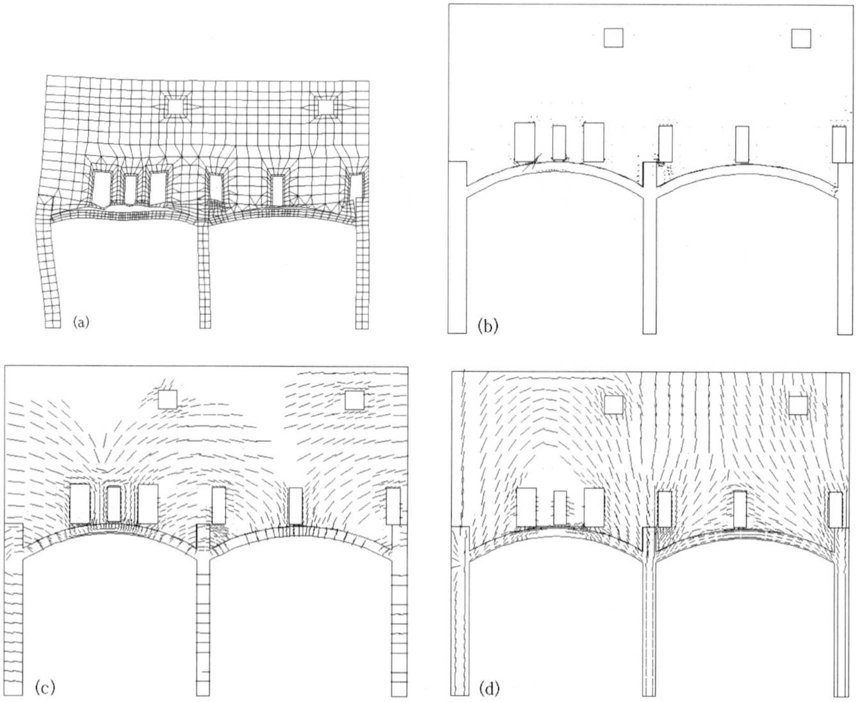

Figure 9.8: Results of the analysis with nonlinear interfaces, Drucker-Prager law: (a) deformed shape; (b) principal plastic strains; (c) isostatic lines of tension; (d) isostatic lines of compression.

Prager model. Both analyses indicate that the arches undergo some "cracking" at their keystone intrados and, in the case of the De Felice model, some "cracking" also at the abutments. Other "cracking" is predicted at the corners of the openings in the wall, but this should be interpreted as a local phenomenon, not significant from the overall stability viewpoint. The other important nonlinearity is of course localised at the interfaces between wall, arches and supporting diaphragms, where significant detachment and sliding is predicted.

The less expensive analysis here is that performed using the De Felice model, which, unlike all others, is a piecewiselinear one. The FEM code used is STRUPL-2 [61], based on the LCP formulation of rate plasticity for discrete structures introduced in [62]. The computer time required by the analysis based on the De Felice model is about 3 percent of that required by all the other analyses. The fact that the obtained results are almost coincident with the others indicates this model as very suited to perform elastic-plastic analyses within this context. The reasonable cost of analyses based on the De Felice model also allows one to perform collapse analyses,

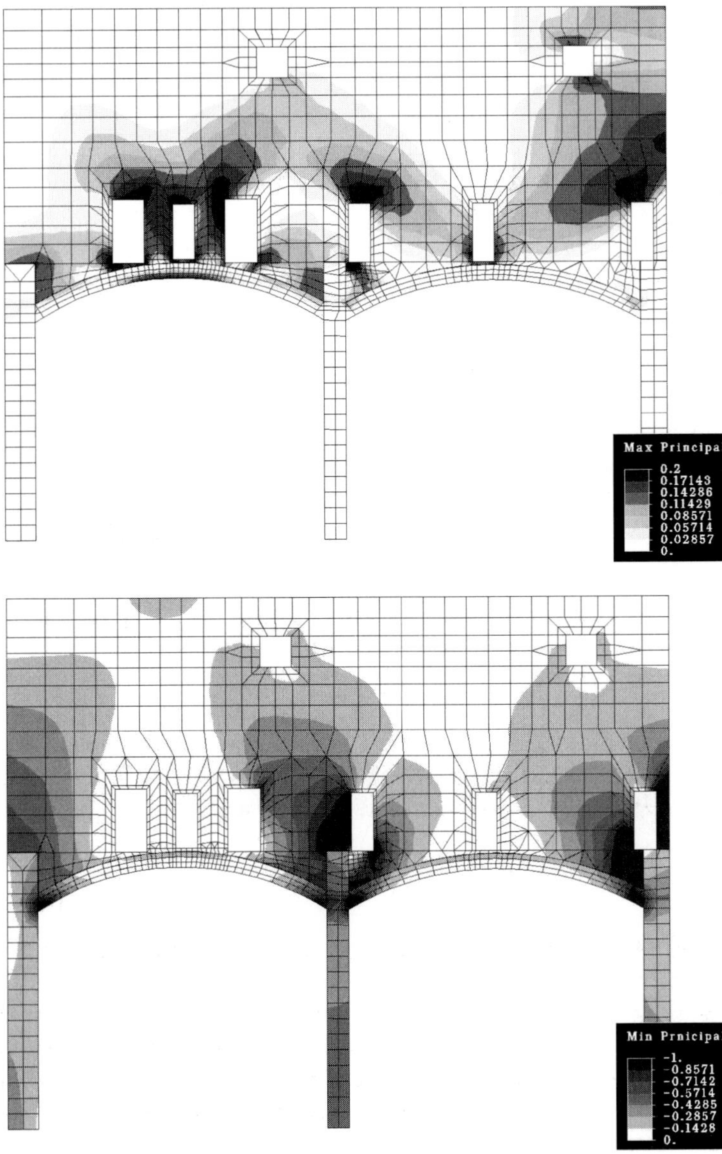

Figure 9.9: (a) Maximum and (b) minimum principal stresses for the analysis with nonlinear interfaces, Drucker-Prager law.

which usually require considerably more time than a standard incremental analysis, owing to the need for detecting both a singularity in the matrix governing the rate

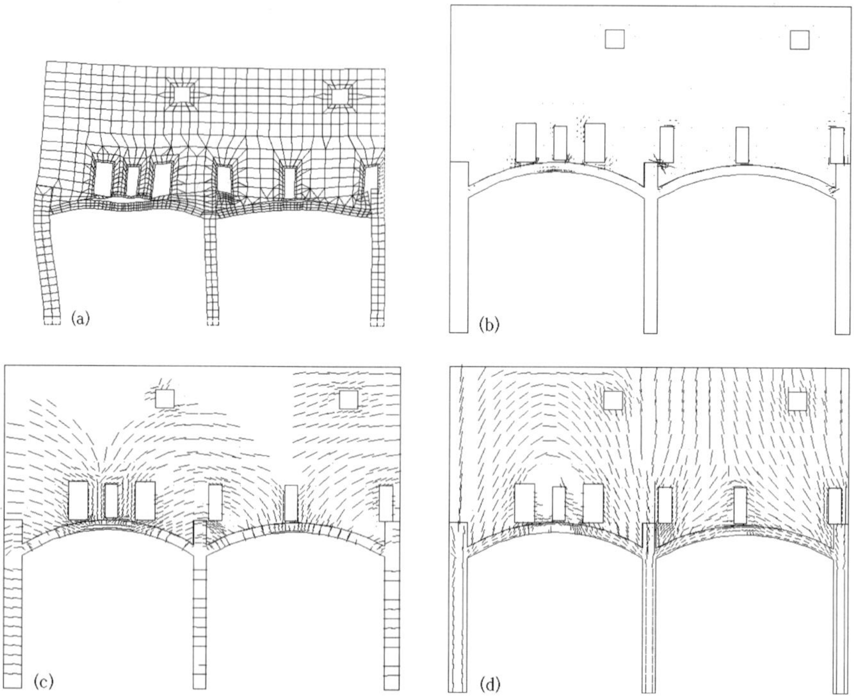

Figure 9.10: Results of the analysis with nonlinear interfaces, De Felice law: (a) deformed shape; (b) principal plastic strains; (c) isostatic lines of tension; (d) isostatic lines of compression.

problem and a suitable collapse mechanism. The computer time needed to bring the analysed wall to collapse, using the De Felice constitutive law, is in fact almost a hundred times that required to study the same problem under the service load. The estimated collapse load factor here is $\alpha_L \approx 12$, referred to the live load only (snow). The collapse mechanism is associated with failure of the left arch due to lack of horizontal support from its left abutment. However, as already said, limit analysis in this framework – associated plasticity – may really mean little, since the actual failure of masonry is due to very different phenomena than what are implied by associated plasticity theory.

To conclude this section, one can observe that the results shown in Figures 9.7-9.10 do actually furnish useful information to the engineer who wants to design a repair for the structure. All the analyses pinpoint problems due to the lack of horizontal restraint at the abutments of the low rise arches, which indicates the need for either placing a tie rod connecting the abutments of the arches or placing a tie rod inside the wall, connecting the top extremities of the supporting diaphragms. The analyses also indicate the need for improving the mechanism of stress diffusion in the proximity of the lower parts of the openings, where significant "cracking" is predicted. Some

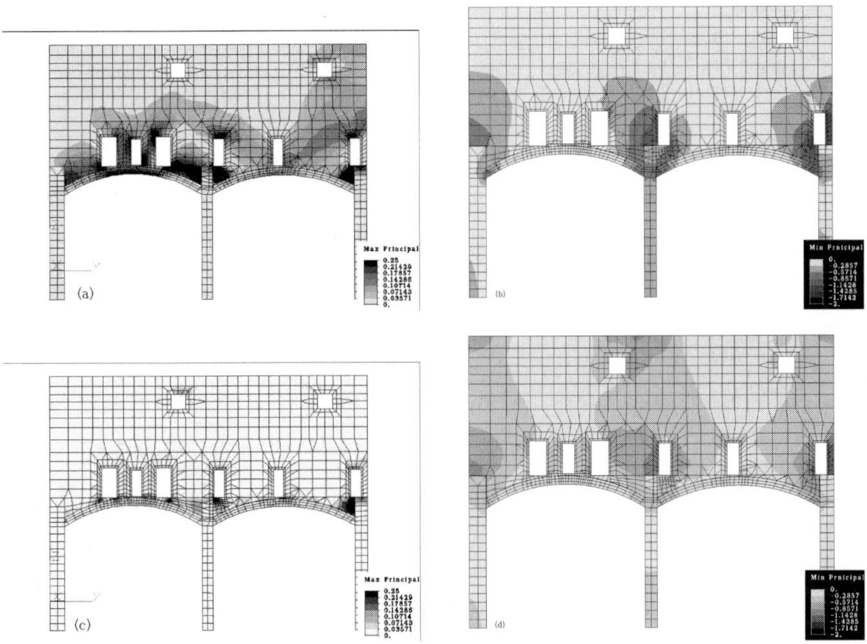

Figure 9.11: Results of the linear elastic analyses: (a) max. princ. stresses (orig. state); (b) min. princ. stresses (orig. state); (c) max. principal stresses (after the consolidation work) (d) min. principal stresses (after the consolidation work).

elastic analyses have been performed in order to check the differences in stress and strain states between the situations before and after the proposed consolidation work, which amounted to inserting both a horizontal tie rod, placed within the wall just below the bottom openings, and several skew tie rods, connecting the top and the bottom of the wall. Also, devices for stress diffusion have been placed below the floor, in correspondence with the openings for the doors in the wall. Figure 11 shows contour plots of the principal stresses: Figures 11a and 11b refer to the situation before the consolidation, Figures 11c and 11d to the situation after the repair work. One can see immediately that the consolidation has the effect of eliminating all the tension peaks, and of bringing all stresses within reasonable limits. This simple comparison indicates that an elastic analysis may prove very useful, especially in terms of comparison between "before" and "after" situations.

9.3.2 A wall of the church "Chiesa della Disciplina" in Verolanuova, Italy

We will here briefly comment on the results of structural analyses performed on a main wall and arch of the Chiesa della Disciplina in Verolanuova (Italy), a rather small church with a single nave along which 5 walls with ogival arches support the ceiling. The arches, probably built in the XIV century, are placed at a distance of 4.5 m from each other, and were initially equipped with tie rods connecting their abutments. In the XVII century, during some remodelling work, the tie rod was removed from the first arch. This has caused severe structural weakness, which became evident in the following decades when extensive cracking, involving both the arch and the wall above it, took place. The geometry of this first arch and wall is shown in Figure 9.12,

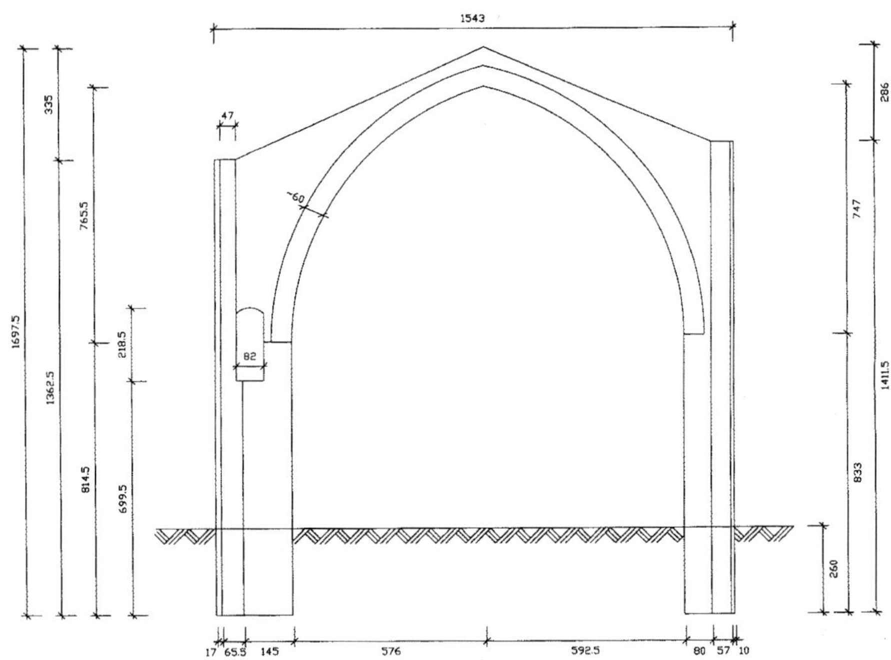

Figure 9.12: Geometry of the first arch of the church "Chiesa della Disciplina" in Verolanuova, Italy.

where all the dimensions are given in centimeters. The thickness of the ogival arch is of 74 cm. The walls which support the arch have thickness of 74 cm on the right side of Figure 9.12, and of 90 cm on the left side. The thickness of the wall above the arch, which transmits the load from the ceiling to the arch itself, is of 74 cm. This wall appears to be simply resting on the arch, with no particular connection devices.

The whole structure goes 260 cm into the ground and does not appear to have any foundation structure. At both sides of the wall and arch the lateral walls of the church exert a small confining effect on the lateral displacements of the arch abutments. More details about this analysis can be found in [63]. Figure 9.13 shows the results of a

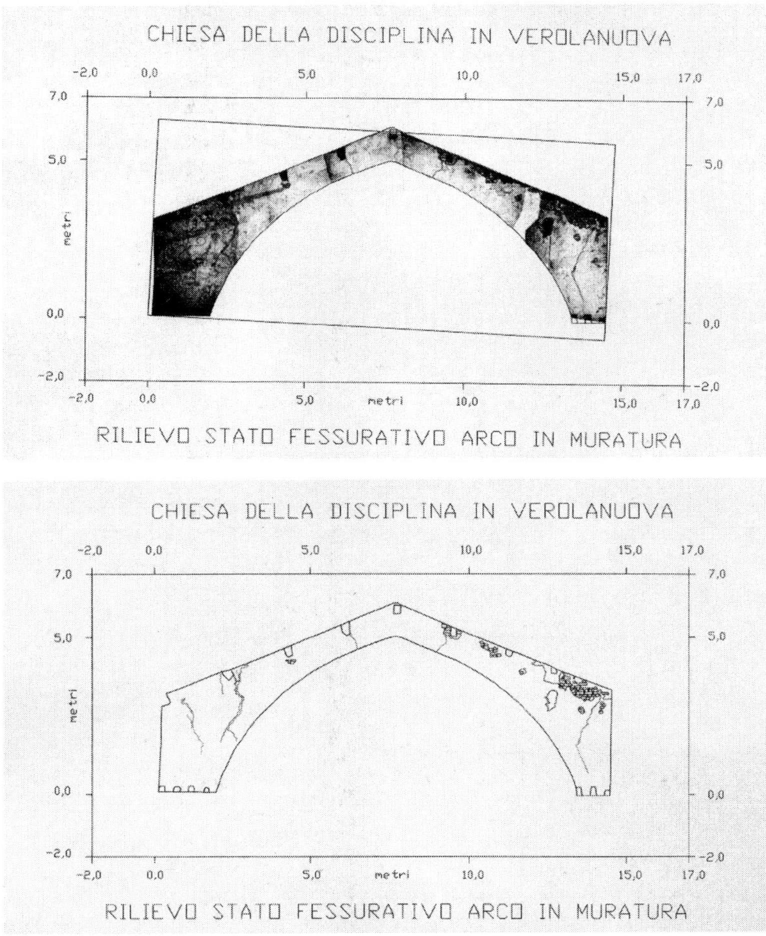

Figure 9.13: (a) Photomosaic of the cracking pattern in the arch (b) Surveyed cracking pattern in the arch.

photographic survey of the cracking pattern in the arch and supported wall. The largest cracks are indicated in Figure 9.13b, and the zones of extensive damage to the plaster have been darkened. The largest cracks are $4 \approx 5$ cm wide. There are four of them, roughly symmetric with respect to the arch keystone. The first two, towards the lateral walls, start from the holes dug in the wall to insert the wood beams which support the roof, initially go down vertically and then deflect following the weak interface between arch and wall. The other central two seem to cross the thickness of the arch. Such an alarming picture suggests that a static collapse of the structure has already happened (the church has in fact been closed to the public for a long time), and that it was certainly due to the lack of restraint of the arch abutments against horizontal displacements. More details about this analysis can be found in [63].

Some flatjack tests have been performed on the external side of the lateral walls (we could not obtain permission for extracting samples of masonry to test in our laboratory). These indicated a stress state, at the ground level, of about 0.2 MPa in compression. The indications about the Young modulus of the masonry were definitely unclear, and affected by permanent strains quite difficult to measure. The average indication for the Young modulus here was $E \approx 500$ MPa, which means a very poor material. The visual inspection of this masonry fully confirmed that it was both very poor from the beginning and now in a very deteriorated state, with several bricks missing owing to erosion and mortar layers likewise corroding.

We have analysed this structure in the linear and in the nonlinear range, both before and after some consolidation work, aimed at restoring its full static effectiveness. The applied load consists of permanent and live load. Again, all the FEM analyses have been performed using the STRUPL-2 code. The numerical model is shown in Figure 9.14. Here, as done for the San Faustino wall, the single structural components (parts of wall and arch) have been kept physically separated in the FEM mesh, by inserting interface elements simulating several possible interface conditions. We stress here the importance of using such a discretisation, which allows us to reproduce some of the most important phenomena occurring in these structures, and at the same time allows us to avoid finding fictitiously high stresses around the interfaces.

There is no space here to describe the problems encountered in trying to define correct boundary conditions for this model, which implied introducing interface elements also between the bases of the wall and the reference system, and trying several stiffness and strength values. We can only say that such details influence, in a really important way, the predictions of the structural behaviour. The material models and parameters used are as follows.

- Masonry has been considered as isotropic elastic-plastic. Young's modulus $E = 750$ MPa, Poisson coefficient $\nu = 0.2$. The yield condition used has been De Felice [25], which proved very effective in the context of the previous analysis. The relevant material parameters are cohesion $k = 0.05$ MPa, friction angle $\Phi = 30°$, tensile strength $\sigma_t = 0.05$ MPa, compressive strength $\sigma_c = 1.0$ MPa, ratio m between brick width and height $m = 5$ for the walls and $m = 0.2$ for the arch. The lateral walls have been assumed as linear elastic.

- Interface elements. Their elastic properties are the same as those of masonry. The nonlinear behaviour has been defined in terms of a Coulomb model with no cohesion and a friction angle of $\Phi = 20°$. A strength limit in compression has also been added, with $\sigma_c = 0.2$ MPa.

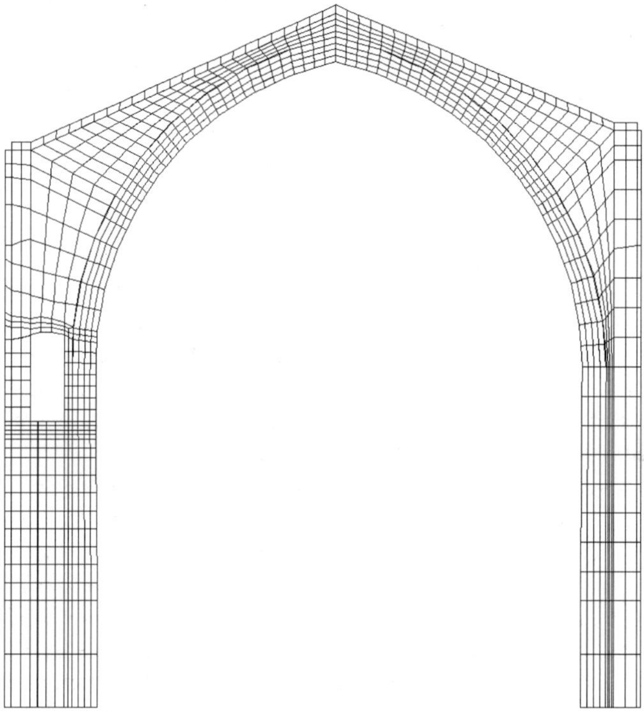

Figure 9.14: Mesh of the wall and arch of the "Chiesa della Disciplina".

The reader will notice that all the material parameters have quite low values. We want to emphasise that this must be a rule to be followed when trying to adapt a plasticity model to follow the nonlinear behaviour of old masonry. This should not, however, lead to the extreme point of choosing zero strength values since, as already pointed out, "no-tension" models are inappropriate for several reasons.

Several analyses have been performed for this structure. Here we show results for only two, concerning the state before the consolidation work, and the state after it. Figure 9.15 shows the deformed shape (9.15a) and the contour plots of the principal stresses (9.15b and 9.15c). Figure 9.16 shows the isostatic lines of tension (9.16a) and compression (9.16b), as well as the distribution of plastic strains (9.16c) in the structure before the repair work, under permanent loading only. If one also applies the live load to the structure, the FEM code predicts the structural collapse at 70% of the design value, caused by rigid motion of the right diaphragm and consequent collapse of the supported arch. The predictions under permanent load only are quite disturbing, with large portions of the structure undergoing severe "yielding". The plastic strain map of Figure 9.16c suggests both detachment between arch and wall, and large dam-

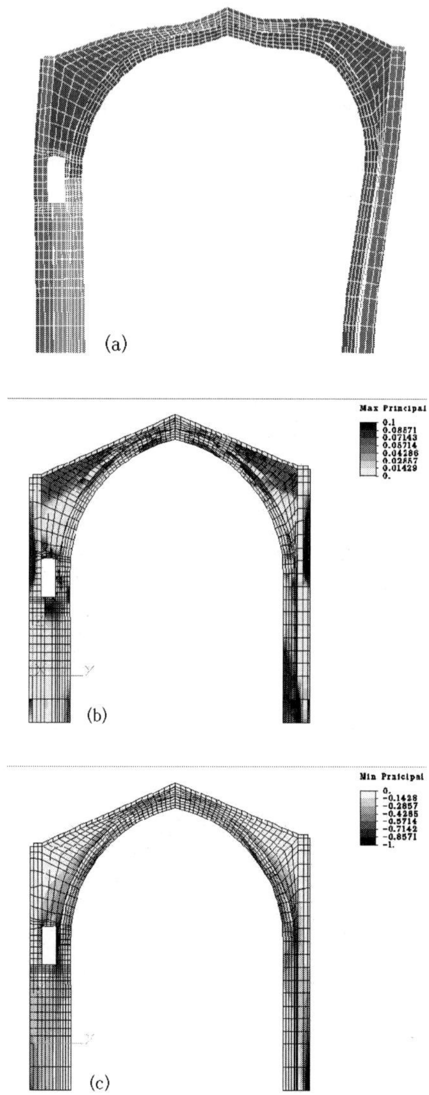

Figure 9.15: Results of the analysis under permanent loads only: (a) deformed shape; (b) maximum principal stresses; (c) minimum principal stresses.

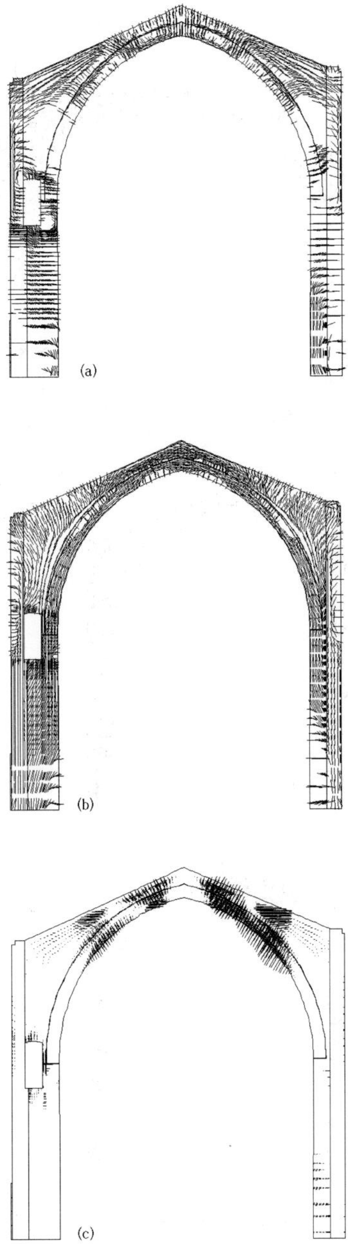

Figure 9.16: Results of the analysis under permanent loads only: (a) isostatic lines of tension; (b) isostatic lines of compression; (c) principal plastic strains.

age occurring in proximity of the zones where actual cracking has occurred. The arch is severely strained, with large zones undergoing bending. The addition of live load worsens the picture which, at global collapse, implies displacements of the order of metres and plastic strains of the order of 100%. These values have no meaning with respect to the real behaviour of a brittle structure, of course, but just indicate the danger of the situation. Figure 9.17 shows the mesh used to study the problem after the consolidation intervention, which consisted of restoring the tie rod connecting the arch abutments and in placing several other steel pins through the upper part of the wall, connecting a concrete girder, added on the top of the wall, with the arch. Figures 9.18 and 9.19 illustrate the numerical results obtained in this new situation under the full live loading. The vertical displacement at the arch keystone is 0.95 cm; the stress in the steel tie rod is 95 MPa. All the stresses in the masonry are limited within the range $-0.5 \leq \sigma \leq 0.07$ MPa; the arch is practically subjected to compressive stresses in its entire span and there is an indication of very little "yielding". All this is obtained without pretension of the tie rod, which, if applied, would probably eliminate any residual nonlinearity in the behaviour of the structure. The safety factor against live load, in the new situation, is greater than 30.

On the basis of these results one can easily conclude that the consolidation work examined is indeed quite effective in restoring sufficient stiffness and strength to a structure which, in the original state, was not able to withstand even a very realistic amount of snow. It is worth concluding this section by remarking that much useful

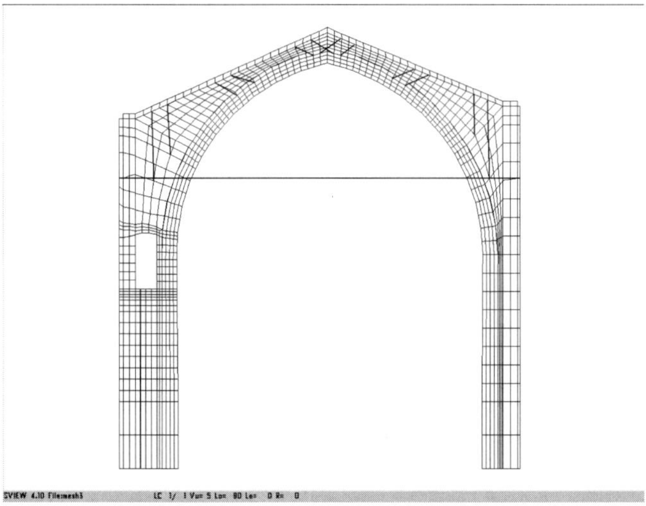

Figure 9.17: Mesh of the wall and arch after the consolidation work.

information could be obtained by simply comparing results of elastic analyses done before and after the consolidation. The work has the effect of limiting most elastic stresses within acceptable values, often reducing them by an order of magnitude with

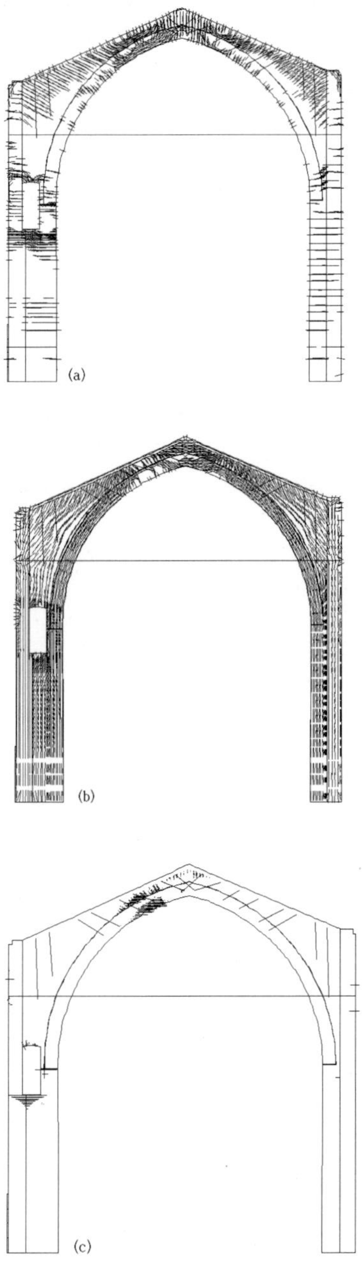

Figure 9.19: Results of the analysis after the consolidation work: (a) isostatic lines of tension; (b) isostatic lines of compression; (c) principal plastic strains.

respect to the analogous values found before the restoration.

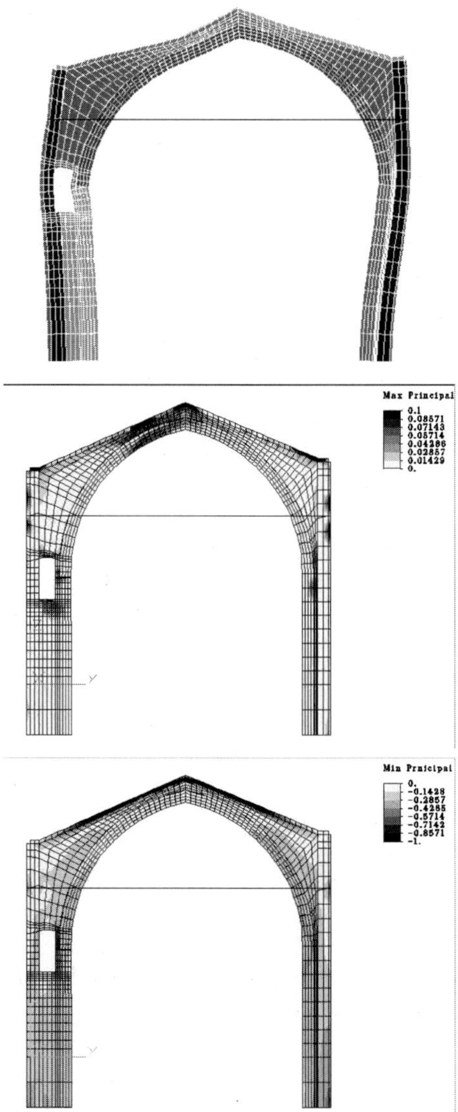

Figure 9.18: Results of the analysis after the consolidation work: (a) deformed shape; (b) maximum principal stresses; (c) minimum principal stresses.

9.4 Analysis of arches and vaults

The field of the analysis of the behaviour of masonry arches and vaults has received considerably more attention, in the past, than that of masonry walls. For this reason we feel that a minimal amount of historical perspective can be given, as an introduction to the specific problems, and possible solutions, arising in this field. Also, in the case of arches and vaults it is easier to classify and describe some of the possible geometry and construction details which, as already mentioned, influence so deeply the local and often the global structural behaviour. We will therefore try and attempt a short summary of these factors, with specific reference to the analysis of vaults.

Historical summaries of the work done at the end of the XVIII and during the XIX centuries on the statics of masonry vaults, arches and domes, treat in a comprehensive way all the effort spent by the researchers to explain by means of mathematical models, and, as a consequence, by means of a general theory, the statics of curved masonry structural elements [64]. Two main research trends are recognisable:

1. the suggestions developed during the XVIII century for the definition of either the "best shape" of the vault or of the correct "failure modalities" [65,66];

2. first attempts at establishing a general theory of masonry structures; general methods for computing the "thrust line" and, most significant, first attempts at extending to masonry the analytical tools developed for general elastic systems [67–71].

By the end of the XIX century, elastic analysis was established as the most reliable tool for the study of long-span masonry vaults, even if the need for a better understanding of the real behaviour was very clear. A new step, at the beginning of the XX century, was the understanding of the three-hinge behaviour of arches, and the proposal of strategies for making it happen in reality; in the meanwhile, a renewed interest in the significance of the intrados shape on the stability of the arch reappeared, almost like a lost heritage of the gothic culture [72,73].

Despite the improvement in the theoretical models, these are still far from being able to give reliable and general formulations. This is due to different reasons, of which the following are of great significance:

(a) the uncertainties about the real behaviour of the material;

(b) the limitations of the analysis methods, based on one-dimensional models used for approximating two- or three-dimensional shapes, often doubly curved;

(c) the difficulty in properly considering the static contribution of each single component of the vaulted structure, which, during the centuries, has undergone substantial changes from the essential gothic concept.

9.4.1 Influence of structural details on the computational model

The establishment of the elastic theory of single- and double-curvature shells gave great enhancement to the analytical description and understanding of the statics of vaulted structures [74]. However, owing to the complexity of the calculations involved, the reference models were necessarily oversimplified. The structural effect of local geometric details, such as sudden changes in the cross-section, or the clamping modality of blocks in the connection zones, or the individual shapes of blocks at the keystones of the arches, can hardly be taken into account by a general model.

The difficulty of analysing any local feature in large structures is obvious, but nowadays the numerical tools of analysis allow, in principle, a very accurate description of the structure, provided that the structural details are correctly included into the model. As a matter of fact, numerical tools are the only way, together with some *in-situ* tests, to try and estimate in a realistic way the structural collaboration among all the different parts of an old structure.

It is well known that historic structures are based on the harmonious development of three main aspects strictly linked together: geometry, statics and construction technology [71, 75]. With specific reference to the analysis methods for vaulted structures, since these three parameters have a profound influence on the static performance, none of them should, in principle, be disregarded in the construction of a correct model. Every typology of vaults is the final result of ideas and construction modalities, so that the model for studying the statics of the gothic ribbed cross-vault should be different from that used to study the barrel, the cloister vault, the flat ceiling, or the ellipsoidal cross-vault. For example, the basic principles and the construction of gothic vaults were based on the main idea of the structural diagonal ribs with webs resting on them (Figure 9.20); today's analyst must decide how to evaluate the collaboration between web and ribbed arches, and how to consider in this system the role of the filling extrados materials.

9.4.2 The computational models of the vault structural details

Both the original strengthening intrados or extrados elements and the confining structural details contribute to improving, in different ways, the resources of the structure, beyond the relatively small limits of the material. Architects were well aware that vaulted roofs with severe limitations in the height, such as in the cloister or "schifo" (flat ceiling) vault, had to achieve their static requirements by means of new different structural details; for this reason they resorted to using a net of ribs or a brick layout, designed in the extrados part of the section of the vault (see Figure 9.21). Thus, what is usually called "filling material" can no longer be considered, as is generally done, to have the minor role of just levelling the upper floor; the significance of the extrados structural components must therefore be taken into serious consideration for performing accurate analyses of the entire vault, which cannot be considered of uniform section, but on the contrary has to be considered as a "flanged vault".

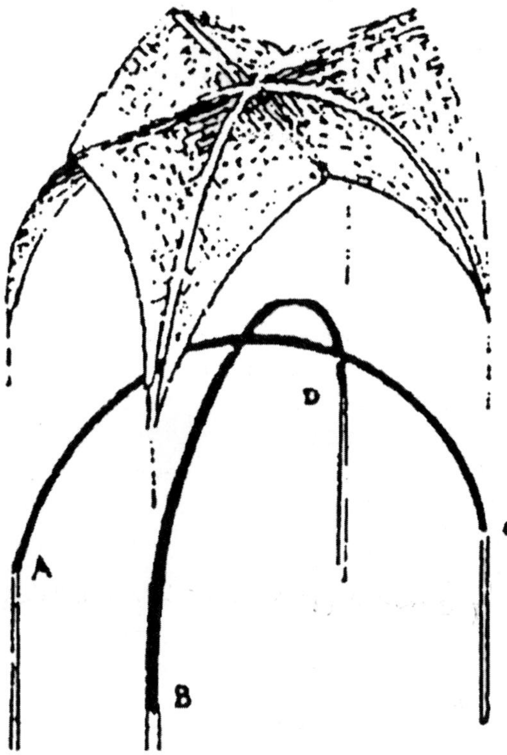

Figure 9.20: Structure made of diagonal ribs with webs resting on them.

9.4.2.1 The role of the extrados buttresses

The actual role of the extrados buttresses, and therefore the correct way to model them in the analysis, is an open question. It is clear that they carry the thin extrados smaller secondary countervaults (Figure 9.22) and therefore most of the live load. But the static equilibrium of the principal vault cannot be satisfied if we do not take into account a structural collaboration between vault and buttresses. Two different assumptions may be realistic:

1. the buttresses just work as connecting compressed rods (Figure 9.23) when the typical deformed shape of a symmetrical arch may occur;

2. the buttresses are made of a well-compacted material (good mortar and bricks) and therefore may be modelled as a structural element sharing the loads between the lateral walls and the vault.

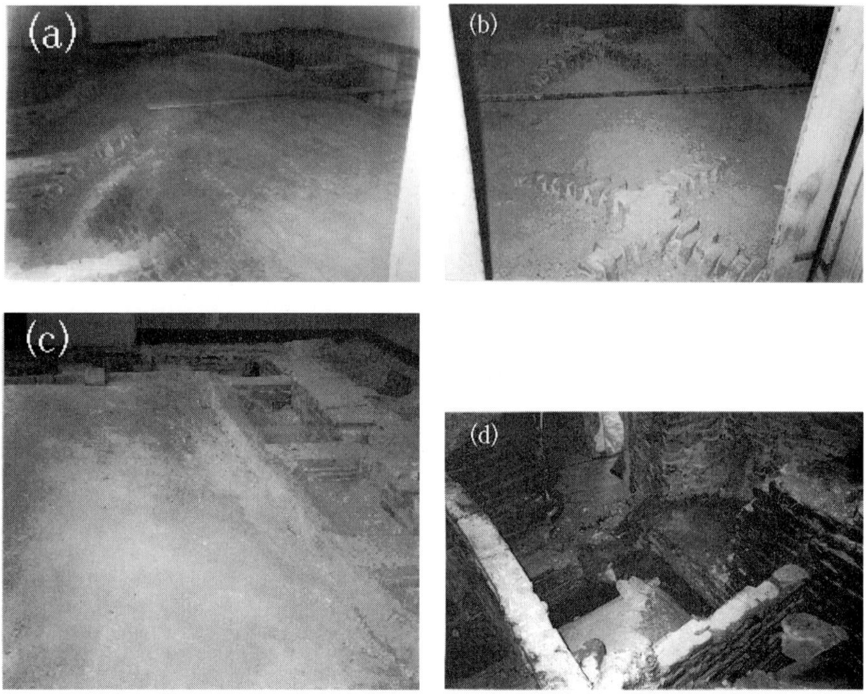

Figure 9.21: (a) extradossal brick wallettes of a barrel vault; (b) cross ribbed extradossal brick reinforcement; (c) wallettes and perimetral extradossal brick cord of a flat ceiling vault; (d) brick wallettes and clamped bricks at a corner vault.

9.4.2.2 The role of the cross-section variable thickness

Another problem is how to evaluate the real thickness of the cross-section of the vault. Often, the intrados of the vault is frescoed, and therefore untouchable; the apparent extrados thickenings, sometimes made after the original construction of the vault, are oriented in many different directions, not easy to observe. All this creates obvious modelling difficulties, sometimes almost insurmountable.

9.4.2.3 The role of supporting walls

The interaction between the supporting walls and the abutments of the vault always requires a high degree of interpretation, often non trivial (Figure 9.23). The vault horizontal thrust has to be carried by the walls which, in many cases, exhibit a series of windows, which weaken the wall stiffness at the vault abutments level.

No analytical approach can disregard the support modality, i.e., the continuity be-

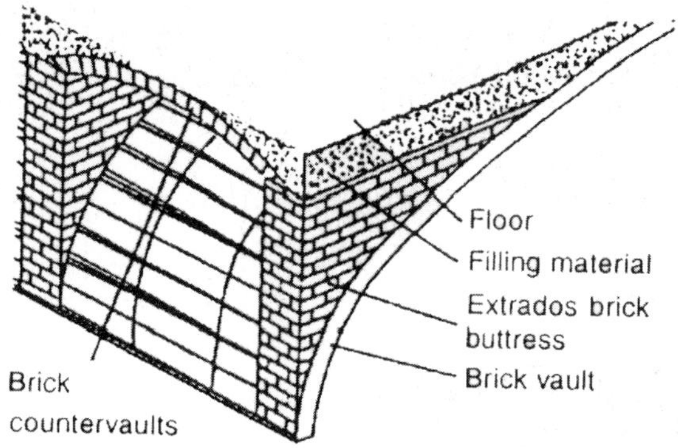

Figure 9.22: Detail of the extradossal thin countervaults.

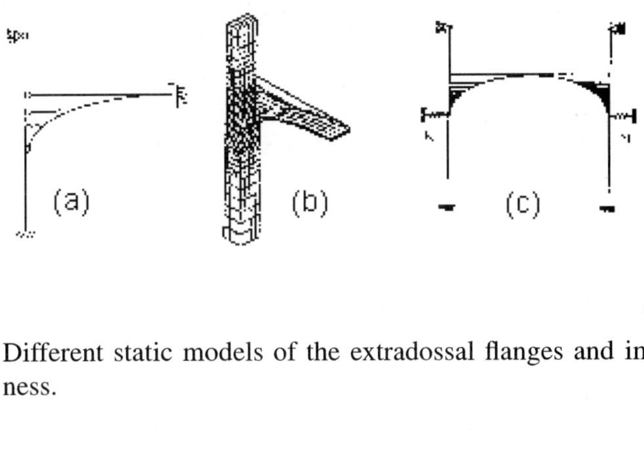

Figure 9.23: Different static models of the extradossal flanges and impost stiffness.

tween vault and wall and the stiffness of the wall. This would imply the definition of a three-dimensional model, able to describe the statics of all the walls supporting the vault. In the case where a full three-dimensional model is intractable, such as, for instance, when trying to perform several complex nonlinear analyses, the analyst is left with the problem of correctly modelling the stiffness and strength of the supporting walls within a two-dimensional model. In this case, as was described earlier, the only reasonable choice is to consider the supporting walls as elastic members only.

9.4.3 Elastic analysis of a cloister vault with frescoes of the XVIII century

An example, considered significant because of the structural detail complexity and static relevance, is presented. It concerns a cloister brick vault of a XVIII century palace in northern Italy, whose three-dimensional finite element model is shown in Figure 9.24. The vault section, of elliptical shape, has a thickness of one brick

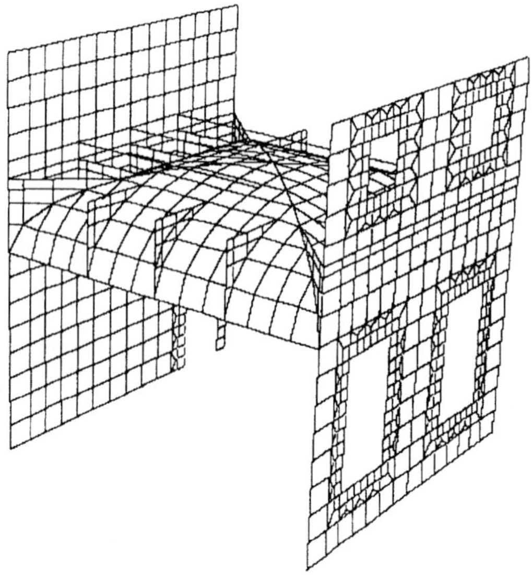

Figure 9.24: Three-dimensional mesh of the cloister vault.

of 12 cm. Both the frescoes and the structure are in a good state of preservation; there is only one crack at one corner, as shown in Figure 9.25, but the crack does not involve the wall. The extrados structural details, quite complex, consist of a system of masonry triangular flanges and thin countervaults resting on them; crossed ribs along the diagonals are present. Photographs, planar representation and the assumed load distribution are shown in Figure 9.25. The wall of the façade has large windows, as apparent from Figure 9.24.

In such a situation a numerical analysis performed on one- or even two-dimensional models cannot properly catch the structural behaviour; in particular, it would be almost impossible to describe all the interactions which contribute to the structural integrity, apparent from its inspection. Only by taking into account all the three-dimensional contributions of all the components of the structure to its stiffness, and by modelling all the restraint conditions due to the confinement given by adjacent structures, as

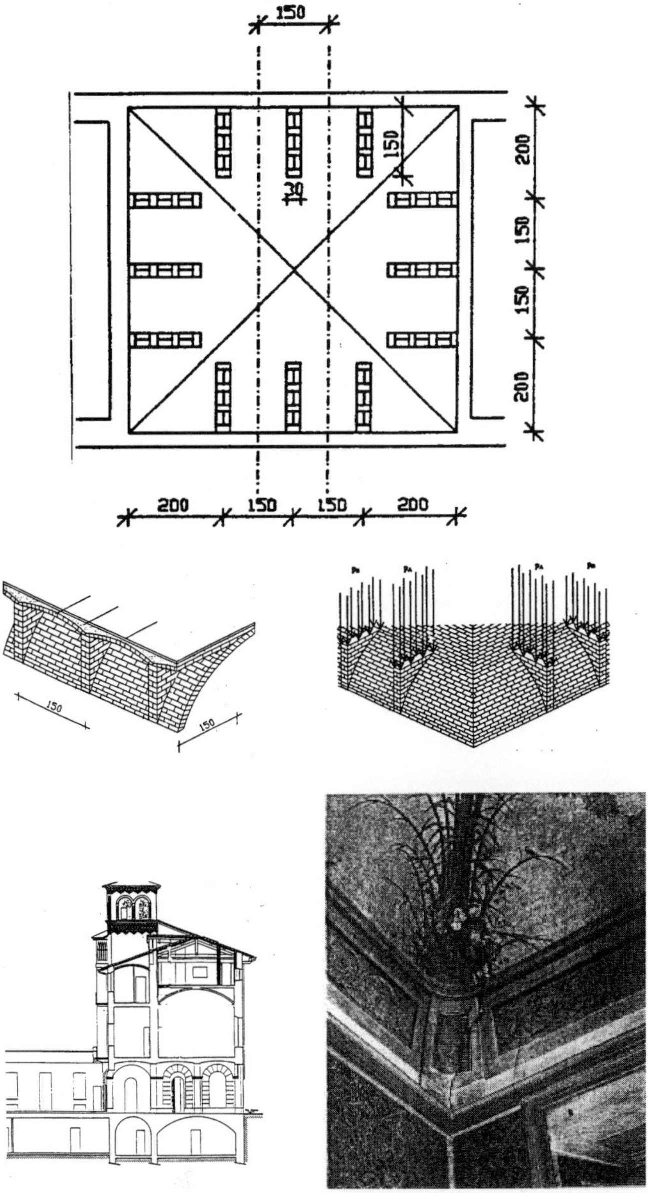

Figure 9.25: The examined vault. Geometry and structural model.

shown by the model of Figure 9.24, is it possible to capture and describe a realistic static behaviour of the actual structure.

Figures 9.26 and 9.27 are displayed to compare the results of a one-dimensional analysis with those given by the three-dimensional complex model of Figure 9.24, in terms of displacements and of thrust line in the vault. With the three-dimensional model also the cracking due to restrained displacements, at the corner of the vault, can be reasonably predicted (Figure 9.28), in terms of concentration of high tensile stresses. Figure 9.28 also shows that the brick extrados wallettes provide a significant

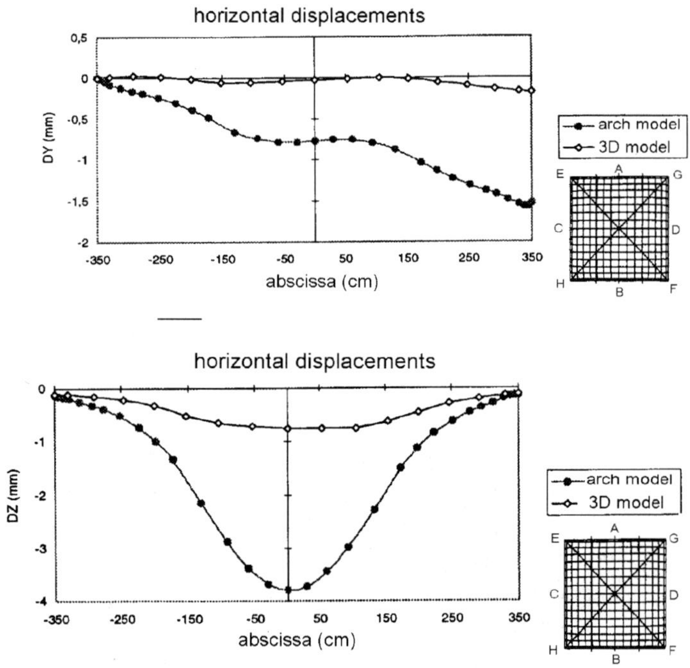

Figure 9.26: A-B line displacements for the two models.

structural contribution, and that the weakest parts of the vault are the sections between two wallettes, where high tensile stresses are present at the extrados.

9.4.4 Limit analysis of vaulted masonry structures subjected to both vertical and horizontal actions

The determination of the ultimate safety coefficient of a vaulted masonry structure is often required when the structure is damaged, or when the structure itself might be required to carry different or heavier loading conditions than the original ones. Dealing with arches and vaults, the interpretation of results given by elastic parametric analyses, as exemplified in the previous paragraph, is useful to indicate the weakest

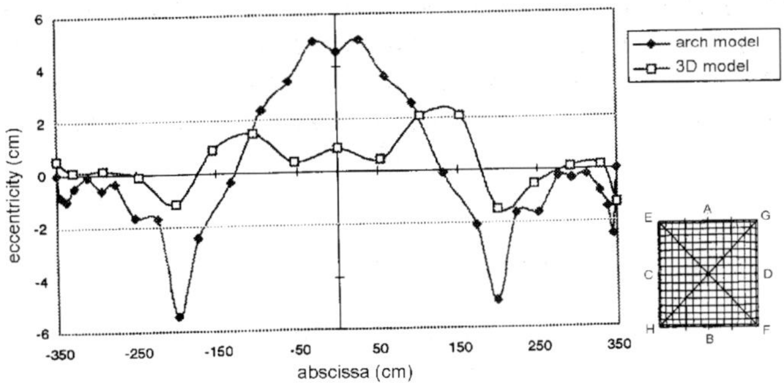

Figure 9.27: The thrust line along the A-B direction.

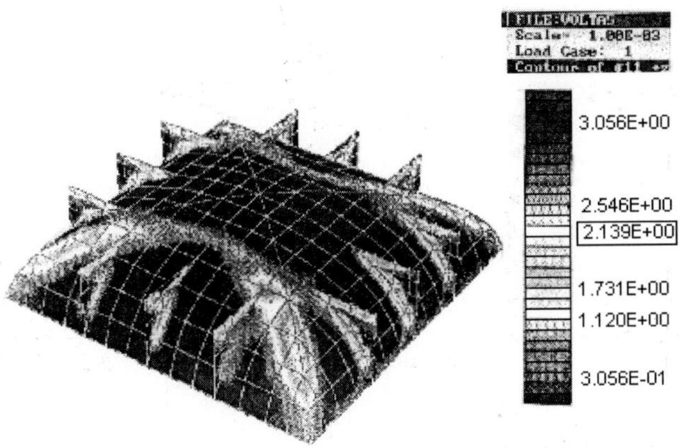

Figure 9.28: Tensile stresses in the intrados vault corner.

parts of the structure, in terms of tensile stresses, and the possible localisation of the first cracks. On the contrary, the use of non elastic models, for studying the ultimate load bearing capacity, may lead to incorrect results, difficult to check. This is because, as mentioned at the beginning of this chapter, the texture of the material and the structural details are often extremely complex, and reflect in a rather unpredictable way on the global structural behaviour. This reason is sufficient to justify the study of simplified numerical methods, based on simple but realistic models, as a tool for checking, at least qualitatively, the more detailed FEM results. One of these methods is here recalled and further developed, with respect to what is already available in the literature, precisely as a tool able to give basic reference solutions for the limit load analysis of vaulted structures in the presence of horizontal loads.

Excluding cases in which particular geometrical conditions lead to failure modes not strictly involving the section capacity of the vault (some of these cases are shown in Figure 9.29), a realistic and conservative procedure for analysing the inelastic behaviour of the structure consists of assuming that the collapse of a masonry arch (assumed also as part of a vault) is the consequence of the formation of a number of hinges sufficient to transform the arch into a kinematically admissible mechanism [4, 76]. The kinematically admissible mechanism procedure, suggested by Heyman [4], is here extended to the case of horizontal loads, such as those simulating the effect of a seismic action.

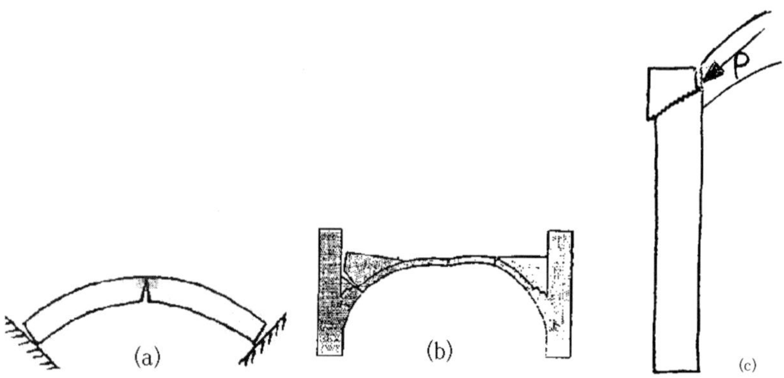

Figure 9.29: Particular collapse mechanisms: (a) crushing failure; (b) detachment of the extradossal structure from the lateral wall; (c) weakness of lateral walls.

The particular behaviour of the hinge, seen as the opening of two adjacent bricks, is illustrated in Figure 9.30, along with the relevant mathematical relationships, where the bending moment M has been assumed coherent with the relative rotation velocity (a superimposed dot indicates rate quantities) and the axial force N positive of traction.

The first relation in Figure 9.30 links the relative rotation of two adjacent sections (corresponding to the adjacent sections of two adjacent bricks) to the centroid opening of the crack; the second relation expresses the limit condition in terms of M and N, illustrated in the same figure where, at the limit, the line of the axial force N reaches the bottom fibers of the section [4]; the third relation states that this class of structures has zero dissipation; the fourth and fifth relations tie this simple mechanical model to the general theory of plasticity (normality rule).

We refer to a generic single degree of freedom (d.o.f.) mechanism illustrated in

Figure 9.31. The hinges A, B, C and D separate four rigid parts of the arch: part AB rotates around A, supposed fixed; part BC rotates around C and part CD around D, supposed fixed to the ground. A simple equation of Principle of Virtual Velocities permits the computation of the load amplifier α (of the external live load q) which ensures equilibrium between external and internal forces:

$$\alpha \int_l [q_x(l)\dot{\delta}_x(l) + q_y(l)\dot{\delta}_y(l)]\, dl + \int_l p(l)\dot{\delta}_y(l)\, dl = 0 \tag{9.1}$$

where $q_x(l)$ and $q_y(l)$ represent the cartesian components of the external load at the

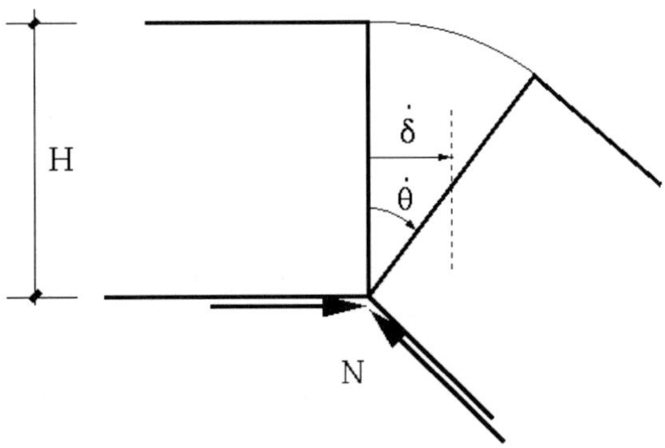

$$\dot{\delta} = \dot{\theta} \times H/2$$
$$M = -N \times H/2$$
$$\dot{D} = M \times \dot{\theta} + N \times \dot{\delta} = -N \times (H/2) \times \dot{\theta} + N \times \dot{\theta} \times H/2 = 0$$
$$\dot{M} = -\dot{N} \times H/2$$
$$\dot{M} \times \dot{\theta} + \dot{N} \times \dot{\delta} = -\dot{N} \times (H/2) \times \dot{\theta} + \dot{N} \times \dot{\theta} \times H/2 = 0$$

Figure 9.30: Plastic behaviour of the hinge.

curvilinear abscissa l; $p(l)$ indicates the self-weight of the arch and the weight carried by the arch at the same abscissa. In order to estimate the safety factor of the arch against the formation of a kinematically admissible mechanism the following operations are required:

- choice of the single d.o.f. mechanism;

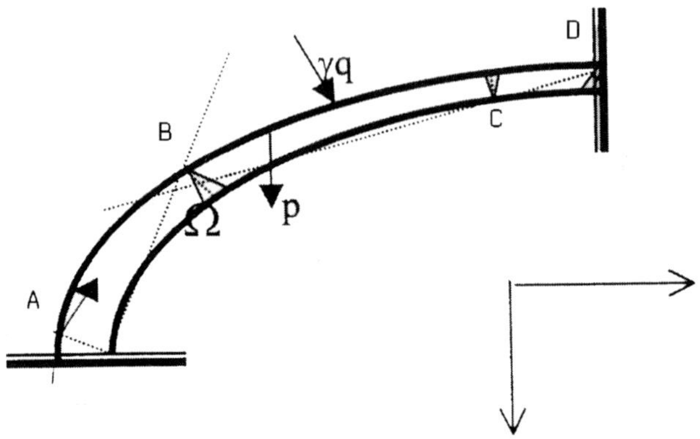

Figure 9.31: Generic single degree of freedom mechanism.

- discretisation of the arch into a set of rigid blocks, in order to concentrate the external loads (permanent and live) into their centroids, transforming the above integral in a suitable summation;
- computation of the vertical and horizontal velocity components of the blocks centroids due to the mechanism formation. These velocities vary linearly inside any rigid part of the mechanism.

In Figure 9.32 the last point of the entire procedure (see Franciosi [76]) is shown to be an easy task, that can be easily implemented. The simplicity of the procedure, which does not require any sophisticated computer program, together with the basic importance of the information which it is able to provide, makes Heyman's approach important and recommendable.

As an illustrative example, the amplifier α of the external loads is computed for the case of the cloister vault analysed in the elastic field with all its extradossal components in the previous subsection. Having recognised that the weakest part of the vault was the vault strip between two flanges, the limit analysis has been applied considering only the section of the vault. The mechanism corresponding to the lowest load multiplier is shown in Figure 9.33 (half span load condition).

9.4.5 Limit analysis of a supporting arch of the Basilica Superiore of Assisi, Italy

This example refers to a typical shape of gothic vault, with ogival arch, as shown in Figure 9.34. The geometry is that of the vault and arch of the first span of the Basilica Superiore of Assisi, Italy, recently damaged by a severe earthquake (1997).

Here three different zones can be recognised: a first part where the extrados of the arch is connected to the lateral walls by means of a structural masonry diaphragm; a

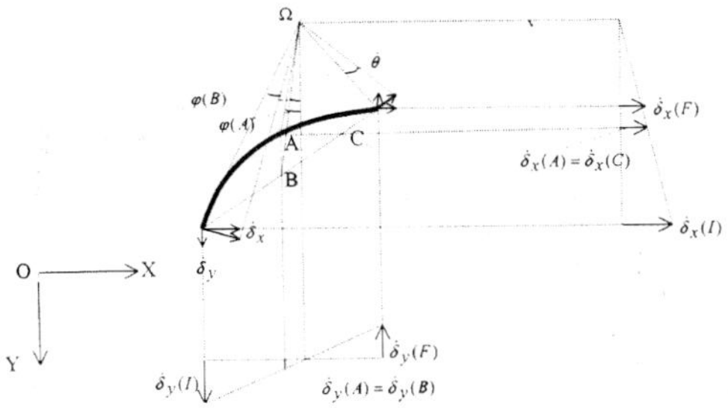

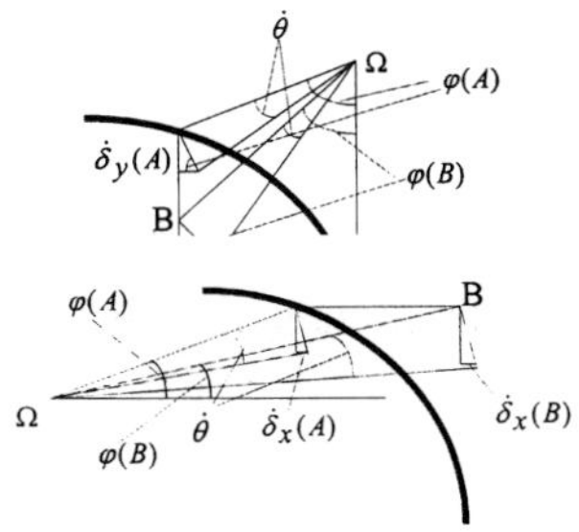

$$
\begin{gathered}
\dot{\delta}_y(A) = (A\Omega)\dot{\theta}\sin\varphi(A) \\
\dot{\delta}_y(B) = (B\Omega)\dot{\theta}\sin\varphi(B) \\
(A\Omega)\sin\varphi(A) = (B\Omega)\sin\varphi(B) \\
\Longrightarrow \dot{\delta}_y(A) = \dot{\delta}_y(B) \\
\dot{\delta}_x(A) = (A\Omega)\dot{\theta}\sin\varphi(A) \\
\dot{\delta}_x(B) = (B\Omega)\dot{\theta}\sin\varphi(B) \\
(A\Omega)\sin\varphi(A) = (B\Omega)\sin\varphi(B) \\
\Longrightarrow \dot{\delta}_x(A) = \dot{\delta}_x(B)
\end{gathered}
$$

Figure 9.32: Velocity components along a rigid arch element rotating around Ω and horizontal velocity component of two points located along a horizontal line.

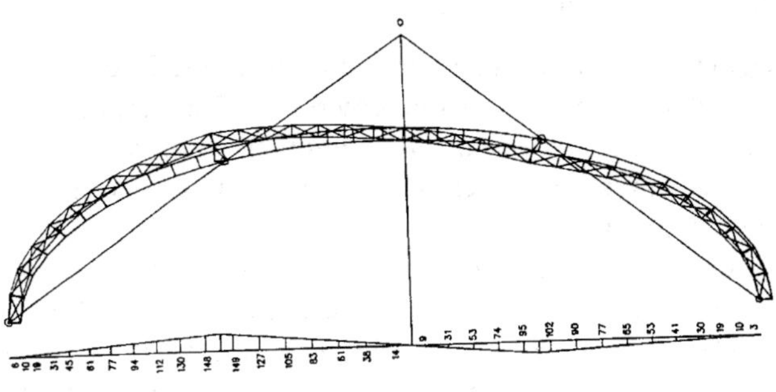

Figure 9.33: The mechanism corresponding to the lowest load multiplier for the arch model.

second part where a non coherent filling material is placed between the vaults and the lateral walls, and a third zone where the vault is uncovered.

The numerical model considers the first zone as a two-dimensional structural element with a well defined strength, able to completely prevent all displacements of the vault and arch abutments. This part has therefore been removed from the model; the arch has been idealised to be fully clamped at the bottom end of this zone.

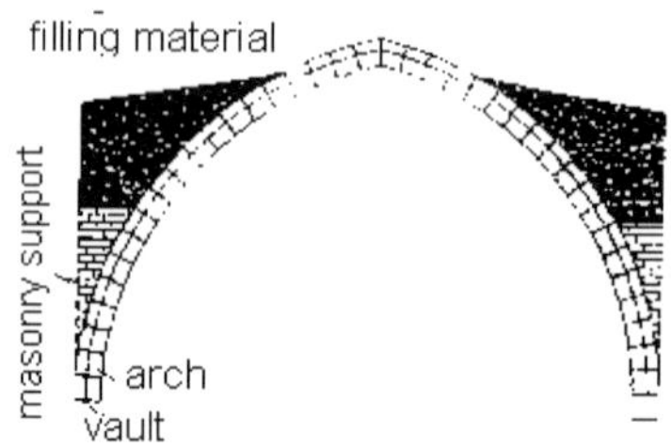

Figure 9.34: Geometry of the filling material masonry support vault and arch.

The second zone, where the filling material is superimposed onto the vaults, is considered to be fixed if, when computing the collapse mechanism, the work of the filling material turns out to be negative; it is considered as a structural vault-arch element if the above-mentioned work turns out to be positive.

The third zone consists of the structural vault-arch element only.

This model has been studied by means of the analytical procedure described in the previous subsection, as well as by using the already mentioned FEM code STRUPL-2 [61]. In the latter case the nonlinearity has been confined only into the interface elements which connect the arch blocks; all the other parts of the mesh have been considered as linear elastic. The interface elements have been described in terms of a "no tension" behaviour coupled with a limited strength in uniaxial compression, thus reproducing the constitutive law used in [4], in terms of generalised variables, to perform limit analysis of masonry frames.

The analysis is aimed at computing the collapse load multiplier under horizontal actions only. Two types of mechanical schemes have been analysed. A first, more conservative one does not consider any bond between arch and vault, and therefore considers the arch (40 cm high) as the only structural element, while the vault acts as a dead weight only. A second scheme considers perfect bond between arch and vault; therefore, the structural model is reduced to that of an arch of height 68 cm: 40 cm of the actual arch plus 28 cm of the vault.

Two types of geometry have been considered: one considers the arch, the vault and the filling, while the second models arch and vault only. The loading consists of the self weight of the arch, the self weight of the vault and that of the filling corresponding to a certain influence area; this is the major simplifying assumption and therefore the major cause of possible poor interpretation of the real behaviour of the three-dimensional structure. The seismic loading has been assumed as an horizontal force proportional to the weight with a common load multiplier. A unit value of the horizontal load multiplier means a force corresponding to an acceleration of 1 g. Preliminary analytical results, in terms of vertical load multiplier (Figure 9.35), con-

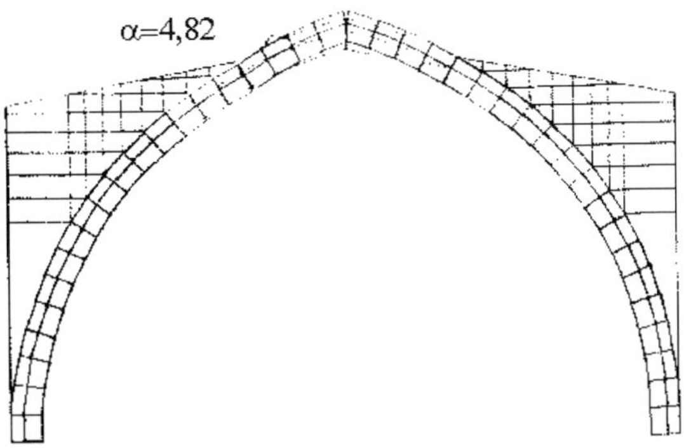

Figure 9.35: The load multiplier $\alpha = 4.82$ corresponds to a dead load on half span equal to 960 Kg/cm^2.

sidering the filling material just like a dead load, confirm the "optimal shape" of the gothic arch.

For the case of live horizontal load, the results of the analytical computations are summarised in Figure 9.36, which gives the different structural schemes, the failure mechanisms and the horizontal load collapse multipliers. Four results are presented: two relevant to the arch with bonded and unbonded vault, and two relevant to the arch and vault with or without filling. These results allow us to draw the following conclusions:

- the collapse acceleration factor ($0.3\ g \div 0.6\ g$) is very close to the maximum acceleration expected by a single degree of freedom simple visco-elastic model subjected to the ground recorded acceleration in this area;
- the filling seems to have negligible influence on the collapse;
- a significant static role can be attributed to the collaboration between arch and vault; it may be assumed that this collaboration, due to the bond strength of the mortar and to the friction between the bricks, has deteriorated during the time because of previous seismic actions.

The above indications, obtained by means of relatively simple analytical tools, have been confirmed by the more refined and expensive FEM nonlinear analyses, not reported here for the sake of brevity [77, 78].

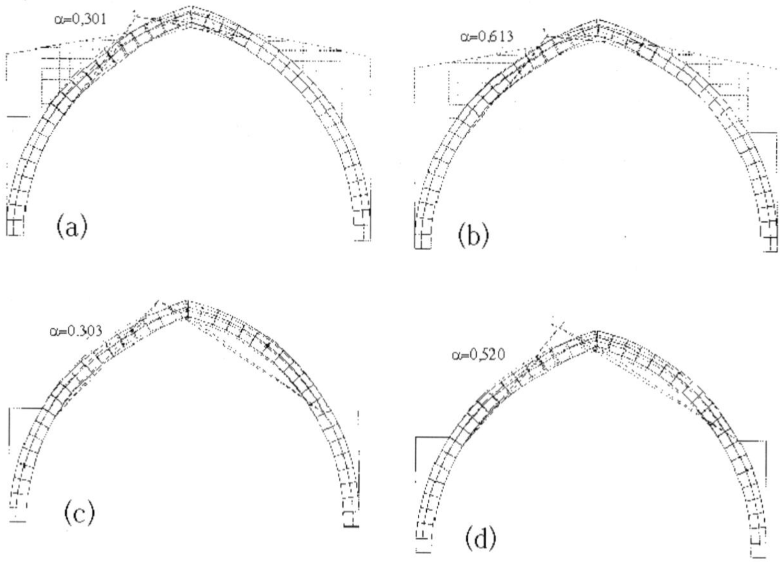

Figure 9.36: Results of the collapse analyses: (a) The arch with non collaborating vault and with filling; (b) the arch with collaborating vault and with filling; (c) the arch with non collaborating vault and without filling; (d) the arch with collaborating vault and without filling.

9.5 Conclusion

In the field of conservation and rehabilitation of old masonry structures the main problem to be faced is the uncertainty about the correct way of studying the safety and stability of structural elements made of such a "difficult" material as old masonry. If, as shown in this chapter, numerical methods are used as the reference techniques to study these structures, some general concluding remarks can be drawn:

1. old masonry can exhibit significant structural resources, behind the apparent weakness of the material. This in turn implies that the structure may reach unpredictable limit states;

2. in the particular case of arches and vaults, the phase of structural assessment – the first step in the rehabilitation process – has to take into consideration every technological and structural detail, most often hidden, and, most of all, it has to face the problem of the stiffness interaction among these details, and their calibration into the analytical or numerical model;

3. as a consequence, when dealing with the task of interpreting the structural behaviour of a masonry building analysed by means of a FEM technique, a mo-

saic of different assumed structural models and analysis methods have to be compared to each other;

4. provided the structural details are realistically calibrated into the model, the linear elastic analysis and result interpretation stands as a guideline for choosing more refined models, to use for assessing the ultimate bearing capacity of the structure;

5. in any case, in order to obtain results useful from a practical viewpoint the simplest non linear models should be used, in view of the difficulty of calibrating the material parameters and recalling that, in practical situations, it may be useless to try and follow the structural behaviour in the post-critical regime.

References

[1] Zienkiewicz O.C., Taylor R.L., "The finite element method", McGraw-Hill, London, 1991.

[2] Hartmann F., "Introduction to Boundary Elements", Springer-Verlag, Berlin, 1989.

[3] Nappi A., Perego U., "Boundary element analysis by linearized nonlinear elastic material models: an application to no-tension systems", Engineering Structures, 10, 146-156, 1988.

[4] Heyman J., "The stone skeleton", Int. J. Solids Structures, 2, 249-279, 1966.

[5] Sacchi Landriani G., Riccioni R. (eds.), "Comportamento statico e sismico delle strutture murarie" (in Italian), CLUP, Collana di ingegneria strutturale, Milano, Italy, 1984.

[6] Lee J.S., Pande G.N., Middleton J., Kralj B., "Numerical modeling of brick masonry panels subject to lateral loadings", Computers & Structures, 62(4), 735-745, 1996.

[7] Ganduscio S., Romano F., "FEM and analytical solutions for buckling of nonlinear masonry members", Journal of Structural Engineering, ASCE, 123, 104-111, 1997.

[8] Wang R., Elwi A.E., Hatzinikolas M.A., "Numerical study of tall masonry cavity walls subjected to eccentric loads", Journal of Structural Engineering, ASCE, 123, 1287-1294, 1997.

[9] La Mendola L., "Influence of nonlinear constitutive law on masonry pier stability", Journal of Structural Engineering, ASCE, 123, 1303-1311, 1997.

[10] Anthoine A., "Derivation of the in-plane elastic characteristics of masonry through homogenization theory", Int. J. Solids Structures, 32(2), 137-163, 1995.

[11] Caddemi S., "The global cracking laws for a finite-element model of no-tension material", Int. J. Non-Linear Mechanics, 27(1), 63-73, 1992.

[12] Maier G., Nappi A., Papa E., "Damage models for masonry as a composite material: a numerical and experimental analysis", in Desai C.S., Krempl E., Frantziskonis G., Saadatmanesh H. (eds.), Constitutive Laws for Engineering Materials, ASME Press, New York, 427-432, 1991.

[13] Pietruszczak S., Niu X., "A mathematical description of macroscopic behaviour of brick masonry", Int. J. Solids Structures, 29(5), 531-546, 1992.

[14] Gambarotta L., Lagomarsino S., "Damage models for the seismic response of brick masonry shear walls. Part II: the continuum model and its applications", Earthquake Engineering and Structural Dynamics, 26, 441-462, 1997.

[15] Lourenço P.B., De Borst R., Rots J.G., "A plane stress softening plasticity model for orthotropic materials", Int. J. Numerical Methods Engineering, 40, 4033-4057, 1997.

[16] Lourenço P.B., Rots J.G., Blaauwendraad J., "Continuum model for masonry: parameter estimation and validation", Journal of Structural Engineering, ASCE, 124, 642-652, 1998.

[17] Andreaus U., "Failure criteria for masonry panels under in-plane loading", Journal of Structural Engineering, ASCE, 122, 37-46, 1996.

[18] Riddington J.R., Naom N.F., "Finite element prediction of masonry compressive strength", Computers & Structures, 52(1), 113-119, 1994.

[19] Lotfi H.R., Benson Shing P., "Interface model applied to fracture of masonry structures", Journal of Structural Engineering, ASCE, 120, 63-80, 1994.

[20] Gambarotta L., Lagomarsino S., "Damage models for the seismic response of brick masonry shear walls. Part I: the mortar joint model and its applications", Earthquake Engineering and Structural Dynamics, 26, 423-439, 1997.

[21] Boothby T.E., Domalik D.E., Dalal V.A., "Service load response of masonry arch bridges", Journal of Structural Engineering, ASCE, 124, 17-23, 1998.

[22] Page A.W., "Finite element model for masonry", Journal of Structural Division, ASCE, 104(8), 1267-1285, 1978.

[23] Chiostrini S., Vignoli A., "Application of a numerical method to the study of masonry panels with various geometry under seismic loads", Structural repair and maintenance of historical buildings, Computational Mechanics Inc., 309-317, 1989.

[24] Ali S.S., Page A.W., "Finite element model for masonry subjected to concentrated loads", Journal of Structural Engineering, ASCE, 114(8), 1761-1784, 1988.

[25] De Felice G., "Analisi limite di strutture a blocchi modellate come mezzi continui" (in Italian), Proc. VIII Convegno Italiano di Meccanica Computazionale, Torino, Italy, 31-36, June 1994.

[26] Baggio C., Trovalusci P., "Limit analysis for no-tension and frictional three-dimensional discrete systems", Mech. Struct. & Mach., 23(6), 287-304, 1998.

[27] Ma M.Y., Pan A.D., Luan M., Gebara J.M., "Arch bridges", Melbourne C. (ed.),

Thomas Telford, London, 247-256, 1995.

[28] Maier G., "Linear flow-laws of elastoplasticity: a unified general approach", Acc. It. Lincei, Rend. Sc. Fis. Mat. Nat., VII, XLVII(5), 266-276, 1969.

[29] Franchi A., Genna F., "A numerical scheme for integrating the rate plasticity equations with an 'a priori' error control", Comp. Meth. Appl. Mech. Engrg., 60, 317-342, 1987.

[30] Franchi A., Genna F., "A stable/neutral equilibrium path for the numerical solution of elastic-plastic softening problems", Comp. Meth. Appl. Mech. Engrg., 90, 921-942, 1991.

[31] Franchi A., Genna F., Corradi L., "On the determination of bifurcation and limit points", Journal of Engineering Mechanics, ASCE, 124(8), 866-874, 1998.

[32] Cohn M.Z., Erbatur F., Franchi A., "Strupl 1 user's manual", University of Waterloo Press, Waterloo, Ontario, Canada, 1982.

[33] Genna F., "An accurate numerical integration scheme for solving structural problems in the presence of a 'no-tension' material", Computers & Structures, 53(2), 253-273, 1994.

[34] Maier G., Nappi A., "A theory of no-tension discretized structural systems", Eng. Struct., 12, 227-234, 1990.

[35] Lucchesi M., Padovani C., Zani N., "Masonry-like solids with bounded compressive strength", Int. J. Solids Structures, 33(14), 1961-1994, 1996.

[36] Binda L., Gatti G., Mangano G., Poggi C., Sacchi Landriani G., "The collapse of the Civic Tower of Pavia: a survey of the materials and structure, Masonry International, 6(1), 11-20, 1992.

[37] Binda L., Anzani A., Gioda G., "An analysis of the time-dependent behaviour of masonry walls", The Masonry Society Journal, 11(2), 1058-1067, 1991.

[38] Anzani A., Binda L., Melchiorri G., "Time dependent damage of rubble masonry walls", Proceedings IV Intl. Masonry Conf., London, 2, 341-351, 1995.

[39] Mirabella Roberti G., Binda L., Anzani A., "Experimental investigation into the effects of persistent and cyclic loads on the masonry of ancient towers", Proceedings VII Intl. Conf. and Exhib., Structural Faults+Repair 97, Edinburgh, 3, 339-347, 1997.

[40] Mehrabi A.B., Benson Shing P., "Finite element modeling of masonry-infilled RC frames", Journal of Structural Engineering, ASCE, 123(5), 604-613, 1997.

[41] Weihe S., Kröplin B., De Borst R., "Classification of smeared crack models based on material and structural properties", Int. J. Solids Structures, 35(12), 1289-1308, 1998.

[42] Carpinteri A., Chiaia B., Bocca P., "Size dependence of strength and fracture properties of brick masonry walls", Journal of Engineering Mechanics, ASCE, 123(8), 816-822, 1997.

[43] Bažant Z.P., "Is no-tension design of concrete or rock structures always safe? – Fracture analysis", Journal of Structural Engineering, ASCE, 122(1), 2-10, 1996.

[44] Ferretti D., Iori I., Riva R., "Un approccio allo studio della stabilità delle antiche torri: il crollo della Torre Civica di Pavia" (in Italian), Studi e Ricerche, Scuola di Specializzazione in Costruzioni in C. A. – Fratelli Pesenti, Politecnico di Milano, Italy, 19, 169-191, 1998.

[45] Tiraboschi G., Binda L., Facchini M., Mirabella Roberti G., "Study of the interaction between bricks and mortar joints through electronic speckle pattern interferometry", Proceedings Intl. Symposium on Non-destructive testing in Civil Engineering (NDT-CE), Berlin, 1, 253-260, 1995.

[46] Falter H., Baronio G., Binda L., Reinhardt H.W., Tedeschi C., "Experimental tests on the mechanical behaviour of reproduced byzantine conglomerates", Proceedings II Intl. Conf. RILEM on Rehabilitation of Structures, Highett, Melbourne, Australia, 498-513, 1998.

[47] Ronca P., "The significance of the gauging system in the flatjack in-situ stress test for masonry: experimental investigation", The Masonry Society Journal, 79-86, August 1996.

[48] Ronca P., "Nuovi sviluppi sull'interpretazione meccanica della prova in situ coi martinetti piatti" (in Italian), Proceedings of the Congress La Meccanica delle Murature tra Teoria e Progetto, Messina, Italy, 18-20 september 1996.

[49] Andreaus U., Maroder M., "Stato dell'arte sui legami costitutivi dei solidi murari" (in Italian), Studi e Ricerche, 2/91, Dipartimento di Ingegneria Strutturale e Geotecnica, University of Roma "La Sapienza", July 1991.

[50] Moriconi G., Castellano M.G., Collepardi M., "Mortar deterioration of the masonry walls in historic buildings. A case history: Vanvitelli's Mole in Ancona", Materials and Structures, 27, 408-414, 1994.

[51] Binda L., Baronio G., "Physic-mechanical characteristics and durability of bricks from some monuments in Milan", Masonry International, 4, 29-35, 1985.

[52] Binda L., Baronio G., "Alteration of the mechanical properties of masonry prisms due to aging", Proceedings VII IBMaC, Melbourne, Australia, 1, 606-616, 1985.

[53] Mezzanotte G. *et al.*, "Percorsi del restauro in San Faustino a Brescia" (in Italian), Il Polifilo, Milano, 1997.

[54] Morbiducci R., "Comportamento non lineare della muratura: modelli costitutivi, risultati sperimentali ed identificazione parametrica" (in Italian), PhD Thesis, Department of Structural Engineering, Politecnico di Milano, 1998.

[55] Cerioni R., Brighenti R., Donida G., "Use of incompatible displacement modes in a finite element model to analyze the dynamic behavior of unreinforced masonry panels", Computers & Structures, 57(1), 47-57, 1995.

[56] Villaggio P., "Stress diffusion in masonry walls", J. Struct. Mech., 9(4), 439-450, 1981.

[57] Di Pasquale S., "Questioni concernenti la meccanica dei mezzi non reagenti a trazione" (in Italian), Proceedings VII Congresso Nazionale AIMETA, Trieste,

227-239, 2-5 October 1984.

[58] Romano G., Sacco E., "Sul calcolo di strutture non resistenti a trazione" (in Italian), Proceedings VII Congresso Nazionale AIMETA, Trieste, 217-226, 2-5 October 1984.

[59] Genna F., Di Pasqua M., "Numerical analysis of masonry structures: a comparison between 'no-tension' and 'softening' constitutive laws", Proceedings VII Convegno Italiano di Meccanica Computazionale, Trieste, June 1993.

[60] Genna F., Di Pasqua M., Ronca P., Veroli M., "Numerical analysis of old masonry buildings: a comparison among constitutive models", Engineering Structures, 20(1-2), 37-53, 1998.

[61] Franchi A., Genna F., "STRUPL-2 theoretical manual", Internal Report, Department of Structural Engineering, Politecnico di Milano, Italy, 1984.

[62] Maier G., "A matrix structural theory of piecewise linear elastoplasticity with interacting yield planes", Meccanica, 5, 54-66, 1970.

[63] Carini A., Genna F., "Indagine sperimentale e numerica sul quadro statico di alcuni elementi strutturali della chiesa della Disciplina in Verolanuova (Brescia)" (in Italian), Technical Report N. 5/1999, Department of Civil Engineering, University of Brescia, Italy, 1999.

[64] Benvenuto E., "A history of statics and structural mechanics", Chapt. 3, Berlin and New York, 1988.

[65] Lamé G., Clapeyron B.P.E., "Mémoire sur la stabilité des voutês" (in French), Annales des Mines, 1823.

[66] Poncelet J.V., "Solutions graphiques des principales questions sur la stabilité des voutês" (in French), Comptes Rendus – XXXV, 1852.

[67] Navier L., "Resumé des Leçons (...) sur l'application de la mecanique" (in French), Bruxelles, 1839.

[68] Méry E., "Memoire sur l'equilibre des voutês en Berceau" (in French), Annales des Ponts et Chaussées, XIX, 1840.

[69] Castigliano A., "Formule razionali ed esempi numerici per il calcolo pratico di archi metallici e delle volte a botte murali" (in Italian), L'Ingegneria Civile e le Arti Industriali, II, 1876.

[70] Winkler E., "Die Lage der Stutzlinie in Gewölbe" (in German), Deutsche Bauzeitung, 1879.

[71] Breymann G.A., "Trattato di costruzioni, costruzioni in pietra" (in Italian), Milano, 1885.

[72] Jorini A.F., "Teoria e pratica della costruzione dei ponti" (in Italian), Milano, 1921.

[73] Campanella G., "Ponti in muratura" (in Italian), Milano, 1928.

[74] Flügge W., "Stresses in shells", Springer-Verlag, Berlin, 1961.

[75] Korkecki Z., Szarlinski J., "Structural problems of the gothic art of building",

Computer Methods in Structural Masonry, Swansea, 1993.

[76] Franciosi V., Augusti G., Sparacio R., "Collapse of arches under repeated loading", J. Structural Division, ASCE, 20, 1964.

[77] Ronca P., Franchi A., Arnetti G., "The vaulted structure of the Basilica of San Francesco in Assisi: static-dynamic analysis and the influence of structural anomalies" (in Italian), Internal Report, Department of Civil Engineering, University of Brescia, 1998.

[78] Ronca P., Franchi A., Paolucci R., "On the structural behavior of the gothic vault under seismic loading: the case of the S. Francis' Basilica in Assisi", Proceedings Workshop on seismic performance of built heritage in small historical centers, Assisi, Italy, 1999.

©2001, Saxe-Coburg Publications, Stirling, Scotland
J.W. Bull, (Editor), Computational Modelling of Masonry, Brickwork and Blockwork Structures,
Saxe-Coburg Publications, Stirling, Scotland 273-312.

Chapter 10

Historic Masonry Structures

E.A.W. Maunder and W.J. Harvey

Department of Engineering, University of Exeter, United Kingdom

Abstract

Computational techniques are considered in this Chapter in the context of structural assessment. Techniques are focused on fundamental levels of investigation which aim to establish feasible states of equilibrium by methodologies which encourage interaction between the engineer and the computer. The computational models presented are based on primitive finite element concepts which are developed with a hierarchy of characteristics relating to equilibrium elements. The techniques described involve the determination of thrust lines for skeletal structures with the aid of spreadsheets, the optimization of general thrust line solutions by limit analyses, and the use of conventional finite element models with adaption of mesh topology to simulate cracking. All these techniques are illustrated by case studies involving actual historic structures.

10.1 Introduction

In this chapter historic masonry structures are considered with two aspects in mind: their structural assessment, and computational modelling appropriate for assessment. The implications of the terms "historic", "masonry", and "structure" in the present context are first discussed, and particular problems associated with structural assessment are introduced.

- **Historic**: structures can be classified as historic when they become of interest as evidence of human activity, or, in other words, they become part of our built heritage. This classification may be officially recognized by a listed building status, or as a scheduled ancient monument [1]. Such a status does not deny a structure from having a useful function, and with or without this status there are many old structures which need to be maintained to enable them to continue to serve a useful purpose other than simply monumental [2]. This purpose may be different from that originally intended, and the timescale for maintenance may be "indefinite", for example cathedrals which have existed for 600 years may be envisaged to last for at least another 600 years [3].

- **Masonry**: traditional forms of masonry construction involve blocks of stone shaped and joined in a variety of ways [4]. Examples range from: dry stone walls of Dartmoor granite, Chinese arch bridges using large pebbles without any mortar [5], to mediaeval castles in Slovakia using irregular stones with mortar (rubblestone) and rendering e.g. Figure 10.1, cathedrals with ashlar masonry laid in courses, and lighthouses such as Eddystone with interlocking granite blocks [6]. The term "masonry" may of course be interpreted in a more general way to embrace assemblages of blocks of material other than stone, e.g. brick, cob, etc.. Perhaps the most significant characteristics of masonry construction in the present context are the predominance of self weight in any loading, and the inability to tolerate tensile stress.

Figure 10.1: Blatnica Castle - abandoned 1790 - Velka Fatra range, Slovakia.

- **Structure**: the variety and evolution of structural forms with masonry has been well documented by, for example, Mainstone [7]. The development of slender skeletal forms from more massive wall forms was a particular feature of the

Gothic period of cathedral reconstruction. A good example where this development can be observed is in the Abbey Church of Saint Sauveur in Redon, Brittany. However, the observation has also been made by Salvadori [8] that although the nominal stress levels were generally low, the articulation of structures developed to a state where all their components were needed for stability. Such a state implies that a skeletal structure is at best isostatic, or at worst hypostatic, and that "if a minor component malfunctions, the entire frame is endangered" [8]. This situation is clearly undesirable, and most existing historic masonry structures are to some degree hyperstatic, and can sustain some local damage, for example Exeter Cathedral did not collapse when it suffered wartime bomb damage which destroyed some flying buttresses and part of a side aisle vault [9]. The question of tolerance to damage is of current concern [10], and it would be of interest to speculate on whether such questions were considered by the mediaeval master masons.

The development of structural forms without any "redundancy" appears to be regarded even now as in some sense optimal. Material based topological optimization [11, 12], based on maximising stiffness subject to constraints on the mass of the structural materials, appears to lead to the development of statically determinate skeletal forms from original continuous forms. These skeletal forms convey load by strut and tie forces, and they are dependent on the load case. Not only will the structure fail if one component fails, but it is also possible for the design to evolve into a form which is unstable for another load case which may not have been foreseen!

Particular problems of structural assessment concern: uncertainties as regards geometry; internal structural forms and materials, for example, columns may appear to be solid masonry, but they may well contain rubble within an ashlar masonry shell; and uncertain structural integrity. A frequently made observation is that structures are seen to stand, but without an obvious explanation!

This Chapter continues in Section 10.2 with a philosophy proposed by the authors for the purposes of structural assessment, Section 10.3 reviews and proposes alternative computational techniques with interactive use in mind, and Section 10.4 illustrates these techniques with some case studies of investigations on actual structures. Concluding remarks are made in Section 10.5.

10.2 Structural Philosophy

Structural assessment may become necessary in a variety of circumstances, for example, as part of a maintenance programme, when a change of use or ownership is anticipated, or after damage has occurred. In the authors' opinion, assessment should begin by developing an understanding of the possible alternatives for the transfer of loads. In the present context, such an understanding may be all that can be reasonably achieved. This philosophy, and its consequences, are elaborated as follows:

- A philosophy for assessment: a fundamental objective in the assessment of old structures is to develop an understanding of the force paths available to transfer loads to the foundations. The availability of alternative force paths must be established, which then raises the questions of which ones are feasible, and which are the most reliable. Once these questions have been addressed, one may continue to consider margins of safety, tolerance to local damage, the effects of subsidence and other forms of movement of supports etc.

Heyman [13–15] has frequently emphasized the shortcomings of linear elastic analyses for masonry structures for the purposes of assessment. Mortar will only tolerate tensile stress at very low levels. Unique solutions to particular load cases require knowledge of ground movements at supports, and other ways for exciting residual stresses which are difficult to quantify. So conclusions based on photoelastic models or linear elastic finite element models may be unreliable.

It is proposed instead to focus on the determination and exploration of alternative equilibrium states, and to seek those which are feasible from the point of view of the material. Masonry structures are normally highly statically indeterminate. They are sensitive to construction sequence and to changes in geometry due to creep of both the masonry and its foundations. We can aim to find viable load paths, and so assure ourselves of continuing structural competence.

- A methodology for assessment: means are proposed to establish the existence of a statical solution (i.e. statically admissible forces and/or stresses), or solutions, which may be judged to be safe, e.g. within the capacity of the materials. The main idea is to gain confidence in a structure through a study of statics. This methodology is based on the lower bound theorem of plasticity, **but** is intended to include those structures where the necessary properties of plasticity may not be completely valid [16, 17]. In such cases the mathematical consequences for safety are not so clear (i.e. "sufficiency" may be absent), however the "necessity" for satisfying equilibrium is vital. Thus this methodology may not guarantee safety, but it is a vital primary consideration which may be sufficient in many cases.

- Engineer/computer interaction: the aim is to enable the complementary use of the engineer's knowledge and skills and the computer's power for 3-D analysis. The challenge to develop similar interaction has also been made in the context of design for modern structures [18].

10.3 Computational Techniques

This section starts by briefly reviewing techniques which already exist and those which are under development by the authors' research programmes. This review is presented under three headings: objectives of modelling; computational models and their associated methods of analysis; and organisations for computing. The Section then continues with more detailed consideration of techniques developed by the authors.

10.3.1 A Review

(i) **Objectives of computational modelling**: there is a spectrum of objectives in the present context. The spectrum ranges from investigating states of equilibrium (e.g. as represented by internal force paths) and their feasibility, to predicting stages in non-linear behaviour *en route* to reaching a limit state (e.g. collapse) [19]. It is the authors's intention to focus on the initial part of this spectrum as it is the most fundamental, and may according to the judgement of the engineer be sufficient for his/her purposes.

(ii) **The computational models**: these are based on finite element concepts which involve some form of discretization of a structure whether it be in the form of a "continuum" such as a wall or dome, or a more recognizable skeletal form. There is a further spectrum of types of element and methods for analysis. Conventional elements have attributes such as nodes, shape functions to describe assumed modes of deformation, and material constitutive relations. Analysis usually involves a stiffness method which may be incremental in order to simulate non-linear behaviour, e.g. due to large displacements and elastoplasticity [20]. However other types of element can also be considered in a hierarchical way according to the objectives of the analysis. Thus equilibrium elements are proposed with the following hierarchy of attributes:

- a region of space occupied by structural material; the element has the sole function of transferring tractions on its boundary through its interior, and in the process equilibrating with the gravitational force representing the self weight of the element. The tractions are more naturally transferred through the sides of an element, as opposed to its vertices.
- internal stress fields which are statically admissible with tractions and self weight. Such fields may be required in limit analyses, although limits may also be restricted to the tractions themselves [21].
- the constitutive relations for the materials of an element; these may be invoked in order to make predictions of deflections. For example, linear elastic assumptions with a tension cut-off may be assumed. These assumptions imply a non-linear analysis to account for changes in contact between elements in order to relieve tensile stresses. This type of analysis has been implemented by Maunder [22,23], where tension crack simulation is made by topological changes to the mesh in an iterative procedure. The procedure involves conventional linear elastic analyses, and post-processing to determine where contact at element interfaces should be broken. This has been made dependent on user interaction, however the procedure may fail to converge to a stable solution, or fail to recognize stability in a complex structure. Choo [24] has also considered an iterative analysis using an automated adaptive scheme for masonry arch structures with a single ring, but based on tapered beam elements and elastoplastic material.

Concepts from force or flexibility methods such as: statical indeterminacy, statical basis, and material compliance or flexibility properties become more natural when exploiting this hierarchy with equilibrium elements.

More complicated finite element models account for separation of elements or substructures into discrete elements [25,26]. These models tend to depend on a complete knowledge of material constitutive relations, are non-interactive, and lead to a very indirect route to satisfying our more pragmatic aims.

(iii) **The organisation of the computations**: the engineering user of software may wish to consider various levels of interaction with a computer [18], (e.g. a desktop PC), and various forms of visualisation. The way to obtain flexibility in use depends on the organisation of the software, e.g. based on routine or object oriented programs (such as Fortran 90 or C++), and/or the use of spreadsheets such as Excel [27].

10.3.2 Thrust lines in skeletal forms

10.3.2.1 Interactive development of thrust lines

The concepts of thrust lines and zones of thrust are more complex than they appear. The thrust line is a very limited representation of a state of stress. It carries with it implied knowledge of the peak normal stress on a section of a skeletal structure, and of the depth of cracking, but it tells us nothing of the principal directions of stress, or of the coexistent shears on the section. Despite these drawbacks, the concepts have two major strengths: to an experienced user they convey a great deal in a simple picture, and they are relatively simple to calculate.

Lines of thrust can be derived from the results of the most complex analysis, and might take into account both geometric and elasto-plastic non-linearity of behaviour. In general, though, the use of this simple visualisation is allied with a simple statical view of behaviour.

Heyman [14] revitalised the line of thrust approach in combination with the plastic theorems. His view was a simple one, and anchored in hand calculation. More complex approaches using micro-computers were quick to follow, particularly in the field of arch bridge assessment, though the Heyman approach continued finding the arch of minimum thickness which would just fail by forming a mechanism. It is perhaps worth stressing that this approach finds a mechanism in order to show that one cannot form in the real structure. In this way the method is used to deliver lower bound solutions.

In more complex structures than a simple arch, the mechanism approach to finding the structure of minimum viable thickness is less feasible. Engineers have become familiar with the idea of simply showing that a zone of thrust can be accommodated within the structure, and that this zone is best found by trial and error. We are aware of only one specialist piece of software that has been designed to allow the engineer to make this exploration freely for multi-span arch bridges [28]. In more complex skeletal masonry structures, this exploratory approach has been used through spreadsheet calculations which effectively modernise Fuller's graphical trial and error approach [29].

In this approach, the skeletal structure is first divided into substructures, whose shapes might be defined by straight segments, or arcs of different radii. The divisions between the substructures should be chosen with care, or the calculations may become excessively complex. A compromise must be sought between choosing dividing planes which lead to sections with simple shapes, and those where the thrust will cross as nearly as possible normal to the section. A second subdivision is generally required in order to model each substructure as a series of primitive elements, that is, elements which have the lowest hierarchical level. In arching substructures, the interfaces of the elements might sensibly include the radii of the arc, while in vertical piers they will be parallel, and probably horizontal. Figures 10.29 and 10.30 in Section 10.4.4.2 show a buttress at Wells cathedral which, though including an arch, is sufficiently vertical to be subdivided into elements by horizontal slicing. The fact that the element interfaces are not normal to the principal stress means that there are additional components of force to deal with on each interface or section.

The process of computing a thrust line proceeds by accumulating all the forces to one side of a particular section, and determining where their resultant crosses that section. Determination of the latter requires consideration of rotational equilibrium about a centre. It should be noted that for 2D models the resultant is a simple force or thrust, whereas for 3D models the resultant may be a wrench which incorporates both a thrust and a twisting moment about its line of action. The simplest parallel occurs in the study of the stability of a retaining wall. In this case the engineer typically checks horizontal, vertical and rotational equilibrium, and finally finds the position of the resultant by dividing the moment about the toe by the vertical component of reaction. In complex three dimensional masonry skeletons the calculations are made more complicated by the need to choose appropriate centres for the calculation of rotational equilibrium. If one centre is chosen for the whole structure the calculations can become quite unwieldy. It is often much simpler to define the substructures so that each one has a convenient centre. For straight substructures, such as piers, it is easier to consider parallel sections and shift the centre in one direction only from section to section.

Most skeletal forms of masonry structure are statically indeterminate, and so the thrust lines depend upon "redundant" forces. This is true of a simple 2D model representing a single arch ring. In this case the redundancies may be defined by three quantities associated with one of the reactions at a springing, e.g. the magnitude and direction of the reactive thrust, and its line of action. The interactive approach allows redundant quantities to be used as variables which are given trial values. Once the structure has been divided into elements, some care must be exercised in deciding what quantities are to be treated as redundant variables. These variables can then be used to compute and control the lines of thrust throughout the structure. In the most recent versions of high level spreadsheets it is possible to include such controls as slider bars with which to input values for selected variables.

10.3.2.2 Thrust lines from limit analysis

The finding of a feasible thrust line that lies within a structure may be sufficient in itself, but then the question may be raised: what is the optimal solution? In considering this question, possible objectives are:

1. find a solution which minimises the eccentricity of a thrust line, or, in similar vein,

2. find a solution with the maximum geometric factor of safety (the ratio of actual to minimum thickness). This factor is more relevant to considerations of damage tolerance, e.g. to what depth can masonry be allowed to deteriorate before the structure reaches an unstable state?

3. find a solution to give the maximum load factors for load distributions which distinguish between dead and live loads.

More complex issues concern sensitivity analyses, for example, how much can a structure tolerate variations in loading, material strengths, and/or dimensions?

In order to make further progress in addressing such objectives, particularly for models with high degrees of statical indeterminacy, it is proposed to consider limit analyses with appropriate optimization procedures. The formulation of limit analyses will be based on simple objectives and constraints on interface tractions in both tension **and compression**, as well as shear. In making structural assessments it would seem important to consider nominal stress levels rather than assume infinite compressive strength!

Thus consider a solid element such as a hexahedral with flat sides to represent a block or blocks of masonry. The tractions on a side will be constrained by the following limits:

$$0 \leq \sigma \leq \bar{\sigma} \quad \text{and} \quad \tau \leq \mu\sigma \tag{10.1}$$

where $\bar{\sigma}$ represents a normal stress limit (compression positive), and τ represents tangential stress with a Coulomb frictional limit. It should be noted that in the context of plasticity, the frictional limit implies a non-associative flow rule, and so as already indicated the lower bound (safe) theorem may not be necessarily valid.

The traction limits imply constraints also on the interface thrust forces transferred between elements. The form of the constraints depends on the assumptions made for the distributions of tractions. The simplest assumption is that of uniform tractions over the contact areas, similar to the use of the rectangular stress block in reinforced concrete. Formulation of these constraints is first illustrated for a 2D model with rectangular sides of elements. With reference to Figure 10.2, a thrust of magnitude T has direction defined by angle θ, the inclination from the normal, and line of action defined by eccentricity e from the centre of the side.

A contact zone of area tz is defined with z from Equation (10.2).

$$z = (d - 2e) \tag{10.2}$$

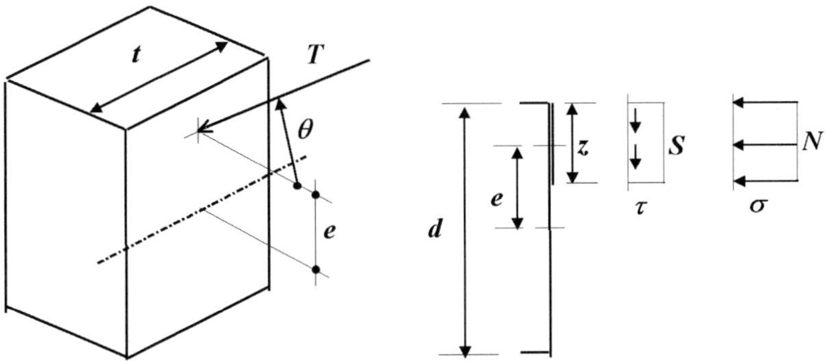

(a) thrust definition (b) contact zone with uniform tractions

Figure 10.2: Thrust for a 2D problem.

Limits on the thrust parameters can now be expressed in non-dimensional form as in Equation (10.3) and illustrated in Figure 10.3.

$$\left|\frac{e}{d}\right| \leq \frac{1}{2}\left(1-\frac{N}{\bar{N}}\right) \quad \text{and} \quad |\theta| \leq \tan^{-1}\mu \quad \text{where} \quad \bar{N}=\bar{\sigma}\cdot td \tag{10.3}$$

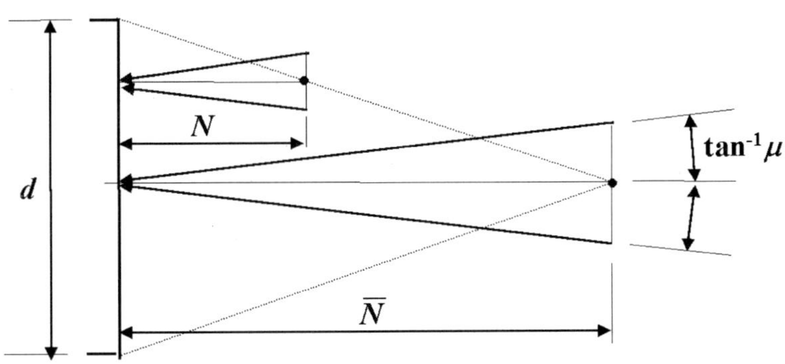

Figure 10.3: Limits on thrust parameters.

However it is more convenient for limit analysis to formulate constraints on the three components of action N, S, and $M = Ne$ which represent the thrust T. These constraints are given in Equations (10.4) to (10.6) in non-dimensional form.

$$0 \leq \frac{N}{\bar{N}} \leq 1.0 \tag{10.4}$$

$$6\frac{M}{d\bar{N}} \leq 3\frac{N}{\bar{N}}\left(1-\frac{N}{\bar{N}}\right) \tag{10.5}$$

$$\frac{S}{\bar{N}} \leq \mu \cdot \frac{N}{\bar{N}} \tag{10.6}$$

The corresponding limit surfaces are shown in Figure 10.4. Clearly constraint (10.5) can be linearised as a convex polygon in many ways, and then optimization through linear programming becomes feasible. Figure 10.4 includes one such scheme which provides a total of eight piecewise linear constraints for each interface. These polygons are used in the Case Study of Bridgemill arch bridge which is described in Section 10.4.1. Limit points 1 to 4 refer to the limits reached in four cross-sections in the analysis illustrated in Figure 10.15.

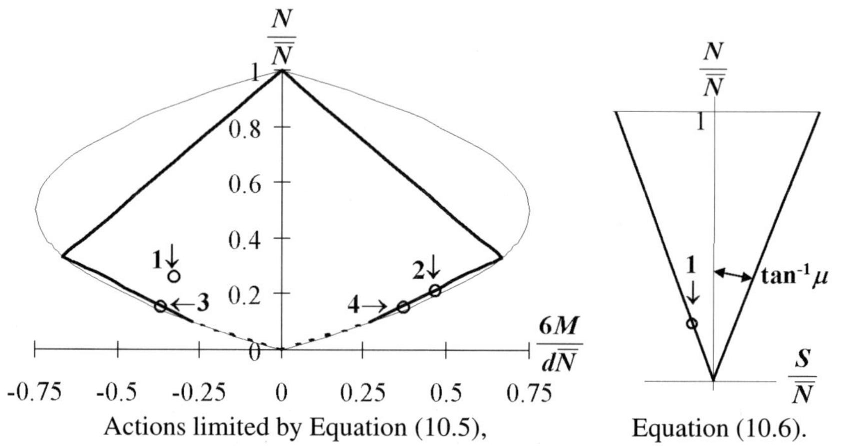

Figure 10.4: Limit surfaces for components N, S, and M representing thrust T.

It would be optimistic to expect constraints for 3D problems to be formulated with similar ease by extending the 2D case! The 3D thrust case is illustrated in Figure 10.5 where the thrust now requires 5 parameters for its specification: magnitude T, inclinations θ_x and θ_y from the normal, and eccentricities e_x and e_y from the principal axes of the section.

For uniform contact stresses, the contact zone ABCD can be delineated with the straight line CD so that the centroid of the zone coincides with the point of application of the thrust T. A three dimensional limit surface could be developed for N, $M_x = Ne_x$, and $M_y = Ne_y$, which would be dependent on the shape of the side of the element. A linearised convex polyhedron could then be defined within it for a linear programme.

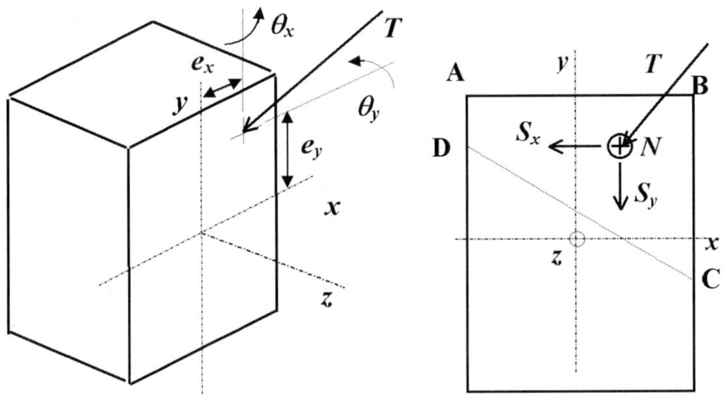

(a) Thrust T (b) contact zone ABCD

Figure 10.5: Thrust in the 3D case.

However, an extra complication arises in 3D systems of forces - the generalised vector representing stress resultants on a side has six components which includes a twisting moment about the normal axis z. From inspection of Figure 10.5(b) this twisting moment is restricted to the expression in Equation (10.7), and it is not in this case a freely variable quantity.

$$M_z = (S_x \cdot e_y - S_y \cdot e_x) \tag{10.7}$$

In this case, the twisting moment about the line of action of the normal component N has been released. A general 3D system of forces requires the concept of a wrench acting on a side, that is a thrust force together with a twisting moment about the line of action of the thrust. If the twist component about a normal to the side is released, then this introduces the possibility of an unstable system for which statical solutions do not exist for some or all load cases. Although each situation could be considered for stability, this would be an unsatisfactory feature [30].

However if the wrench is included for completeness and stability, then the uniform traction assumption is no longer sufficient. An additional field of shear stresses could be incorporated to transfer the wrench as shown in Figure 10.6, where the contact zone is subdivided into a perimeter ring for uniform torsional stress, and an inner core for uniform shear stress.

The subdivision of the contact zone with further dimensional parameters h_1, h_2, etc. complicates the formation of limit surfaces for N, S_x, S_y, and M_z, and further work is required to judge whether such an approach for a linear programme for 3D is tractable.

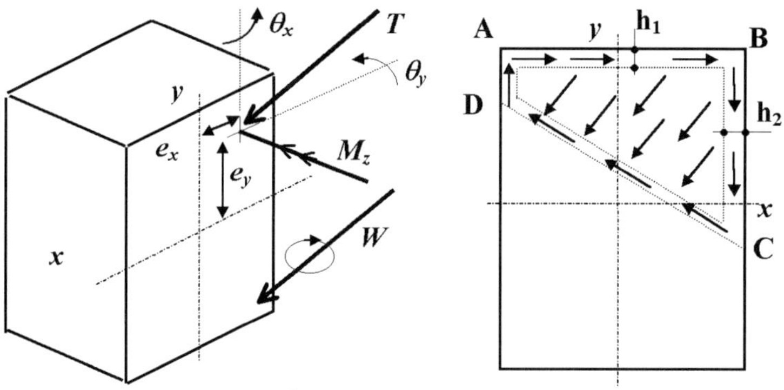

(a) Wrench $W \equiv$ thrust $T+$ twist M_z (b) subdivided contact zone for shear and torsional stresses

Figure 10.6: Wrench interaction.

10.3.3 Thrust lines in continuous forms

10.3.3.1 Force paths

For skeletal forms of structure, an adequate model may be formed by a single element through the depth, e.g. to represent a series of voussoirs. However an arch with multiple rings where delamination is possible, or a structural form such as a wall where a skeletal form is not identifiable, leads to models with complex possibilities for internal force paths. Although such models may have high degrees of statical indeterminacy, investigations of force paths can be simplified by combining concepts from force methods of analysis and graph-theoretic concepts [31].

In a force method of analysis a statical solution for internal forces is defined as a combination of particular and complementary solutions, the latter being self balancing. A complete set of complementary solutions forms a statical basis for the vector space of complementary solutions. The dimension of this space is the degree of statical indeterminacy. The formulation of a statical basis has been advocated as an algebraic process using Gauss-Jordan elimination on a complete set of equilibrium equations for elements or interfaces [30,32]. However the use of graph-theoretic concepts leads to a direct definition of a statical basis without the need for matrix elimination procedures.

Figure 10.7 serves to illustrate a typical hexahedral element with a tree graph having 6 vertices and 5 edges. The vertices represent the potential interfaces, or "terminals", of the element, and the edges represent paths for the transmission of generalised forces via a "root" vertex. There are 3 or 6 generalised force components per side or "terminal" in 2D or 3D cases respectively.

When the elements of a model are interconnected, the separate tree graphs are connected at their vertices to form a single graph for the structural system with n vertices and e edges. So long as no forces are released, the degree of statical indeter-

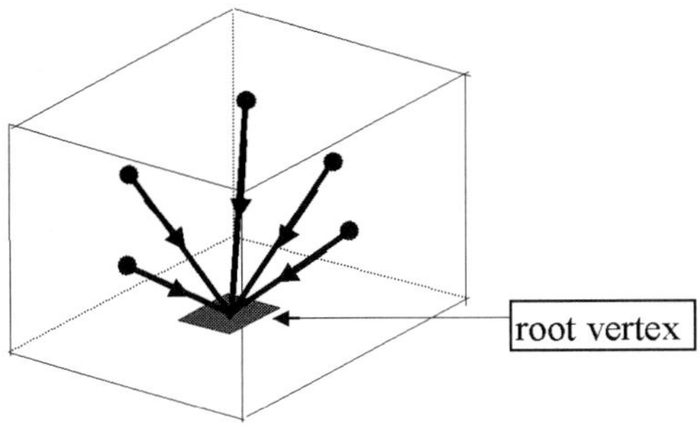

Figure 10.7: Hexahedral element and its tree graph.

minacy is $3\times$ or $6\times\ (e-n+1)$ [31]. Structural supports are represented by a single "ground" element. A statical basis is constructed by associating 3 or 6 self balancing force systems with each circuit of a set of $(e-n+1)$ independent circuits of the graph. A particular solution can be formed by association with a spanning tree of the graph. The main computational problems reduce to the selection of circuits and trees from graphs, these problems are relatively simple although the circuit problem for the general 3D form is more complex [33].

The circuit selection problem is greatly simplified for 2D models since then a suitable set of circuits frame the "windows" formed by embedding the graph into a plane, refer to Figure 10.8 for an example. In this case the components of the corresponding statical basis become very localised.

10.3.3.2 Particular solutions from conventional finite element models

An alternative derivation of a particular solution can be made by transforming nodal forces obtained from an analysis of a conventional finite element model, e.g. using isoparametric displacement elements and a linear elastic type of analysis. A thrust line solution derived in this way accounts for material deformation and may serve as a better starting solution for subsequent variation in an investigation. The concepts for this derivation [34] are again explained with reference to an 8-noded hexahedral element E illustrated in Figure 10.9.

The gravitational loads applied to the element E are represented by a set of consistent and statically equivalent node loads f_n^E where n is a node number. As a result of a finite element analysis, element nodal forces F_n^E are derived from nodal displacements, and these forces are self balancing for an element. It follows that nodal forces $\hat{F}_n^E = \left(F_n^E - f_n^E\right)$ maintain an element in equilibrium under its own weight. The analysis also enforces the equilibrium of each node under the action of loads and forces from adjacent elements, i.e.

$$\sum_E \left(-F_n^E + f_n^E\right) = 0 \quad \text{or} \quad \sum_E \hat{F}_n^E = 0 \tag{10.8}$$

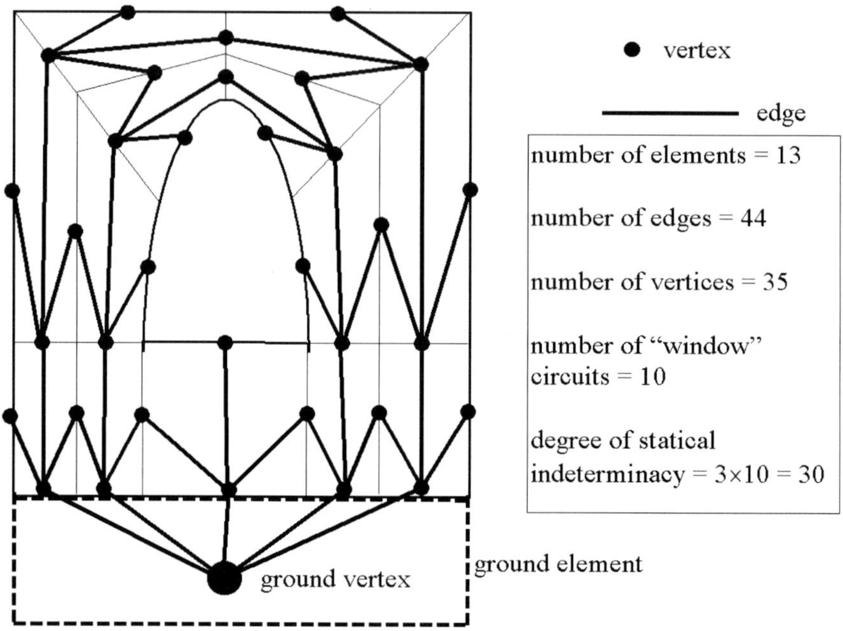

Figure 10.8: 2D model, graph, and circuits.

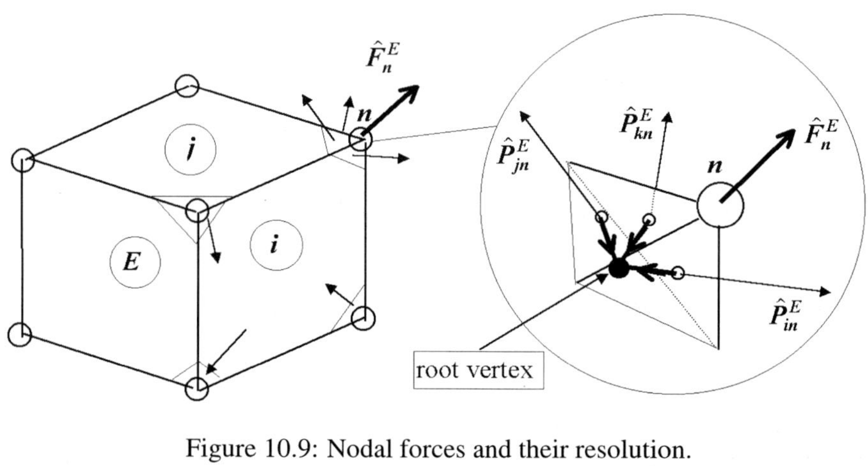

Figure 10.9: Nodal forces and their resolution.

Thus node n, when considered as a free body, is in equilibrium if subjected to forces $-\hat{F}_n^E$ from adjacent elements. Each force $\hat{F}_n^E$ is resolved into three interface components such as $\hat{P}_{in}^E$ where i refers to a side which contains node n. These components are illustrated in Figure 10.9. The aim is to resolve all element nodal forces so that interface components are transferred between adjacent elements in an equal and opposite way. For example, if elements E and M share interface i, then:

$$\hat{P}_{in}^{E} + \hat{P}_{in}^{M} = 0 \tag{10.9}$$

Assuming this aim is achieved, side i is then acted on by four corner forces with a resultant wrench $\hat{W}_i^E$. The adjacent element M is acted on by an equal and opposite wrench, and all elements are maintained in equilibrium by their side wrenches. In this way a particular solution for thrusts (and twisting moments) is derived.

In fact, not only are such resolutions possible, but in general they are not unique. A resolution procedure again makes use of graph-theoretic concepts. In Figure 10.9 an "infinitesimal" tetrahedral element is cut as a free body from a corner of element E, and is shown with its tree graph consisting of four vertices and three edges. The edges correspond to the three component forces $\hat{P}_{in}^{E}$ with varying i. A typical mesh of hexahedral elements having an interior node n in common gives rise, for example, to 8 infinitesimal tetrahedra enclosing the node. The graph associated with this substructure then has the topology indicated in Figure 10.10. If the substructure is loaded with forces $-\hat{F}_n^E$ applied at the root vertices, then the interface forces between the tetrahedra define suitable resolutions of $\hat{F}_n^E$. Clearly the substructure, or shell, has a statical indeterminacy of $3\times$ number of independent circuits, in this case, $3 \times 5 = 15$.

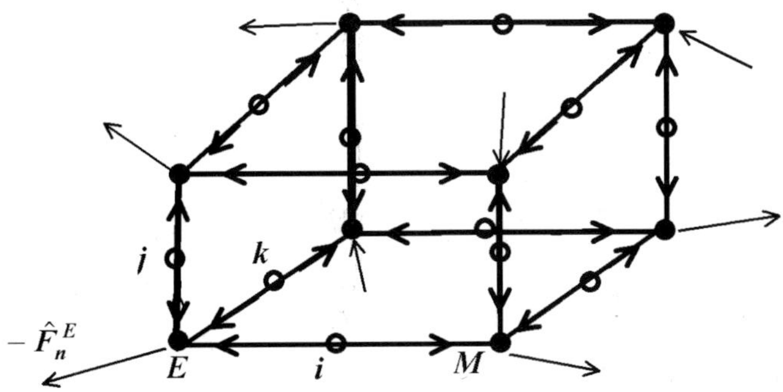

Figure 10.10: Oriented graph associated with resolutions of forces at node n.

How should the interface forces be determined? The approach advocated in [35] is based on a least squares procedure which, for the present context, is outlined as follows. Corresponding to each force $\hat{P}_{in}^{E}$, there exists the force P_{in}^{E} which is consistent with the surface tractions on side i as derived from the deflections of the isoparametric element. We then seek to minimize $\sum_{E,i}\left(\hat{P}_{in}^{E} - P_{in}^{E}\right)^2$ subject to the equilibrium conditions between $\hat{P}_{in}^{E}$ and $-\hat{F}_{in}^{E}$. In the present example, the sets of forces $\hat{P}_{in}^{E}$ and P_{in}^{E} form vectors $\hat{P}_n$ and P_n with a total number of 3D cartesian components equal to $8 \times 3 \times 3 = 72$. Since there is no coupling between the cartesian components, the minimization can be determined for each component separately. Thus using "x" as a subscript to denote components parallel to a reference x axis, the x-components of $\hat{P}_n$ are expressed in Equation (10.10).

$$\left\{\hat{P}_{nx}\right\} = [C]^\top \{R_{nx}\} + \left\{\hat{P}^0_{nx}\right\} \tag{10.10}$$

where $\hat{P}^0_{nx}$ represents a particular solution obtained from a tree, R_{nx}represents bi-actions corresponding to each independent circuit, and $[C]$ represents the incidence matrix of graph edges in the independent circuits.

The least squares minimization is given by the solution to Equation (10.11), with a coefficient matrix for the example in Figure 10.10 as in Equation (10.12).

$$[C][C]^\top \{R_{nx}\} = [C]\left\{P_{nx} - \hat{P}^0_{nx}\right\} \tag{10.11}$$

$$[C][C]^\top = 2\begin{bmatrix} +4 & -1 & 0 & -1 & -1 \\ -1 & +4 & -1 & 0 & -1 \\ 0 & -1 & +4 & -1 & -1 \\ -1 & 0 & -1 & +4 & -1 \\ -1 & -1 & -1 & -1 & +4 \end{bmatrix} \tag{10.12}$$

The coefficient matrix is symmetric and positive definite, and is repeated for all components and all similar arrangements of element connections. Thus the solution by minimisation requires only a minor computational effort.

The resolution and transformation of nodal forces becomes much simpler for 2D models. In this case, nodal forces $\hat{F}^E_n$ are resolved into only two components, the substructure around node n is formed from infinitesimal triangular elements, and the associated graph has only one circuit when node n is an internal node of a mesh. In fact the least squares solution then corresponds to a very simple construction of vectors within a Maxwell force diagram [34]. The forces $-\hat{F}^E_n$ form a closed polygon, and if the element forces are drawn in a sequence corresponding to a path taken around the circuit, then their resolution is achieved by inserting a pole point P into the Maxwell diagram as illustrated in Figure 10.11. The least squares solution of Equation (10.11) corresponds to locating P at the centroid of a set of unit weights associated with the vector components of P_n.

The resolution is also shown in Figure 10.11 by the string polygon around the node, which has the same form as the associated graph.

10.3.3.3 Admissible solutions by adapting mesh topology

A particular solution derived from conventional finite element models may be reliant on thrusts or wrenches which are not feasible, e.g. they imply tensile tractions or shear slip. A general scheme for investigating variations to a solution may involve mathematical programming or other schemes for optimization. However a particular solution may also be improved more directly by the simpler expedient of disconnecting elements to simulate tension relieving cracks. It is not suggested that such changes

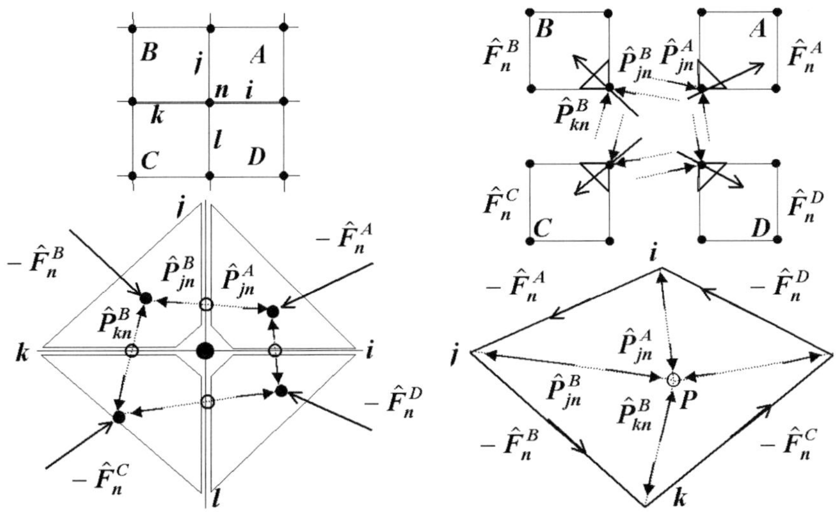

Figure 10.11: Resolution of nodal forces.

to mesh topology offer a panacea, but there are situations where it is worth considering.

For simplicity of presentation this option is discussed for 2D models. As is often the way, the extension to 3D models is possible, but it becomes much more challenging to formulate a general algorithm! For 2D models using elements with corner nodes only, the resolution of nodal forces leads to a pair of side forces denoted by $\hat{P}^E_{in}$ and $\hat{P}^E_{im}$ acting on side i of element E at the positions of corner nodes n and m. The resultant thrust $\hat{T}^E_i$ is not feasible if one or both of these forces has a tensile normal component. This situation may be alleviated by changing the positions of the pole points in the corresponding Maxwell diagrams. However if this fails to relieve tension, then the next option is to partially disconnect one or more elements from their original nodes, and so release thrusts from some interfaces. In order to maintain corner nodes for element definition, distinct duplicate nodes are generated having the same coordinates as the original nodes. Typical arrangements of disconnections are illustrated in Figure 10.12, where (a) represents the common case of tension and crack initiation at a surface. The continuation of the disconnection process may involve one or more nodes at a time according to judgement, in a trial and error procedure.

The adaptive changes in mesh topology are followed by reanalyses in an iterative sequence which aims to relieve tension and maintain a stable system of elements. It may become appropriate to reconnect elements and nodes when renewed contact is indicated by the relative displacements determined by linear elastic analyses.

Relief of shear tractions at interfaces when shear friction cannot be mobilised is more problematic. In this case simple separation of interfaces may be too drastic a

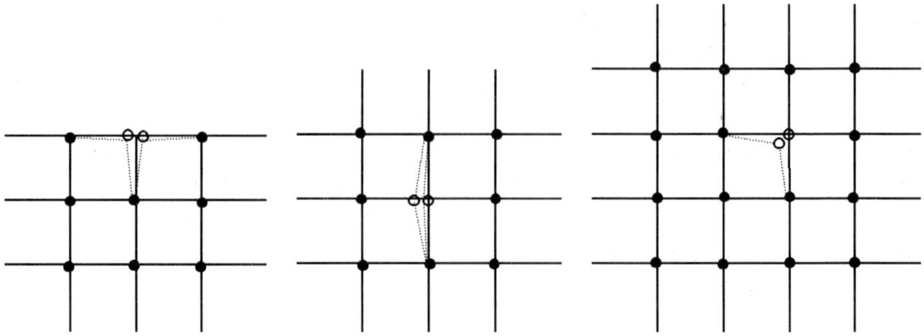

Figure 10.12: Crack simulation by element disconnections.

way of removing the problem!

10.3.3.4 Statically admissible stress fields

Once a feasible set of thrusts or wrenches has been determined for a model, it may be of interest to carry out more detailed investigations into local stress fields within selected elements. As an example, consider the voussoir block in Figure 10.13 which transfers a 2D thrust between opposite sides. A qualitative description of the internal stresses can be developed by "strut and tie" models [36], with the assumption that the thrusts are introduced through localised contact zones. This type of model of a stress field is included in Figure 10.13, and it can lead to a reasonable estimate of the tensile stresses which are developed by the spreading of the compressive stresses.

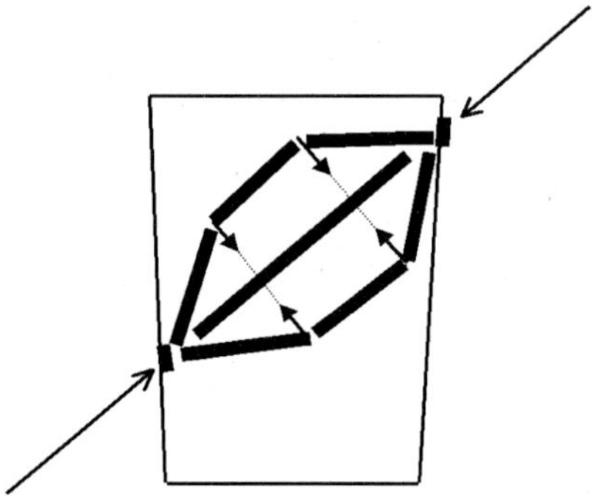

Figure 10.13: Strut and tie model of thrust transfer.

Establishing a system of strut and tie forces extends thrust line concepts to a more general structural form. A complete description of statically admissible stress fields can be made with stress trajectories. These trajectories can be regarded as the most detailed form of a strut and tie system.

The derivation of stress fields and their trajectories is proposed by using hybrid equilibrium elements [37–40], whose stress attributes provide a natural way to increase the hierarchical level of a model. These elements are based on internal statically admissible stress fields, and dual traction/displacement fields defined independently for each side of an element. Elements can be of arbitrary shape, but continuity of tractions is more readily satisfied when elements have straight or flat sides, and the degrees of traction polynomials are not greater than the degrees of the polynomial stress fields.

Stress recovery can be carried out for individual elements (or substructures) with each element subjected to tractions which are statically equivalent to the thrusts or wrenches acting on its sides. For simplicity the tractions may be assumed to have constant or linear distributions over the contact zones. In order to ensure that tractions can be transmitted by assumed stress fields, the stress fields are defined in a piecewise sense in macro-elements [39] with the possibility of using polynomials of variable degree. For degrees greater than one, the stress fields become hyperstatic, and more freedom is available for stress recovery. In these cases the stress fields may be determined using linear elastic assumptions. Alternatively the stress fields may be redistributed with the aim of keeping within material stress limits [41].

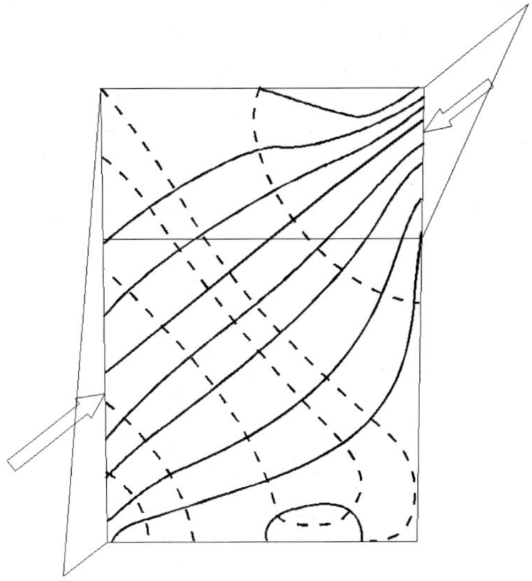

Figure 10.14: Stress trajectories from macro-elements of degree 2.

Complete stress fields for the voussoir block of Figure 10.13 are illustrated in Figure 10.14 as stress trajectories. In this case the tractions representing the thrusts are distributed in a linear way over contact zones which extend as far as possible over the sides. The lines of action of the thrusts penetrate these zones at the extremities of their middle thirds. The voussoir is modelled by two macro-elements of stress degree 2, and the stone is assumed to be elastic with a Poisson's ratio of 0.15.

10.4 Case studies of historical masonry structures

10.4.1 Bridgemill Bridge

Bridgemill bridge was a road bridge constructed in 1869 at Girvan in Strathclyde. It was constructed with dressed sandstone voussoirs, and although described as being in "superb condition" it became redundant due to road re-alignments. The bridge was selected for load testing to destruction in 1984 in a programme of tests for TRRL. The arch ring had a span of 18.29m, a rise of 3.195m, and a thickness of 0.71m. It has been the subject of several types of analysis designed to assess the strength of the bridge, and compare with the actual load tests [24, 42–44]. The main reason for its inclusion here is to serve as an example of limit analysis based on a linear programme. The objectives are to illustrate how a simplified analysis can assist in investigating the sensitivity of the capacity of the structure (i.e. its partial load factors) to parameters such as material strength, arch thickness, and friction coefficients [21].

The arch ring was modelled as a 2D structure with 40 uniform voussoir block elements arranged to form a parabola. The linear programme has 3 static variables (equal to the degree of statical indeterminacy) and 328 constraints (8 constraints per interface as indicated in Figure 10.4) on local thrust values. The dead loads consist of the arch self weight (2650 kg/m^3), and fill material (2040 kg/m^3) assumed to raise the road level to 3.75m above the centre of the springing. The fill is assumed to be non-structural with no capacity for load dispersion. Live load is represented by a single axle placed at any position along the span. The arch is illustrated in Figure 10.15, which includes one limiting solution corresponding to an axle force of 255 kN/m when the material compressive strength is limited to 5 N/mm^2, and the friction coefficient is 0.3.

A summary of the key conclusions made from investigations with this model are as follows:

- With dead load alone (the total dead weight is estimated to be 845 kN/m) the strength of the structure is limited by the material strength. For a material strength of about 15 N/mm^2, the load factor is about 11, but the geometric safety factor is about 2.2. Thus the minimum arch thickness for stability is $\frac{0.71}{2.2} = 0.32$m.

- With the addition of an axle live load at the quarter point of the span, the maximum axle load is estimated to be 265 kN/m sensibly independent of the compressive strength when this is above about 15 N/mm^2. For lower material strengths the axle load is reduced, e.g. to about 190 kN/m for 5 N/mm^2, and zero for 1.4 N/mm^2. At this lower strength the dead load is just supported.

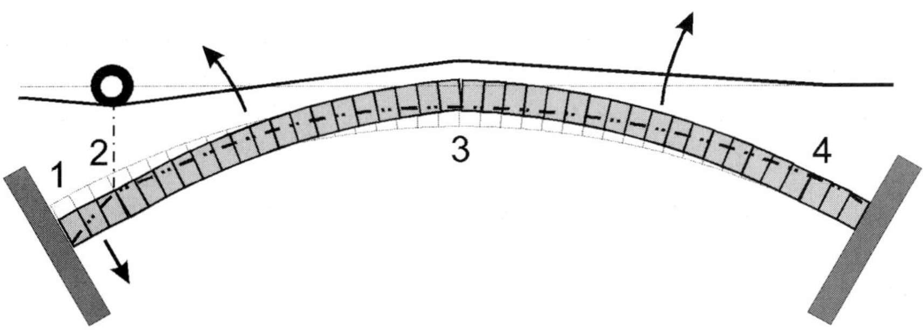

Figure 10.15: Bridgemill bridge - a thrust line solution derived from a linear programme, and the corresponding collapse mechanism.

- The frictional limit is activated when the axle moves closer to an abutment. Figure 10.15 illustrates the case when the axle is 1.21m from the top of the abutment. The limiting axle load corresponds to compressive stress limits being reached at sections 2, 3, and 4, whilst a friction limit is reached at section 1. The corresponding limit points are shown in the piecewise linear limit polygon in Figure 10.4. In this case the corresponding mode of collapse can be predicted to be as shown in Figure 10.15 since positive work is done at the sliding section. The dashed line indicates the thrust line which is based on the original shape of the arch.

The test loading to destruction involved a line load across the bridge at the quarter point of the span. The maximum test load was reported to be 373 kN/m, and it was considered that collapse would have occurred if this load had been maintained [42]. A more refined mechanism analysis which also accounted for support from the fill predicted 520 kN/m [43], and an incremental cracked elastoplastic analysis which further accounted for changes in the arch shape predicted a collapse load of 353 kN/m [24].

10.4.2 Horrabridge

This road bridge, in the Devon village of Horrabridge north of Plymouth, is a Grade II listed structure and a Scheduled Ancient Monument [6]. It is a 14th century 3-span arch bridge spanning the river Walkham, with a central span of 4.27m and a rise of 1.83m. The spandrel walls enclose a road between them of width 3.5m. One of the main structural features of interest is the relatively steep and pointed geometry of the arches which are formed by double rings of granite blocks, giving a total thickness of about 0.55m. The shape of this arch, as shown in Figure 10.16, is such as to place it outside the normal range for assessment by the MEXE method [45,46].

The central arch was investigated as a single 2D structure without the spandrel walls, but including the fill as a medium to disperse a concentrated axle load. The

Figure 10.16: Horrabridge - a view from downstream.

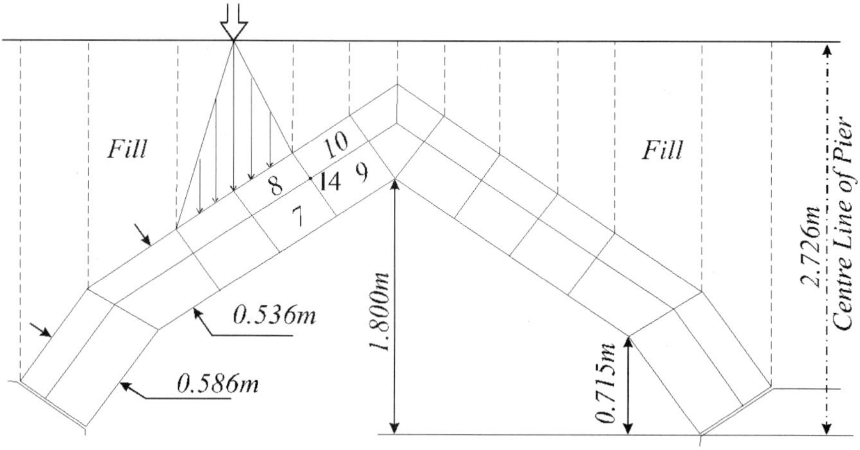

Figure 10.17: Horrabridge - a 2D finite element model of an arch ring.

fill also applies dead load to the arch extrados, but any other interaction, e.g. leading from uplift of the arch ring, was not accounted for. The investigation was based on a conventional finite element model [23, 47], with adaption of the mesh topology, to study the ability of the structure to support "standard" loads representing traffic. Two load cases were considered, the first to represent normal HA loading, and the second to represent the load from a single axle of a 40 tonne vehicle.

A simplified finite element model which represents the double ring construction of the arch is shown in Figure 10.17, and this contains 23 4-noded isoparametric elements. This Figure also includes the first load case with the axle load positioned approximately at the quarter point of the span. The material properties were assumed as follows:
for the stone: density = 2200 kg/m^3; Young's modulus = 10 kN/mm^2; Poisson's ratio = 0.15; for the fill: density = 2000 kg/m^3.

- The first load case includes a 40 kN/m axle load and an UDL of 10 kN/m^2. The results of the iterative finite element analyses are illustrated in Figures 10.18 and 10.19 for deflections and thrust lines respectively. The deflected view includes the disconnections made to simulate cracks for the relief of tension. The nodal disconnections only involved nodes on the extrados or the intrados. The maximum deflection normal to the original arch alignment is only 0.1 mm, and hence no account need be taken of the change in geometry for the analysis of the internal thrusts.

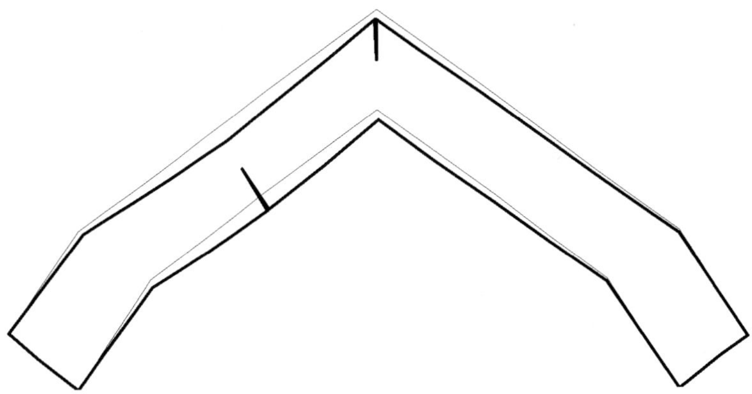

Figure 10.18: Horrabridge - deflections ($\times$1000) of a cracked model with normal HA loading.

- The maximum thrust is about 150 kN/m, and the maximum tractions based on uniform stresses on the contact zones are very low, e.g. 0.75 N/mm^2 in compression, and 0.25 N/mm^2 in shear. However this solution depends on high friction coefficients where indicated in Figure 10.19, e.g. 0.66 on a cross-section, and 2.62 on a longitudinal-section. Such friction is unlikely to be mobilised, and sliding or delamination may occur. The ability to transfer shear tractions would also depend on the cohesion and bond that could be developed in mortared joints.

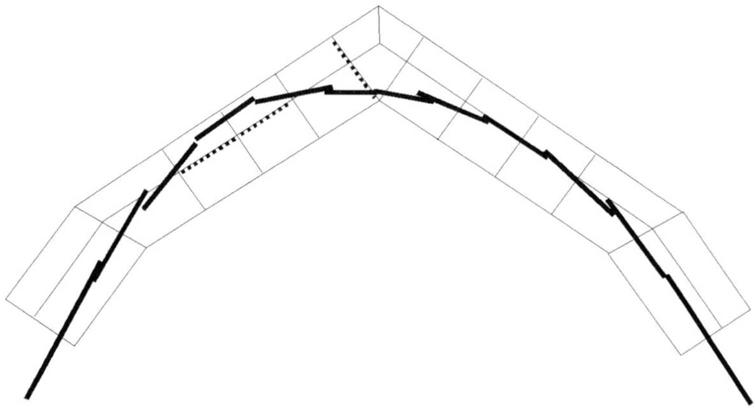

Figure 10.19: Horrabridge - a derived thrust line solution, including potential slip planes.

- In the second load case the UDL is removed and the axle load is increased to 130 kN/m. The results are illustrated in Figures 10.20 and 10.21. The mode of deflection changes to one where the right hand side of the arch rises by 0.2mm, and the maximum deflection normal to the arch increases to 0.4 mm on the left hand side. No passive interaction with the fill was considered in this simplified model.

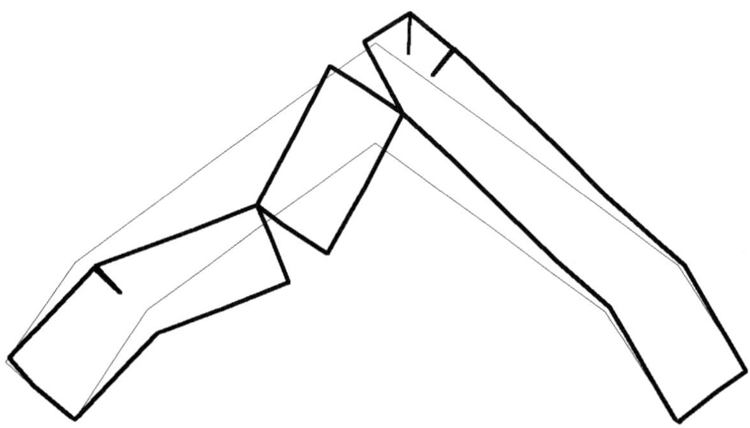

Figure 10.20: Horrabridge - deflections ($\times 1000$) of a cracked model with a high axle load.

- The maximum thrust is about 230 kN/m, but the thrust line just leaves the extra-

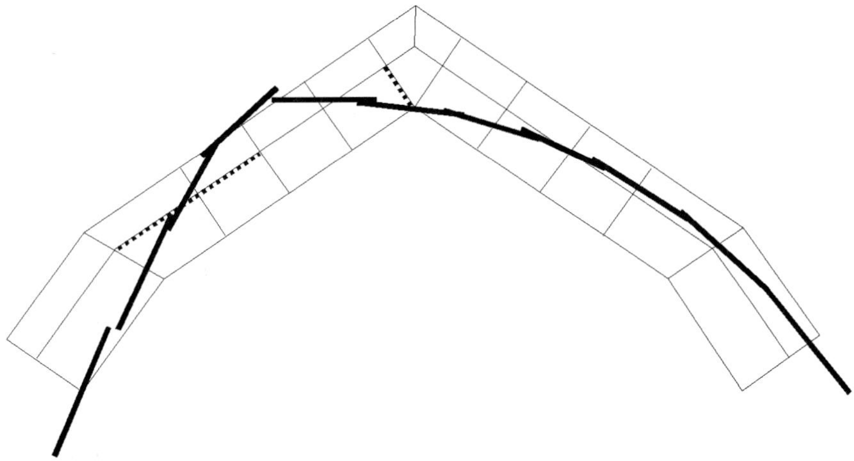

Figure 10.21: Horrabridge - a derived thrust line solution, including potential slip planes.

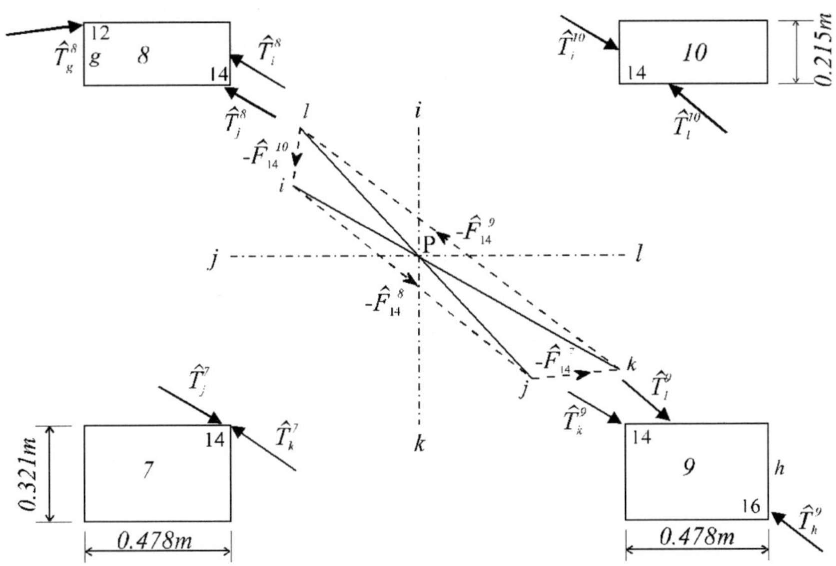

Figure 10.22: Horrabridge - a Maxwell diagram for node 14.

dos under the axle load, and just reaches the intrados at two critical sections at the springing and at the crown. Thus the iterative solution just failed to confirm stability, and again the need for high friction coefficients is indicated, e.g. 0.85

Thrust	Magnitude kN/m	Inclination
$\hat{T}_g^8$	117.6	7.5°
$\hat{T}_g^9$	87.8	40.4°
$\hat{T}_i^8, \hat{T}_i^{10}$	47.2	32.8°
$\hat{T}_j^7, \hat{T}_j^8$	40.3	59.9°
$\hat{T}_k^9, \hat{T}_k^7$	40.1	34.8°
$\hat{T}_l^{10}, \hat{T}_l^9$	47.6	48.8°

Table 10.1: Local thrusts on elements adjacent to node 14.

on a cross-section, and 1.96 on a longitudinal-section.

- For the second load case a typical resolution of nodal forces at node 14 is illustrated in Figure 10.22, together with the Maxwell force diagram for that node. It should be noted that the pole point P occurs at the intersection of the diagonals of the force polygon. This position was chosen to minimise the maximum friction coefficient required by the resolved nodal forces acting as components of local thrusts on the sides of elements. The resultant local thrusts on the sides of the elements in Figure 10.22 are tabulated in Table 10.1, where the inclinations are relative to the normals to the interfaces.

This example has shown that details of force paths are recoverable from conventional finite element analyses, and potentially weak aspects of a structure can be identified. A finer mesh would allow for greater precision in locating tension relieving cracks, and a limited compressive strength. A more complete investigation could be continued with a limit analysis.

10.4.3 Exeter Cathedral

10.4.3.1 The investigation

The present form of the cathedral stems mainly from the Gothic reconstruction in the 13th and 14th centuries [48, 49]. This resulted in one of the longest Gothic structures with a continuous pattern of vaulting extending over some 100 m. The western arm of the nave was reconstructed in the comparatively short period between 1328 and 1342, and a cross-section taken between the columns is shown in Figure 10.23. A typical bay of this part of the nave has been subjected to a structural study involving an investigative type of analysis with the aims of understanding its stability, establishing levels of stress, and sensitivity to local damage [22].

Only loading due to self weight has so far been considered, and a number of 2D and 3D internal force systems have been derived. The examples presented here are based on 2D systems to gain an understanding of how the major 2D components such as buttresses and columns transfer the self weight to ground level. The first system

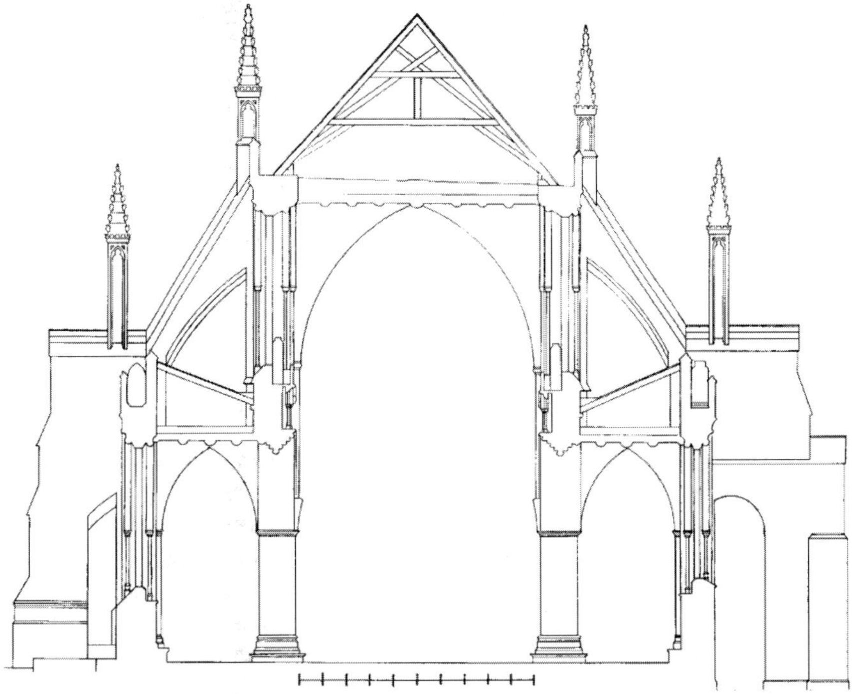

Figure 10.23: Exeter cathedral - a cross section through the western nave (10 m scale bar).

derived takes into consideration a possible construction sequence, the main stages of which are illustrated in Figure 10.24. The main assumptions made for the sequence are as follows:

- The side aisle vaults were constructed before the clerestory walls in order to stabilise the columns of the nave;
- The ribs of the high vault were constructed after the flying buttresses and the timber roof, and were then used as supports for the construction of the webs of the vaulting. The function of vault ribs has long been a controversial subject in general [50]. However in the case of Exeter Cathedral, with its dominant forest of ribs, the authors are inclined to agree with the view [48] that "every rib really supports part of the vaulting, and none is purely ornamental". This assumption implies that the thrust lines of the vault can be contained within the ribs, and preliminary 3D analyses indicate that there is a range of such thrust lines which are feasible. This range can then be extended for the completed structure by allowing thrust lines to pass into the webs.

Figure 10.24: Exeter cathedral - a construction sequence.

10.4.3.2 The computational model

Although Figure 10.23 indicates some lack of symmetry in the external buttressing, the model was assumed symmetric about the east-west axis based on the northern part. This part was modelled by the 2D finite element mesh shown in Figure 10.25. This mesh is composed of 370 6- and 8-noded isoparametric elements of different thicknesses. Since the primary load is the self weight of the masonry, priority was given to modelling the mass distribution rather than the stiffness. Element thicknesses vary between 0.5 m and 6.2 m, where the latter dimension is the spacing of the columns along the nave. A state of plane stress was assumed throughout, together with the following uniform material properties:
Density = 2200 kg/m^3; Young's modulus = 10 kN/mm^2; Poisson's ratio = 0.15.

Nodes at ground level were assumed to be fixed. The conventional boundary conditions on the vertical axis of symmetry would be to fix all nodes in a horizontal direction. However these conditions were released for selected nodes as a means of controlling the horizontal thrust in the high vault. This is another example of user interaction controlling the redundant horizontal thrust introduced across an assumed plane of symmetry. One node was loaded by the timber roof, and the force exerted was determined by another preliminary analysis of the roof structure assuming it to be fixed at its bearings on the masonry walls.

The investigation of internal forces proceeded with a sequence of linear elastic analyses of the finite element model with adaption of the mesh topology. In order to simplify this procedure for interactive analyses, the mesh was partitioned into 11 super-elements as shown in Figure 10.26. The interface between super-elements 1 and

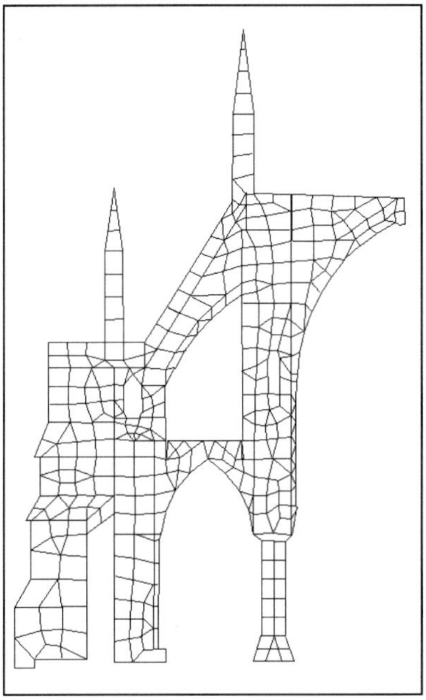

Figure 10.25: Exeter cathedral - a 2D finite element model.

2 excludes the top three elements in Figure 10.25 in order to avoid vault thrusts being transferred to the buttresses via routes which would imply horizontal bending of the clerestory wall. The resultant thrusts on the interfaces of super-elements are generally determined by vector addition of the nodal forces on one side of an interface. Nodal resolution with the aid of a Maxwell force diagram was then only required for one node where three super-elements are connected, i.e. the node connected to super-elements 1, 2, and 3.

A column was represented by super-element 5, and this was dimensioned to represent both the mass and the uncracked inplane stiffness of the column. However in order to demonstrate overall stability, it was necessary to establish column thrust lines which pass through its actual plan form. This form is approximately a square of side length 1.06 m oriented at $45°$ to the plane of the model.

Cracks were generally initiated by disconnecting boundary nodes on super-element interfaces when a thrust line was observed to lie outside the interface. The decision to disconnect was left as a matter for the engineer, and an additional section for potential cracking was identified in the clerestory wall at the level of the top of the walkway opening.

Although in this example the investigations were not continued in any optimal sense with the use of a linear programme, it is to be noted that the super-element model has, effectively, a degree of statical indeterminacy in terms of the resultant

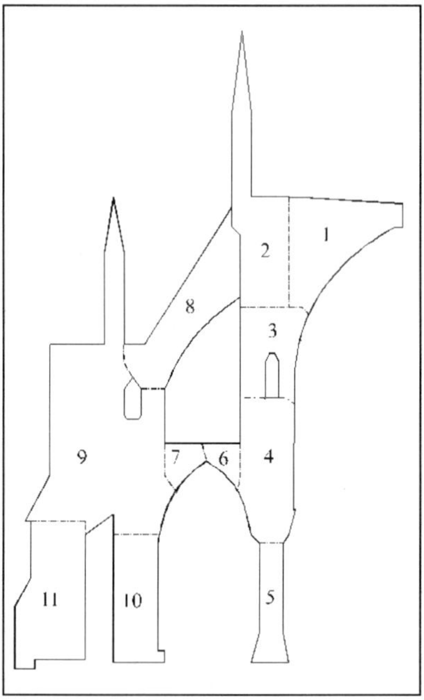

Figure 10.26: Exeter cathedral - a super-element model.

thrusts across interfaces of $(3 \times 7 - 1) = 20$.

10.4.3.3 Summary of results based on the assumed construction sequence

- The final solution shown in Figure 10.27 is the superposition of solutions representing the different stages of construction. The last analysis involved loading from the high vault, the roof, the high pinnacle, and removal of the concentrated reactions from the centering for the flying buttresses. The high vaulting was supported entirely by the *tas de charge* represented by the interface between super-elements 1 and 3. The interface between super-elements 1 and 2 was thus completely disconnected.

- The only parts of the structure which appeared to warrant further disconnections to simulate cracking occurred in the side aisle vaults. Without disconnections, these vaults acted in a similar way to the flying buttresses but at a lower level. The steep direction of the resulting thrust, as indicated by the dotted line in Figure 10.27, could not be maintained within the vault, and disconnections at the bottom of the interface between super-elements 4 and 6, and at the top of the interface between super-elements 6 and 7 became necessary. The extent of the simulated cracks was decided with the aim of stabilising the column,

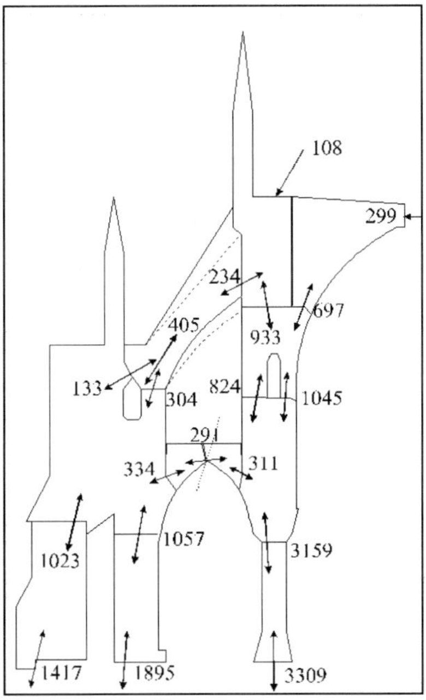

Figure 10.27: Exeter cathedral - a thrust line solution based on a construction sequence.

i.e. minimising the eccentricity of thrust in the column. However the solution obtained led to the thrust in the top of the column lying just outside the kern of the section. At this section the average stress in the column is about 3.0 N/mm^2, and the maximum value based on a linear distribution in compression is about 6.3 N/mm^2.

10.4.3.4 Summary of results for an alternative solution based on allowing vault thrust lines to depart from the ribs

- For this solution the thrust line in the high vault is allowed to rise above the *tas de charge* and pass through the web and/or fill material. This is achieved by connecting the interface between super-elements 1 and 2, and resolving the nodal forces adjacent to the *tas de charge*. The analysis in this case was simplified in that the entire self weight of the structure was applied as a single load case for the complete structure. The consequence on the vault thrust is shown in Figure 10.28 where the horizontal component is increased by some 70%, and the vertical load transferred down the clerestory wall is reduced by some 20%. The flying buttress consequently has a significantly higher thrust, but its alignment is improved.

- Although the load in the clerestory wall is reduced, the local thrust line to the left of the opening implies the need for local relief of tension in the north face. Thus only a single node connection is retained in this section within super-element 3.
- The thrust in the aisle vault is also reduced, but further disconnections are required in the top of the interface between super-elements 7 and 9, making this part similar to a three hinged arch.
- The column thrust is well within the kern, and also reduced in magnitude by some 14%. The maximum stress is then estimated at about 3.0 N/mm^2.

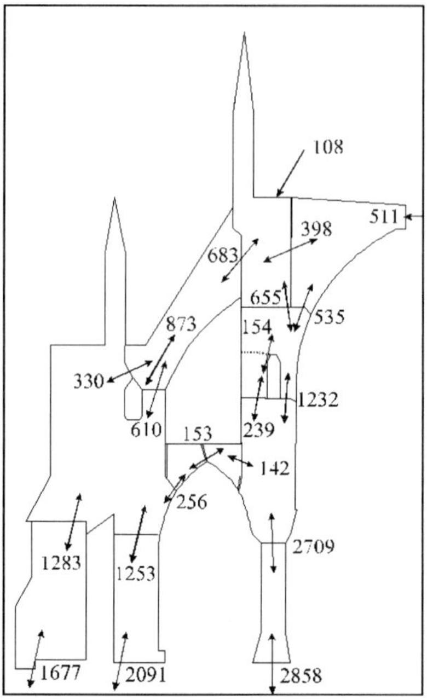

Figure 10.28: Exeter cathedral - an alternative thrust line solution.

10.4.3.5 Some concluding observations on the Gothic designs

The above results presented in conjunction with Figures 10.27 and 10.28 prompt some interesting comparisons between the eastern and western arms of the nave. The most significant differences between the earlier eastern design and the later western design from the structural point of view concern the position of the flying buttress and the opening in the clerestory wall. In the eastern arm there is no walkway at the clerestory level and hence no openings there. The flying buttress is of a shallower construction and at a lower angle of inclination as indicated by the dashed line in Figure 10.27. If these two aspects are considered with the solution in Figure 10.27, then it appears that for the eastern arm:

- The flying buttress is better aligned to buttress the upper wall and vaulting when the assumptions of the construction sequence are made, and;
- The higher thrusts from the vault transferred through the *tas de charge*, and from the wall adjacent to the vault, are more readily transferred down through the clerestory wall when the openings are not present.

These observations raise the question: were the changes in design made as a result of advances in engineering concepts in dealing with new functional requirements, or were they merely fortuitous?

10.4.4 Wells Cathedral

10.4.4.1 The investigation

All cathedrals have elements of structure which are surprising to present day investigating engineers. Often, the most complex elements result from redesign either during or after construction, or from changes which were forced by the need to bring a partially complete design into use. At the east end of Wells cathedral, the great East window is flanked by two flying buttresses which support the thrust from the arches above the Choir windows and the longitudinal thrust from the Choir vault. The buttresses are clearly visible from outside the cathedral as illustrated in Figure 10.29, but there is no obvious sign of them in the space below. The space between the high altar and the Lady Chapel is airy and open with four slender columns supporting an unusual pattern of vaults.

A longitudinal section through the building shows that the buttresses are supported on two of these columns with a heavy block of dressed but square masonry providing the link between the shapely buttress above the roof and the slender column below the vault. Figure 10.30 illustrate the form of one of these linking blocks. A cross section provides a less comfortable view. In this view the buttresses appear to step sideways onto the columns. Here is a situation where the "wrench" described in section 10.3.2.2 is very obvious. The structure is showing no sign of distress, so any analysis must begin from the comfortable view that the structure is safe.

10.4.4.2 The model

The buttress was modelled as a simple arch, with the mass represented with some care. The shape at every section is represented by an octagon, though at some sections this reduces to a rectangle. Because the flying buttress has a large vertical component in its geometry, it was sensible to model the whole structure by dividing it into horizontal slices. Each element was trapezoidal, with front and back faces inclined to the vertical at different angles. In the central section, between roof and vault, all four faces were inclined or stepped. The elements were therefore treated as truncated skew

Figure 10.29: Wells cathedral - buttress above the roof (from the north).

Figure 10.30: Wells cathedral - buttress below the roof (from the south and the east).

pyramids for which the volume and centroid are easily calculated. All calculations were based on the vertical line on the outside face of the east wall through the centre of the buttress.

User interaction with the model involved the “redundant” thrusts at the top of the buttress and from the vaults. Each of these is limited by gravity and the geometry of a complex structure, and limits are not easy to calculate. In the event, the forces required proved to be modest and the engineers confidence in the structure was not challenged by the results. A more detailed exposition of this problem will shortly be offered to the journal: The Structural Engineer.

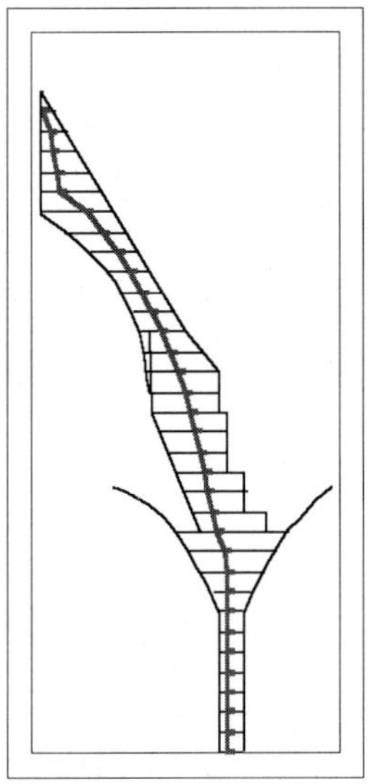

Figure 10.31: Wells cathedral - thrust line in elevation.

Figure 10.31 shows in elevation a thrust line in one of these structures. Figure 10.32 shows the same result in an oblique projection. To understand the action at a particular level, it is useful to extract a plan of that level as in Figure 10.33. Here we can see the centre of thrust with two rings round it. The inner core is calculated to support the direct and shear stresses from the thrust, and the outer zone between the rings is calculated to support the shear stresses from the twist component of the “wrench” described above. The material within the outer ring will thus sustain all the forces applied. Also shown in the view is a vector representing the shear force.

10.5 Closure

The principal aim of this Chapter has been to address structural assessment at the fundamental level of static equilibrium. Non-conventional finite element concepts have been proposed for the purposes of exploring “admissible force paths” in structural forms where such paths are generally not unique, and/or may be far from obvious. In

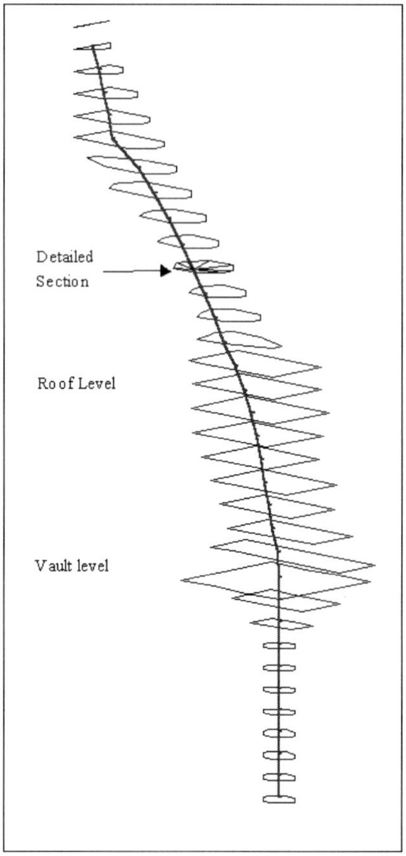

Figure 10.32: Wells cathedral - an oblique projection of a thrust line.

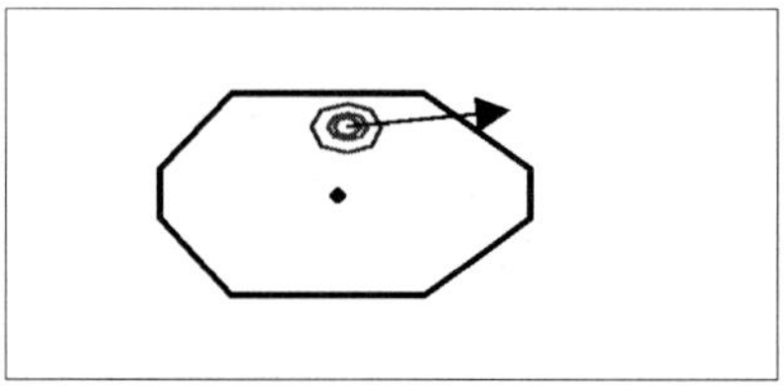

Figure 10.33: Wells cathedral - thrust and wrench at detailed section.

this context "admissible" implies not only the satisfying of equilibrium, but also the satisfying of constraints set by material and joint strengths.

Of course, this fundamental level of assessment is not necessarily the whole story, and it does not guarantee safety - there is too much uncertainty in material properties and connections in ancient structures for that, and mathematical assumptions for a safe solution (e.g. for plasticity theorems) may not be valid. However, this fundamental level of investigation forms an essential part of an assessment, and coupled with the experience of the engineer, it may be deemed sufficient.

Much more developmental work is needed to transform concepts into practical and useful tools for the engineer. In particular, the authors propose further work as regards:

- 3D modelling;
- Visualization with computer graphics;
- Human/computer interaction (HCI).

Acknowledgements

The authors would like to acknowledge with thanks the following contributions: From E.M.Faithfull, formerly an undergraduate project student at the University of Exeter, the original surveyed drawing of the cross-section of Exeter Cathedral reproduced in Figure 10.23; From Professor J.P.Moitinho de Almeida, Technical University of Lisbon, the stress trajectories in Figure 10.13.

References

[1] Harrison R., (Editor), "Manual of Heritage Management", Butterworth-Heinemann, 1994.

[2] Hume I.J., "The structural engineer in conservation", The Structural Engineer, 75, No 3, 33-37, 1997.

[3] Alexander D., "Ancient and trodden", New Civil Engineer, 21-23, 30 May 1991.

[4] Clifton-Taylor A., Ireson A.S., "English Stone Building", Victor Gollancz, 1983.

[5] Mao Yi-sheng, "Bridges in China-Old and New", Foreign Language Press, Peking, 1978.

[6] Otter R.A., (Editor),"Civil Engineering Heritage Southern England", Thomas Telford, 1994.

[7] Mainstone R.J., "Developments in Structural Form", 2nd edition, Architectural Press, 1998.

[8] Salvadori M., "Why buildings stand up - the Strength of Architecture", W.W. Norton & Co., New York, 1990.

[9] Drury H.M.R., "Exeter Cathedral, repair of war damage", The Builder, 330-332, 1948.

[10] Onate E., Hanganu A., Barbat A., Oller S., Vitaliani R., Saetta A., Scotta R., "Structural analysis and durability assessment of historical constructions using a finite element damage model", in "Structural analysis of historical constructions. Possibilities of numerical and experimental techniques", Roca P., Gonzalez J.L., Mari A.R., Onate E., (Editors), CIMNE, 189-224, 1997.

[11] Ramm E., Maute K., "Topology optimization - a general tool in structural design", in "Computational Modelling of Concrete Structures", Mang H., Bicanic N., de Borst R., (Editors), Pineridge Press, Swansea, 805-824, 1994.

[12] Ramm E., Maute K., Schwarz S., "Conceptual design by structural optimisation", in "Computational Modelling of Concrete Structures", de Borst R., Bicanic N., Mang H., Meschke G., (Editors), Balkema, Rotterdam, 879-896, 1998.

[13] Heyman J., "The Stone Skeleton", CUP, Cambridge, 1995.

[14] Heyman J., "The Masonry Arch", Ellis Horwood, Chichester, 1982.

[15] Heyman J., "Arches, vaults and buttresses: masonry structures and their engineering", Variorum, Aldershot, 1996.

[16] Livesley R.K., "Limit analysis of structures formed from rigid blocks", Int. J. Num. Meth. Eng., 12, 1853-1871, 1978.

[17] Calladine C.R., "Plasticity for Engineers", Ellis Horwood, Chichester, 1985.

[18] Blaauwendraad J., Schneider J., "Structural Analysis: Elastic or Plastic?", Structural Engineering International, 3, 165-168, 1993.

[19] Croci G., "The Conservation and Structural Restoration of Architectural Heritage", Computational Mechanics Publications, Southampton, 1998.

[20] LUSAS Bridge, FEA Ltd., Kingston, 1999.

[21] Maunder E.A.W., "Limit analysis of masonry structures based on discrete elements", in "Structural Repair and Maintenance of Historical Buildings III", Brebbia, C.A., Frewer, R.J.B., (Editors), Computational Mechanics Publications, Southampton, 367-374, 1993.

[22] Maunder E.A.W., "Some structural studies of Exeter Cathedral", The Structural Engineer, 73, No. 7, 105-110, 1995.

[23] Maunder E.A.W., "Thrust line solutions for masonry arches derived from finite element models", in "Arch Bridges", Melbourne C., (Editor), Thomas Telford, 215-224, 1995.

[24] Choo B.S., Coutie M.G., Gong N.G., "Finite element analysis of masonry arch bridges using tapered elements", Proceedings I.C.E. Part 2, 91, 755-770, 1991.

[25] Barbosa R., "Discrete finite element method for multiple deformable bodies", in "FEM in the Design Process", Robinson J., (Editor), Robinson and Associates, Okehampton, 70-77, 1990.

[26] Ma M.V., Pan A.D., Luan M., Gebara J.M., "Stone arch bridge analysis by the DDA method", in "Arch bridges", Melbourne C., (Editor), Thomas Telford, 247-256, 1995.

[27] Goodchild C.H., Lupton J., "Spreadsheets and the structural engineer", The

Structural Engineer, 77, No 3, 21-22, 1999.

[28] Harvey W.J., http://www.obvis.com, 2001.

[29] Harvey W.J., "Thrust line analysis in three dimensions using spreadsheets", presented at the First International Conference on the Restoration of Gothic Cathedrals, Vittoria, May, 1998.

[30] Livesley R.K., "A computational model for the limit analysis of three-dimensional masonry structures", Meccanica, 27, 161-172, 1992.

[31] Maunder E.A.W., Savage G.J., "A graph-theoretic model for finite elements with side variables", Civil Engineering Systems, 11, 111-141, 1994.

[32] Meek J.L., "Computer methods in structural analysis", E & FN Spon, London, 1991.

[33] Kaveh A., "Structural Mechanics: Graph and Matrix Methods", Research Studies Press Ltd., Taunton, 1992.

[34] Ladeveze P., Maunder E.A.W., "A general method for recovering equilibrating element tractions", Comp. Meth. Appl. Mech. Eng., 137, 111-151, 1996.

[35] Ladeveze P., Maunder E.A.W., "A general methodology for recovering equilibrating finite element tractions and stress fields for plate and solid elements", Comp. Ass. Mech. Eng. Sci., 4, 533-548, 1997.

[36] Muttoni A., Schwartz J., Thurlimann B., "Design of concrete structures with stress fields", Birkhauser, Basel, 1997.

[37] Hajdin R., Thurliman, B., "Computer-aided design of reinforced concrete walls using stress fields", in "Computer Aided Analysis and Design of Concrete Structures", Bicanic N., Mang H., (Editors), Pineridge Press, Swansea, 515-526, 1990.

[38] Almeida J.P.B.M., Freitas J.A.T., "Alternative approach to the formulation of hybrid equilibrium finite elements", Comput. Struct. 40, 1043-1047, 1991.

[39] Maunder E.A.W., Almeida J.P.M., Ramsay A.C.A., "A general formulation of equilibrium macro-elements with control of spurious kinematic modes", Int. J. Num. Meth. Eng., 39, 3175-3194, 1996.

[40] Almeida J.P.M., Pereira O.J.B.A., "A set of hybrid equilibrium finite element models for the analysis of three dimensional solids", Int. J. Num. Meth. Eng., 39, 2789-2802, 1996.

[41] Temin-Gendron P., Laurent-Gengoux P., "Calculation of limit loads for composite materials via equilibrium finite elements", Comput. Struct. 45, 947-957, 1992.

[42] Hendry A.W., "Test on stone masonry arch at Bridgemill-Girvan", Department of Transport TRRL Contractor Report 7, Crowthorne, 1985.

[43] Harvey W.J., "Application of the mechanism analysis to masonry arches", The Structural Engineer, 66, 77-84, 1988.

[44] Harvey W.J., "Stability, strength, elasticity and thrustlines in masonry structures", The Structural Engineer, 69, 181-184, 1991.

[45] BA 16/97, "The Assessment of Highway Bridges and Structures", DoT, 1997.

[46] BD 21/97, "The Assessment of Highway Bridges and Structures", DoT, 1997.

[47] Maunder E.A.W., "The masonry arches at Horrabridge - thrust lines from a finite element model", Finite Element News, No 4, 27-33, 1995.

[48] Hope V., Lloyd J., Erskine A., "Exeter Cathedral", Dean and Chapter of Exeter Cathedral, Exeter, 1988.

[49] Swanton M., (Editor), "Exeter Cathedral - A Celebration", Dean and Chapter of Exeter, Exeter, 1991.

[50] Courtenay L.T., (Editor), "The Engineering of Medieval Cathedrals, Studies in the History of Civil Engineering - Volume 1", Ashgate Publishing Ltd., Aldershot, 1997.

Index

Author Biographies

John W. Bull (Editor) is a lecturer in Structural Engineering in the Engineering Design Centre and in the Department of Civil Engineering at the University of Newcastle upon Tyne. He has published widely in computational engineering and has published over one hundred and twenty books and papers. He is also an acknowledged expert in the field of soil-structure interaction, concrete pavements, aluminium design and timber design. His current area of research is in Self-Designing Structures and in the use of computational modelling for the determination of the effects of underground voids on above ground structures.

John W. Bull was awarded his BSc (1971) and PhD (1974) by the University of Wales (Cardiff) then gaining the professional qualifications, CEng (1976), MICE (1976), MIMunE (1977), FIWSc (1989), Eur Ing (1990), MIStructE (1991) and FIHT (1993). Between 1974 and 1979 he was employed by Durham County Council designing and building bridges and other civil engineering structures. In 1979 he moved to the Department of Civil Engineering at the University of Newcastle upon Tyne. In 1994 he moved to the Engineering Design Centre for Marine and Other Made to Order Products, within the University of Newcastle upon Tyne from where he developed research into Elastic and Ultimate State Design Tools for Stiffened Shell Structures and into Self-Designing Structures. He presently holds an EPSRC research grant entitled EDO-NET; Engineering Design Optimisation Network which brings together major groups in industry and in academia with the aim of setting up a university based optimisation research centre.

He is the editor or author of a number of books dealing with computational engineering including: 'Numerical analysis and modelling of composite materials' (Blackie Academic and Professional), 'Soil structure interaction: numerical analysis and modelling' (Chapman and Hall), 'Finite element applications to thin walled structures' (Elsevier Science), 'Finite element analysis of thin walled structures' (Elsevier Science) and 'Engineering analysis using PAFEC finite element software' (Blackie and Sons).

He is a member of the Editorial Board of the International Journal of Computers and Structures and has been a member of the editorial board of Structural Engineering Review. Since 1993 he has been a member of the editorial board of seventeen international conferences. He also acts as referee for journals and publishing houses.

Takayoshi Aoki is currently Associate Professor of Structural Engineering in the School of Design and Architecture in Nagoya City University in Japan.

Educated initially at Toyohashi University of Technology he was later awarded his doctorate from Tokyo University. Professor Aoki has held posts in Nagasaki University, Miyagi National College of Technology, and Nagoya City University, and latterly has been visiting Associate Professor at the Politecnico di Torino in Italy.

His principal research interests are in theoretical research on masonry structures, structural restoration of historical masonry structures, experimental research of mortar used for historical masonry structures and experimental research on a non-destructive testing method for concrete.

Both the International Scientific Advisory Committee of STREMA '91 and the WG-17 (Historical Structures) of the International Association for Shell and Spatial Structures have benefitted from his expertise, and he is a regular contributor to conferences worldwide with several published Papers to his credit.

He has been the recipient of awards from the Japan Concrete Institute in various sectors on several occasions.

Charles Augarde is a Lecturer in Civil Engineering at the School of Engineering, Durham University. His research interests are split between topics in structural and geotechnical engineering which are linked by the use of complex numerical modelling. At present he is involved with the development of methods for the prediction of damage to structures affected by tunnelling, analysis of tunnel stability and the mechanics of masonry arch bridges.

Dr Augarde obtained his BSc in 1986 from Lancaster University. He then spent six years in industry, working for Manchester City Engineer's Department, spending time on site and in design offices. His final three years were spent in the Bridge Assessment group as an Assistant Engineer. In 1993 he was awarded an MSc in Structural Engineering Computational Technology at Heriot-Watt University and in 1994 went on to a three-year research assistantship at the Department of Engineering Science at Oxford University. During this period he developed his current interest in tunnelling, which formed the subject of his doctoral thesis (1997). After a brief period back in industry, in London, Dr Augarde returned to academia as a lecturer in structural engineering at the School of Architecture and Engineering at the University of Westminster. He moved on from there, to a Departmental Lecturership at Oxford (1997) and to his current post in 2001.

He is the author of a number of papers concerned with numerical models of tunnelling and general finite element techniques. Dr Augarde has also pursued his interest in the failure behaviour of masonry arches through laboratory testing, an approach which led to the Institution of Structural Engineers' Model Analysis Award in 1999 for undergraduate project work.

Nenad Bicanic is Regius Professor and Head of Department of Civil Engineering, University of Glasgow. Dipl Ing (Zagreb), PhD (Wales), Fellow ICE, Fellow IACM (International Association for Computational Mechanics). Joined the University of Glasgow in 1994, after nine years at the Department of Civil Engineering in Swansea. Member of ASCE/ACI Committee on Finite Element Analysis of Concrete Structures.

Founding member and past chairman of ACME-UK, the UK Association for Computational Mechanics in Engineering. Principal organiser of an established EURO-C series of international conferences on computational modelling of concrete structures in 1984, 1990, 1994, 1998 in collaboration with universities in Delft and Vienna. Research interests include the behaviour and constitutive modelling of concrete at extreme conditions (high temperature, transient dynamics) and the development of computational modelling frameworks for problems involving continua and discontinua (rock blasting, structural masonry, flow in discontinuous fractured rock). Involved in a number of structural integrity assessments and computational modelling of safety critical structures. Author of over 150 papers in international journals, edited books and conference proceedings.

Francesco Genna is Professor of Structural Engineering at the University of Brescia, Italy. Born in Verona, Italy, in 1957, he obtained his degree in structural engineering at the Politecnico of Milan in 1980. Appointed Assistant Professor at the Department of Structural Engineering, Politecnico of Milan, in 1983, he worked in that position until 1991. He then moved as Associate Professor to the University of Brescia, where he was appointed Full Professor in 1994.

In 1984-1985 he visited the Division of Engineering at Brown University as a research assistant; in 1989 he spent several months in the same position at the Materials Department of the University of California, Santa Barbara. He has also collaborated with Hibbitt, Karlsson & Sorensen, Inc., developing software for the finite element code ABAQUS.

His research interests range from the nonlinear stress analysis by finite elements through the study of composite materials to the implementation of finite element codes in parallel processing environments. Recently his interests have included biomechanics topics, such as the stability of osteointegrated dental implants.

He has received research grants from both the Italian Ministry of University and Scientific and Technologic Research and from several private companies.

He has taught or is currently teaching classes on strength of materials, finite elements and plasticity.

He is author or co-author of about eighty technical papers and of four books on strength of materials, finite element analysis and elastic-plastic constitutive laws.

Matthew Gilbert is currently the John Carr Lecturer in Construction at the University of Sheffield. He has a number of years of experience developing novel analysis methods for concrete and masonry structures. Although particularly interested in the application of limit analysis methods, he also has experience of developing realistic constitutive relationships for modelling brittle fracture in masonry. For more than a decade he has also been involved in devising and conducting static and dynamic load tests on full-scale masonry structures.

He graduated with a first class degree in Civil and Structural Engineering from the University of Sheffield in 1989. Following a brief period spent with the civil engineering contractor Henry Boot, he joined Bolton Institute in 1990, where he worked with Clive Melbourne on various aspects of the behaviour of masonry arch bridges. In 1993 he was awarded a University of Manchester PhD for this work. He then joined the University of Sheffield, employed first as a Research Associate, and subsequently as a Lecturer (since 1996).

He is the author or co-author of over 30 research papers on various aspects of the response of masonry structures. He is a recent (1999) recipient of the Institution of Structural Engineers' Husband Award for a paper published in their journal (on the behaviour of multispan masonry arches). He is currently working as a principal investigator on an EPSRC funded research project concerned with developing methods of improving the impact resistance of masonry walls. He has acted as a reviewer of journal papers, grant applications and engineering software, has served on the scientific committees of major international conferences and, in 2000, became a member of the EPSRC peer review college.

Bill Harvey began his career as a bridge builder, and moved into the academic world in 1977. He spent eighteen years at the University of Dundee where he became director of the Wolfson Bridge Research Unit, and Head of Department. His main research interest was masonry arches. From 1995 to 2000 he was Professor of Civil Engineering at the University of Exeter. He recently established a consultancy and a software company, the latter primarily to serve as a vehicle for ARCHIE, the well known arch bridge assessment software.

Edward Maunder is a Senior Lecturer at the University of Exeter in the School of Engineering and Computer Science, and a chartered structural engineer. His main research is in the area of computational structural mechanics, and in particular the development of stress based equilibrium finite element models, visualisation techniques for stress fields, and design optimisation. The main applications for these developments concern reinforced concrete and masonry structures.

He graduated with a BA degree in Mechanical Sciences from Cambridge University in 1964, and this was followed by a DIC from Imperial College in 1965 and a PhD from London University in 1971. He spent some seven years with the Pell Frischmann Group of Consulting Engineers before returning to academia at Exeter University in 1976 in what was then the Department of Engineering Science.

He has been the author or co-author of some 65 papers in his research field, many of them presented at international conferences in Europe and North America. He was invited as a Visiting Research Professor at the University of Waterloo and Memorial University in the summers of 1990 and 1992. He has developed widespread European research collaboration in computational structural mechanics with the support of the EU, CNRS, the British Council, the Royal Society, and the Royal Academy of Engineering. He is currently a member of three International Scientific Committees for international conferences, and has acted as a reviewer for a number of international engineering journals including: International Journal for Numerical Methods in Engineering, Finite Elements in Analysis and Design, and Applied Mathematical Modelling.

He has been a member of the Association for Computational Mechanics in Engineering in the UK since its first conference in 1993, and has acted as its vice-chairman and chairman in 1997/98 and 1998/99 respectively.

Clive Melbourne is Professor of Structural Engineering in the Division of Civil and Environmental Engineering at the University of Salford. His main research is in the area of masonry arch bridge behaviour. He was awarded his BEng Honours (1968) and his PhD (1971) by the University of Sheffield. He had twelve years experience as a bridge engineer with Salford City Engineers and the Greater Manchester County Council before returning to academia at the Bolton Institute, where he ultimately became Head of the School of Civil Engineering and Building. He moved to his present post in 1995.

He is author and co-author of many publications in the field of masonry arch bridges. Most of his work has been involved with large-scale laboratory experiments and field tests for which he has received support from EPSRC, Railtrack, TRL, Marshalls and many others.

He is a Fellow of both the Institution of Structural Engineers (ISE) and the Institution of Civil Engineers (ICE) and is convenor at the Informal Study Group on Arch Bridges for the ISE and was honorary Editor for the ICE's Structures and Building Journal (1994-99).

Martin O'Gara is a graduate in Civil Engineering from the University of Ulster at Jordanstown, Belfast. After graduating in 1996 he joined David O'Connor at the University of Ulster to commence his research into the Numerical Modelling of Masonry Walls in Fire Situations.

Through his research he created a finite element model which simulated the structural behaviour of single leaf masonry walls at fire temperatures. The work was carried out in conjunction with a fire wall testing program conducted by Danny Laverty. The model was then applied to identify the influences of primary wall parameters. He is the author of a number of technical papers concerning the high temperature behaviour of masonry. In December 1999 he completed his research and was awarded his PhD the following year.

He is currently working within the structural engineering software industry developing the use of finite elements in structural design; however his research interests continue in the fire environment area with further publications in preparation.

Enrico Papa is a graduate "cum laude" in Structural Civil Engineering from Politecnico di Milano, Italy. He obtained a PhD graduation in Structural Engineering at the Politecnico of Milano in 1990, discussing a thesis on masonry damage mechanics. During the PhD, he spent a year at the "Laboratoire de Mécanique et Technologie" in Cachan (Paris). From 1990, he started his work as research associate at the Department of Structural Engineering of Politecnico of Milano, where currently he has the Associate Professor position in Structural Mechanics. His research is principally devoted to damage mechanics including: damage models for brittle materials, specially concrete and masonry; experimental tests on brittle materials subject to sustained and cyclic loads; data acquisition by laser interferometry (ESPI technique); parameters identification of nonlinear models; computational nonlinear structural mechanics.

He is author of several publications concerned with these topics. He is a Member of the Italian Mechanics Association (AIMETA), of European Mechanics Society (EUROMECH) and of New York Academy of Sciences.

David Ponniah is presently a Senior Lecturer in the Department of Civil and Environmental Engineering, Edinburgh University. He has extensive experience both in industry (1973-1979), and at the Universities of Sheffield (1972-1973), Glasgow (1979-1984) and Edinburgh (1984-1999) in the fields of geotechnical engineering and construction management. The time in industry has encouraged research into areas of immediate interest and of benefit to practicing engineers.

Working widely in the field of geotechnical engineering, particularly in the transition from theory and research to practice. He has specialised in two areas: piling, and soil-structure interaction in arch bridges. The research in piling has included the design of a novel piling system for offshore structures, the development of equipment for testing of model piles, the planning, design and execution of research field test programmes, and the design and installation of geotechnical instrumentation. The soil-structure interaction research in masonry arch bridges was developed at the University of Edinburgh which has a strong tradition in masonry. His work and subsequent publications have been significant in the understanding of the fill (soil)-structure interaction in arch bridges. This work has involved two field bridge tests, large and small scale model tests in the laboratory, and finite and discrete element based theoretical studies. This research experience has enabled the understanding of a wide range of topics and David has over 40 publications.

Justin Robinson graduated from Edinburgh University with a first class BEng Honours degree in Civil and Environmental Engineering. He subsequently obtained a PhD from Edinburgh University in 2000, researching for his thesis "The Analysis and Assessment of Masonry Arch Bridges". The main areas researched during the PhD course were: an investigation into current masonry arch assessment methods; a series of patch and line load tests on an instrumented double span arch bridge, with corresponding finite element analysis and comparison with previous tests; various discrete element analysis programmes were also investigated for their ability to model masonry arch failure. He is the co-author of several publications in the area of masonry arch bridge analysis and assessment. Justin is about to start work with a civil/structural engineering consultants in order to gain further practical experience.

Paola Ronca is Associate Professor of Structural Analysis and Design at the Structural Engineering Department of the Politecnico of Milan, where she graduated at the Faculty of Architecture in 1971. She took a Master Degree in Civil Engineering at the University of Waterloo, Ontario, Canada in 1976, discussing a thesis on non-linear analysis of reinforced concrete arch bridges.

She is currently teaching Structural Analysis and Design for the Civil and Environmental Engineering Degree at the Politecnico of Milan and Rehabilitation of Structures at the University of Brescia, Faculty of Engineering.

Her current research interests range from numerical modelling of old-damaged masonry structures, in particular arches and vaults in seismic areas, to laboratory setting up and calibration for *in-situ* tests, for both masonry and timber structures.

She is author or co-author of more than seventy technical reports and lecture notes and one book on structural conservation problems of ancient monuments.

She has received research grants from both the Italian Ministry of University and Scientific and Technologic Research, and the National Research Center, and consulting assignments from public and private administrations in the field of structural assessment of ancient monuments.

Alan Selby is a Reader in Civil Engineering in the School of Engineering, University of Durham. His research involves several themes related to actions on buildings, including settlements due to tunnelling in soils, ground vibrations caused by construction activities, and bell forces on bell towers. Inevitably, analysis of such problems requires computational methods to incorporate ground-structure interaction, or transient dynamic effects. Such analyses must be calibrated against *in-situ* records at every available opportunity. He was awarded his BSc (1st class, 1968) and his PhD (1972) by Bristol University. He had several years of industrial experience with Scott, Wilson, Kirkpatrick, (London) and the NE. Road Construction Unit before joining Durham University. He was chartered, MICE, in 1974.

He is co-author of a book on Tunnelling in Soils and Effects on Buildings, and has published some 80 papers, a number of which have been presented at International Conferences around the world. He has collaborated with British Steel Piling Division in seminars, and in writing a brochure on vibration and noise control.

Aris A. Sophocleous is Dr Civil Engineer and Research Collaborator at the Institute of Structural Analysis and Aseismic Research of the National Technical University of Athens. Author of scientific publications in the area of modern methods for the analysis and design of structures under static and dynamic loads. Researcher in more than 35 research projects regarding the analysis, design and redesign of structures and methods of rehabilitation-strengthening of historical masonry and monumental structures. Recent areas of work: the conservation of historical structures by applying Innovative Seismic Protection Techniques, Vibration Reduction Techniques and Structural Control. Member of research committee for the investigation of response of new masonry structures made of autoclaved aerated concrete in seismic areas. Member of the National Scientific Committee for the Eurocodes.

Costas A. Syrmakezis is Dr Civil Engineer and Professor at the National Technical University of Athens. Author - co-author of many scientific publications, textbooks, text notes, scientific books, codes and recommendations, review publications, course notes etc. Convenor of the Task Group 10 of the European Association on Earthquake

Engineering on "Preservation on Historical Monuments". Member of the Committee for the Preservation of the Acropolis of Athens monuments. Member of the Committee for the erection of the new Acropolis Museum. Participation in multilateral ERASMUS projects on the protection of historical buildings against earthquake. Invited lecturer on the topic of masonry response, at more than 50 national or international meetings (seminars, conferences, symposia, etc.). Author of two books on subjects related to the rehabilitation of historical structures. Simple participation in 11, principal investigator to 42 research projects. Responsible for more than 20 research projects of the National Technical University of Athens, on the rehabilitation of masonry structures, with the main target being the establishment of a systematic methodology for the investigation / rehabilitation of all types of historical structures, combining safety (civil engineering) requirements, with other (archeological, architectural, etc) restrictions.

John Wilson graduated from Cambridge University in 1965 with a first class B.A. Honours degree in the Mechanical Sciences Tripos. After graduating he worked as an Assistant Engineer for Scott Wilson Kirkpatrick & Partners mainly on the design of prestressed concrete pressure vessels for Dungeness B Nuclear Power Station and the Hong Kong Cross Harbour Road Tunnel. In 1968 he joined the Central Electricity Research Laboratories, Leatherhead, Surrey as a research officer where he worked mainly on the development of computer programs to analyse the dynamic behaviour of reinforced concrete chimneys and cooling towers. In 1973 he became a lecturer at the University of Durham where to date he has been teaching a variety of undergraduate degree courses in Engineering and the postgraduate Advanced Course in Engineering Geology. He has published several papers based on research of the forces exerted on church towers by bell ringing and tower vibration. His other main interest is the computer simulation of rockfall.

SAXE-COBURG
PUBLICATIONS
mmi